SAFER ELECTRONIC HEALTH RECORDS
Safety Assurance Factors for EHR Resilience

SAFER ELECTRONIC HEALTH RECORDS
Safety Assurance Factors for EHR Resilience

Edited by
Dean F. Sittig, PhD and Hardeep Singh, MD, MPH

Apple Academic Press Inc.
3333 Mistwell Crescent
Oakville, ON L6L 0A2
Canada

Apple Academic Press Inc.
9 Spinnaker Way
Waretown, NJ 08758
USA

©2015 by Apple Academic Press, Inc.

Exclusive worldwide distribution by CRC Press, a member of Taylor & Francis Group

No claim to original U.S. Government works

Printed in the United States of America on acid-free paper

International Standard Book Number-13: 978-1-77188-117-3 (Hardcover)

This book contains information obtained from authentic and highly regarded sources. Reprinted material is quoted with permission and sources are indicated. Copyright for individual articles remains with the authors as indicated. A wide variety of references are listed. Reasonable efforts have been made to publish reliable data and information, but the authors, editors, and the publisher cannot assume responsibility for the validity of all materials or the consequences of their use. The authors, editors, and the publisher have attempted to trace the copyright holders of all material reproduced in this publication and apologize to copyright holders if permission to publish in this form has not been obtained. If any copyright material has not been acknowledged, please write and let us know so we may rectify in any future reprint.

Trademark Notice: Registered trademark of products or corporate names are used only for explanation and identification without intent to infringe.

Library and Archives Canada Cataloguing in Publication

SAFER electronic health records : safety assurance factors for EHR resilience / edited by Dean F. Sittig, PhD and Hardeep Singh, MD, MPH.

Includes bibliographical references and index.
ISBN 978-1-77188-117-3 (bound)
1. Medical records--Data processing--Safety measures. 2. Patients--Safety measures--Data processing. I. Sittig, Dean F., author, editor II. Singh, Hardeep (Physician), author, editor

RA976.S23 2015   610.285   C2014-907729-7

Library of Congress Cataloging-in-Publication Data

Sittig, Dean F., editor.
SAFER electronic health records : safety assurance factors for EHR resilience / Dean F. Sittig and Hardeep Singh.

p. ; cm.
Includes bibliographical references and index.
ISBN 978-1-77188-117-3 (alk. paper)
I. Singh, Hardeep (Physician), author. II. Title.
[DNLM: 1. Electronic Health Records--standards. 2. Patient Safety--standards. 3. Medical Record Linkage--standards. WX 175]

RA976   610.285--dc23   2014045339

Apple Academic Press also publishes its books in a variety of electronic formats. Some content that appears in print may not be available in electronic format. For information about Apple Academic Press products, visit our website at **www.appleacademicpress.com** and the CRC Press website at **www.crcpress.com**

ABOUT THE EDITORS

DEAN F. SITTIG, PhD

Dean F. Sittig, PhD, is a Professor at the School of Biomedical Informatics at The University of Texas Health Science Center at Houston and a member of the UT Houston-Memorial Hermann Center for Healthcare Quality and Safety. Dr. Sittig's research interests center on the design, development, implementation, and evaluation of all aspects of clinical information systems. In addition to Dr. Sittig's work on measuring the impact of clinical information systems on a large scale, he is working to improve our understanding of both the factors that lead to success, as well as the unintended consequences associated with computer-based clinical decision support and provider order entry systems.

HARDEEP SINGH, MD, MPH

Hardeep Singh, M.D., M.P.H. is Chief of the Health Policy, Quality & Informatics program at the Houston Veterans Affairs Center for Innovations in Quality, Effectiveness and Safety, and Associate Professor of Medicine at Baylor College of Medicine. He is a practicing internist and conducts multidisciplinary research on patient safety improvement in electronic health record-based clinical settings. Dr. Singh received the AcademyHealth 2012 Alice S. Hersh New Investigator Award for high-impact research of international significance. In April 2014, he received the prestigious Presidential Early Career Award for Scientists and Engineers from President Obama for his pioneering work in the field of diagnostic errors and patient safety improvement.

CONTENTS

Acknowledgment and How to Cite... *xi*
List of Contributors... *xiii*
Introduction...*xvii*

1. **The Context of EHR Safety and the Need for Risk Assessment**............. 1

    Defining Health Information Technology-related Errors:
    New Developments Since *To Err Is Human*... 1
    Dean F. Sittig and Hardeep Singh
    Eight Rights of Safe Electronic Health Record Use 10
    Dean F. Sittig and Hardeep Singh
    Electronic Health Record–Related Safety Concerns:
    A Cross-Sectional Survey... 17
    Shailaja Menon, Hardeep Singh, Ashley N.D. Meyer,
    Elisabeth Belmont, and Dean F. Sittig

2. **Analysis of EHR Safety** 33

    Review of Reported Clinical Information System Adverse Events in
    US Food and Drug Administration Databases.. 33
    Risa B. Myers, Stephen L. Jones, and Dean F. Sittig
    Exploring the Sociotechnical Intersection of Patient Safety and
    Electronic Health Record Implementation 51
    Derek W. Meeks, Amirhossein Takian, Dean F. Sittig,
    Hardeep Singh, and Nick Barber
    An Analysis of Electronic Health Record-Related Patient
    Safety Concerns... 69
    Derek W. Meeks, Michael W. Smith, Lesley Taylor, Dean F. Sittig, Jeanie
    Scott, and Hardeep Singh

3. **User Context of Safe and Effective EHR Use** 89

    Rights and Responsibilities of Electronic Health Record Users 89
    Dean F. Sittig and Hardeep Singh
    Rights and Responsibilities of EHR Users Caring for Children 98
    Dean F. Sittig, Hardeep Singh, and Christopher A. Longhurst

4. **Conceptual Foundation of SAFER Guides** 105

   A New Socio-technical Model for Studying Health Information
   Technology in Complex Adaptive Healthcare Systems 105
   Dean F. Sittig and Hardeep Singh
   Electronic Health Records and National Patient Safety Goals................. 124
   Dean F. Sittig and Hardeep Singh

5. **SAFER Guide Development Methods** 137

   Safety Assurance Factors for Electronic Health Record Resilience
   (SAFER): Study Protocol 137
   Hardeep Singh, Joan S. Ash, and Dean F. Sittig

6. **Overview of SAFER Guides** 153

   The SAFER Guides: Empowering Organizations to Improve the
   Safety and Effectiveness of Electronic Health Records 153
   Dean F. Sittig, Joan S. Ash, and Hardeep Singh
   A Red-Flag Based Approach to Risk Management of EHR-Related
   Safety Concerns.. 162
   Dean F. Sittig and Hardeep Singh
   High Priority Practices for EHR Safety..................................... 174
   SAFER Guides

7. **Mitigating EHR Downtimes** 187

   Contingency Planning for Electronic Health Record-based
   Care Continuity: A Survey of Recommended Practices......................... 187
   Dean F. Sittig, Daniel Gonzalez, and Hardeep Singh
   Downtime 202
   SAFER Guides

8. **Safely Configuring and Maintaining EHRs and System-to-System
   Interfaces** 209

   Field Study of the System Interfaces SAFER Guide............................. 209
   Rodney E. Howell, Hardeep Singh, and Dean F. Sittig
   System-System Interfaces 224
   SAFER Guides
   Hardware/Software Configuration... 233
   SAFER Guides

9. **Assessment of Patient Identification Related Practices** 245

   Matching Identifiers in Electronic Health Records: Implications for
   Duplicate Records and Patient Safety 245
   *Allison B. McCoy, Adam Wright, Michael G. Kahn,
   Jason S. Shapiro, Elmer V. Bernstam, and Dean F. Sittig*

Contents ix

    SAFER Self-Assessment: Patient Identification 256
    SAFER Guides

10. Assessment of Computer-based Provider Order Entry with Clinical Decision Support 267

    Development and Field Testing of a Self-Assessment Guide for Computer-Based Provider Order Entry 267
    Carl V. Vartian, Hardeep Singh, Elise Russo, and Dean F. Sittig
    Computerized Provider Order Entry with Clinical Decision Support 285
    SAFER Guides

11. Assessment of Diagnostic Test Result Reporting and Follow-Up 301

    Improving Follow-Up of Abnormal Cancer Screens Using Electronic Health Records: Trust but Verify Test Result Communication 301
    Hardeep Singh, Lindsey Wilson, Laura A Petersen, Mona K. Sawhney, Brian Reis, Donna Espadas and Dean F. Sittig
    Improving Test Result Follow-up through Electronic Health Records Requires More than Just an Alert 319
    Dean F. Sittig and Hardeep Singh
    Ten Strategies to Improve Management of Abnormal Test Result Alerts in the Electronic Health Record.. 326
    Hardeep Singh, Lindsey Wilson, Brian Reis, Mona K. Sawhney, Donna Espadas, and Dean F. Sittig
    SAFER Self-Assessment Guide: Test Result Reporting and Follow-up ....
    SAFER Guides

12. Assessment of Clinician-to-Clinician E-Communication 341

    Improving the Effectiveness of Electronic Health Record-Based Referral Processes.. 341
    Adol Esquivel, Dean F. Sittig, Daniel R. Murphy, and Hardeep Singh
    SAFER Self-Assessment Guide: Clinician Communication 358
    SAFER Guides

13. Assessment of Handheld Computing Devices 367

    Sociotechnical Evaluation of the Safety and Effectiveness of Point-of-Care Mobile Computing Devices: A Case Study Conducted in India.. 367
    Dean F. Sittig, Kanav Kahol, and Hardeep Singh

14. **Increasing Resilience in an EHR-Enabled Healthcare Organization** 383

   Resilient Practices in Maintaining Safety of Health Information Technologies.. 383
   Michael W. Smith, Joan S. Ash, Dean F. Sittig, and Hardeep Singh
   SAFER Self-Assessment Guide: Organizational Activities and Responsibilities for Electronic Health Record (EHR) Safety 416
   SAFER Guides

15. **Creating an Oversight Infrastructure for EHR Safety** 429

   Creating an Oversight Infrastructure for Electronic Health Record-Related Patient Safety Hazards.. 429
   Hardeep Singh, David C. Classen, and Dean F. Sittig
   Patient Safety Goals for the Proposed Federal Health Information Technology Safety Center 443
   Dean F. Sittig, David C. Classen, and Hardeep Singh

Author Notes.. 457

Index.. 467

ACKNOWLEDGMENT AND HOW TO CITE

Developing the SAFER (Safety Assurance Factors for EHR Resilience) Guides was an enormous team effort (see list of research team members below). The concept of developing the SAFER guides grew out of the "Anticipating the Unintended Consequences of Health IT" project that was funded by the Office of the National Coordinator for Health Information Technology (ONC).

The following people played key roles in both developing the guides as well as helping analyze the data and writing the manuscripts that became this book. First, Joan Ash, PhD, led a team from the Oregon Health and Science University (OHSU) in Portland, OR. Without her and her team's support, this project never would have finished. Lois Olinger led the team from Westat that was responsible for all project management, including final formatting of the guides and development of the MarCom gold medal winning promotional video (see: https://www.youtube.com/watch?v=LxQE6MdDZwY). Kathy Kenyon from the Office of the National Coordinator for Health IT (ONC) served as both our project officer as well as a key confidant. Without her faith, support, and knowledge, this project never would have seen the light of day.

Finally, we thank the members of our expert panel who helped keep us on the right track, along with all of the individuals at the heathcare organizations that we visited. These people gave both their time and knowledge, which greatly helped us to develop the SAFER guides. Without them, this project would not have been nearly as successful.

Research Team Members:

- Charmydevine Bean, MHA
- Arwen Bunce, MA
- Dian Chase, MSN, MBA
- Deborah Cohen, PhD
- Donna Espadas, BS
- Eric Gebhardt, MBI

- Carmit McMullen, PhD
- Vishnu Mohan, MD, MBI, MBCS, FACP
- Daniel Murphy, MD, MBA
- Elise Russo, MPH
- Michael Smith, MS, PhD
- Ana Stanescu, BA
- Colleen Tercek
- Joseph Wasserman, BA
- Adam Wright, PhD

LIST OF CONTRIBUTORS

Joan S. Ash
Department of Medical Informatics and Clinical Epidemiology, School of Medicine, Oregon Health & Science University, Portland, OR

Nick Barber
Department of Practice and Policy, The UCL School of Pharmacy, London, UK

Elisabeth Belmont
Ms. Belmont serves as Corporate Counsel for MaineHealth.

Elmer V. Bernstam
School of Biomedical Informatics, The University of Texas Health Science Center at Houston (UTHealth), Houston, TX and Department of Internal Medicine, Medical School, The University of Texas Health Science Center at Houston (UTHealth), Houston, TX

David C. Classen
University of Utah and Pascal Metrics, Inc, Salt Lake City, UT

Donna Espadas
Health Services Research & Development Center of Excellence, Michael E. DeBakey VA Medical Center, Houston, USA

Adol Esquivel
St. Luke's Episcopal Health System (SLEHS), Texas Medical Center, Houston, TX, USA

Daniel Gonzalez
Department of Clinical Effectiveness and Performance Measurement; St. Luke's Episcopal Health System, Houston, TX

Rodney E. Howell
System Integration - Information Technology Houston Methodist Hospital, Houston Texas, USA

Stephen L. Jones
University of Texas School of Biomedical Informatics, Houston, TX and The Methodist Hospital, Houston, TX

Michael G. Kahn
Department of Pediatrics, University of Colorado Denver, Aurora, CO

Kanav Kahol
Public Health Foundation of India, New Delhi, India

Christopher A. Longhurst
Division of Systems Medicine, Department of Pediatrics, Stanford University School of Medicine, Stanford, CA, USA

Derek W. Meeks
Baylor College of Medicine, Department of Family and Community Medicine, VA HSR&D Center of Excellence, Michael E. DeBakey Veterans Affairs Medical Center, Houston, Texas, USA

Allison B. McCoy
School of Biomedical Informatics, The University of Texas Health Science Center at Houston (UTHealth), Houston, TX

Shailaja Menon
AHRQ Fellow in Patient Safety and Quality at the Houston VA Center for Innovations in Quality, Effectiveness and Safety

Ashley N. D. Meyer
Dr. Meyer is a cognitive research psychologist at the Houston VA Center for Innovations in Quality, Effectiveness and Safety.

Daniel R. Murphy
Baylor College of Medicine, Department of Internal Medicine, VA HSR&D Center of Excellence, Michael E. Debakey Veterans Affairs Medical Center, Houston, Texas, USA

Risa B. Myers
University of Texas School of Biomedical Informatics, Houston, TX and University of Texas M.D. Anderson Cancer Center, Houston, TX

Laura A. Petersen
Center Director, Center for Innovations in Quality, Effectiveness and Safety, Associate Chief of Staff for Research (ACOS/R), Michael E. DeBakey VA Medical Center and Department of Medicine, Chief, Section of Health Services Research, Baylor College of Medicine

Brian Reis
The Center of Inquiry to Improve Outpatient Safety Through Effective Electronic Communication and the Houston VA HSR&D Center of Excellence at the Michael E. DeBakey Veterans Affairs Medical Center

Elise Russo
Houston VA HSR&D Center of Innovation, Michael E. DeBakey VA Medical Center, and the Section of Health Services Research, Department of Medicine, Baylor College of Medicine, Houston, Texas

List of Contributors

Mona K. Sawhney
The Center of Inquiry to Improve Outpatient Safety Through Effective Electronic Communication and the Houston VA HSR&D Center of Excellence at the Michael E. DeBakey Veterans Affairs Medical Center

Jeanie Scott
Informatics Patient Safety, Office of Informatics and Analytics, Veterans Health Administration, Ann Arbor, MI and Albany, NY, USA

Jason S. Shapiro
Department of Emergency Medicine, Mount Sinai School of Medicine, New York, NY

Hardeep Singh
Houston VA Health Services Research and Development Center of Excellence and The Center of Inquiry to Improve Outpatient Safety Through Effective Electronic Communication, Michael E. DeBakey Veterans Affairs Medical Center and Section of Health Services Research, Department of Medicine, Baylor College of Medicine, Houston, TX

Dean F. Sittig
University of Texas – Memorial Hermann Center for Healthcare Quality & Safety, School of Biomedical Informatics, University of Texas Health Sciences Center, Houston, TX

Michael W. Smith
Houston VA Center for Innovations in Quality, Effectiveness and Safety, Michael E. DeBakey Veterans Affairs Medical Center, Houston, Texas, USA and Section of Health Services Research, Department of Medicine, Baylor College of Medicine, Houston, Texas, USA

Amirhossein Takian
Division of Health Studies, School of Health Sciences and Social Care, Brunel University London, Uxbridge, UK

Lesley Taylor
Informatics Patient Safety, Office of Informatics and Analytics, Veterans Health Administration, Ann Arbor, MI and Albany, NY, USA

Carl V. Vartian
School of Biomedical Informatics, University of Texas Health Science Center at Houston

Lindsey Wilson
The Center of Inquiry to Improve Outpatient Safety Through Effective Electronic Communication and the Houston VA HSR&D Center of Excellence at the Michael E. DeBakey Veterans Affairs Medical Center

Adam Wright
Brigham and Women's Hospital, Harvard Medical School, Boston, MA

INTRODUCTION

Electronic health records (EHRs) have the potential to improve the quality and safety of health care [1]. Since the enactment of the Health Information Technology for Economic and Clinical Health Act (HITECH) [2], organizations are adopting EHRs at an unprecedented rate [3]. While the challenges of rapid EHR implementation can be numerous and disruptive, most clinicians prefer EHRs over paper records [4] in the hopes of improving care with better access to information at the point-of-care [5], advanced clinical decision support [6], and more reliable mechanisms for provider-to-provider communication [7]. Clinicians' willingness to adopt EHRs is reassuring, especially in these early stages of an EHR-enabled health system where benefits thus far have been difficult to achieve on a broad scale. However, implementation of EHRs and other new technologies carries unintended consequences that need to be addressed [8]. Clinicians have also experienced safety concerns from EHR design and usability features that are not optimally adapted for the complex workflow of real-world practice settings [9,10,11]. To respond to these challenges, the Office of the National Coordinator for Health Information Technology (ONC) commissioned the 2012 Institute of Medicine Report *Health IT and Patient Safety: Building Safer Systems for Better Care* [12] and recently released the *Health Information Technology Patient Safety Action and Surveillance Plan* that lays out their proposed response to these issues [13].

National initiatives needed to improve the safety of EHRs must be accompanied by practical and helpful strategies for clinicians on the frontlines of EHR-enabled care delivery. Although organizations are accustomed to developing and using practice standards, clinical guidelines, and evidence-based medicine to provide the best possible care for their patients, they are often unaware of best practices for safe EHR implementation and use. For example, they often have minimal guidance to handle problems such as too many alerts [14,15], an EHR that is too slow, or an

EHR that requires an excessive number of "clicks" to complete simple tasks. These are not skills routinely expected of healthcare providers in the past [16]. Clinicians are also not privy to other safety concerns embedded in flawed interfaces between the various components of the EHR and in the way the EHR system is configured. Solutions to these problems are often multifaceted, involving analysis and redesign of workflow and organizational processes and procedures that cannot be addressed through improvements in technology alone. Addressing EHR-related safety concerns is thus inherently complex and involves a comprehensive and multifaceted systems-based approach. Organizations must be active in finding and demanding solutions, but they need practical and useful guidance for EHR safety.

With support from the ONC, we used a rigorous, iterative process to develop a set of nine self-assessment guides to optimize the safety and safe use of EHRs (see Table 1) [17]. These guides, referred to as the Safety Assurance Factors for EHR Resilience (SAFER) guides, are designed to help organizations self-assess the safety and effectiveness of their EHR implementations, identify specific areas of vulnerability, and change their cultures and practices to mitigate risks.

The goal of this book is to provide EHR designers, developers, implementers, users, and policy makers with the requisite historical context, clinical informatics knowledge, and real-world, practical guidance to enable them to utilize the SAFER Guides to proactively assess the safety and effectiveness of their EHR implementations. The first five chapters are designed to provide readers with the conceptual knowledge required to understand why and how the guides were developed. The next nine chapters consist of 1–3 articles that focus on the underlying informatics concepts, key research activities, or methods used to develop each of the guides. Each of these chapters concludes with a copy of the guide itself. The final chapter provides a vision for the future of how we can create the required socio-technical infrastructure necessary to oversee the work required to ensure that future generations of EHRs are designed, developed, implemented, and used to improve the overall safety of the EHR-enabled healthcare system.

Introduction   xix

TABLE 1: Electronic Health Record-Related Structures and/or Processes addressed by SAFER guides.

Name of Guide	Description of each guide
High Priority Practices	The subset of processes determined to be "high risk" and "high priority" meant to broadly cover all areas that have a role in EHR safety
Computerized provider order entry (CPOE) with clinical decision support	Processes pertaining to electronic ordering of medications and diagnostic tests and aiding the clinical decision making process at the point of care
Test result reporting and follow-up	Processes involved in delivering test results to the appropriate providers
Communication between providers	Communication processes in three high-risk areas: consultations or referrals, discharge-related communications, and patient-related messaging between clinicians
Patient identification	Processes related to creation of new patients in the EHR, patient registration, retrieval of information on previously registered patients, and other patient identification processes
Contingency planning for EHR-based care continuity	Processes and preparations that should be in place in the event that the EHR experiences a hardware, software, or power failure
EHR customization and configuration	Processes required to create and maintain the physical environment in which the EHR will operate, as well as the infrastructure related to the hardware and software that are required to run the EHR
System-system data interfaces	Processes that enable different hardware devices and software applications to be connected both physically and logically so they can communicate and share information
Organizational activities and responsibilities	The organizational activities, processes, and tasks that people must carry out to ensure safe and effective EHR implementation and continued operations

DETAILED OVERVIEW OF EACH CHAPTER

Chapter 1 describes the context of EHR Safety and the need for proactive risk assessment. It begins with an article that defines health information technology-related errors. Interestingly, while many EHR researchers and users commonly discuss EHR-related safety concerns, there is still no widespread agreement on exactly how these concerns are defined. The second article provides a high-level overview of our 8-dimension socio-technical model while discussing what must be done within each of these dimensions if we are to achieve the safe and effective EHR-enabled healthcare system that will transform healthcare from a cottage industry into the evidence-based, high-reliability, scientifically-sound, interoperable healthcare ecosystem that is required to address the healthcare needs of our modern society. The final article in this chapter describes the results of a cross-sectional survey that focused on EHR-related safety concerns. It provides the background to help explain why there is such a widespread interest in improving the safety and effectiveness of current EHRs.

Chapter 2 consists of three articles that focus on various methods for, and results of, analyzing EHR-related safety events. The first article focuses on an analysis of the EHR-related safety events that have been reported through the US Food and Drug Administration's (FDA) Manufacturer and User Facility Device Experience (MAUDE) database. The second article explores the socio-technical intersection of patient safety and EHR implementation as experienced in twelve National Health Service (NHS) hospitals in the United Kingdom. The final article analyzes 100 consecutive EHR-related safety events that were extracted from a non-punitive, voluntary reporting system maintained by the US Veterans Health Administration (VA).

Chapter 3 provides an overview of the user context required to ensure safe and effective EHR use. The first article focuses on the rights and responsibilities of physician users of EHRs. The second article focuses on the additional rights and responsibilities of the sub-set of EHR users that care for children. Taken together, these two articles describe specific EHR features, functions and user privileges that are critical for physicians if they are to provide the highest quality, safest and most cost-effective care. Each of these "rights" is also accompanied by a corresponding responsibility of physicians, without which the ultimate goal of improving the quality of health care might not be achieved.

Chapter 4 describes the conceptual foundation of the SAFER guides. It begins with an in-depth review of an eight-dimension socio-technical model that we have used extensively to study various aspects of health information technology. The second article presents a three-phase approach to ensure that EHRs are implemented and used safely and effectively that focuses on: 1) Ensuring that the EHR itself is working appropriately; 2)

Ensuring that the EHR is uses correctly and completely; and 3) Ensuring that the EHR is used to improve the safety of the healthcare delivery system.

Chapter 5 describes the research methods we used to develop the SAFER guides. It includes a description of how we solicited input from various subject matter experts and relevant stakeholders and created the first iteration of the guides that focused on nine specific risk areas. It goes on to describe how we pilot and beta tested the guides with individuals representative of likely users. It concludes with the methods our multidisciplinary team used to assess the content validity and perceived usefulness of the draft SAFER guides, including interviews, naturalistic observations, and document analysis.

Chapter 6 provides an overview of the high priority items from each of the SAFER guides. It begins with an overview of how these guides can be used to empower organizations to improve the safety and effectiveness of their EHRs. The second article describes a red-flag-based approach that we developed based on the SAFER recommendations to help organizations identify various safety issues. The final section of this chapter includes the High Priority SAFER guide. This self-assessment guide is intended to increase awareness of characteristics that can improve the safety of EHRs and support the proactive evaluation of selected risk areas. It helps organizations identify and evaluate where breakdowns may occur in their healthcare delivery system. This assessment focuses on processes determined to be "high risk" and "high priority" and is meant to broadly cover all areas that have a role in EHR safety. Thoughtful use of this assessment by EHR users is intended to stimulate implementation of the recommended practices, as well as sustain those that are already present. When assessing EHRs at repeated intervals, (such as initially, annually and when changes are made), the assessment can be used to establish a baseline for measuring the effect of interventions designed to improve EHR safety. The assessment works for small, ambulatory physician practices and larger outpatient settings as well as for hospitals.

Chapter 7 provides an overview of a survey of recommended practices that we developed to help organizations assess their contingency planning activities for EHR downtimes. Failures in Electronic Health Record (EHR) software and the hardware infrastructure that supports them, not to mention both natural and man-made disasters are inevitable. The potential consequences of EHR-related failures becomes of increasing concern as large-scale EHR systems are deployed across multiple facilities within a health care system, often across a wide geographic area . This chapter concludes with the self-assessment guide that focuses on processes and preparations that should be in place in the event that the EHR experiences a downtime or if a power outage occurs. It helps organizations proactively identify and evaluate if their practice or organization is prepared to deliver safe health care when the EHR is not available and can help manage downtime procedures adequately. Thoughtful use of this assessment by EHR users is intended to stimulate implementation of the recommended practices, as well as sustain those that are already present. The assessment guide works for ambulatory physician practices and other outpatient settings as well as for hospitals, although the patient safety risks in these settings might vary.

Chapter 8 discusses how an organization can learn how to safely configure and maintain the system-to-system interfaces required to implement a state-of-the-art EHR. It begins with a report of a field study that we conducted to assess the utility of the system-to-system interface SAFER guide. It concludes with both the "System interfaces and System configuration guides." Briefly, the System Interfaces SAFER Guide identifies recommended safety practices intended to optimize the safety and safe use of system-to-system interfaces between EHR-related software applications. Many healthcare organizations are involved in planning, implementing, or maintaining enterprise- or community-wide clinical information systems that require integration [18]. Such integration occurs most often via interfaces between software applications, often from different system developers. These interfaces send and receive information, enabling disparate systems to operate on the same data. System interface projects are complex because they involve many stakeholders (e.g., clinicians, administrators, and information technologists) in various departments, often with differing agendas. Stakeholders must work with hardware devices and software applications that are developed independently, while integrating them flawlessly with complex clinical work processes. Well-designed and well-developed system interfaces enable reliable physical and logical connection of different systems. System interfaces require physical equipment (e.g., hardware such as plugs, cables, and cards), software that controls the data and information that is exchanged, and concepts (e.g., data protocols and controlled vocabularies) that control the interactions between systems. In addition to these technical issues, interfaces involve social and organizational factors, such as agreements to provide data in a consistent format and to use data to refer to concepts in a consistent manner (i.e., multiple systems must manage and coordinate any change to the meaning of a data item). Processes and preparations must be in place to ensure appropriate configuration and maintenance of interfaces [19]. For example, a mapping error between the order entry system and the pharmacy can cause dispensing of the wrong drug [20]. Similarly, researchers have identified errors in the transmission of free-text comment fields between the order entry application and the pharmacy system [21].

The second SAFER guide in this chapter describes how configuration of an EHR includes creating and maintaining the physical environment in which the EHR will operate as well as the infrastructure related to the various aspects of the hardware and software which are required by the EHR. Configuring EHRs and their hardware and software components into their associated environment is complex and vulnerable to errors. It is a continuous process that must be sustained and reliable over time. In the EHR-enabled healthcare environment, we rely upon technology to support and manage many complex clinical and administrative processes. EHR safety and effectiveness can be improved by establishing proper configuration policies and practices and then embedding these concepts within the EHR to ensure that they are carried out. This self-assessment guide is intended to increase awareness of characteristics that can improve the safety of EHR configuration and support the proactive evaluation of selected risk areas. It helps you identify and evaluate where configuration issues may occur in your healthcare delivery system. Thoughtful use of this assessment guide by EHR users is intended to stimulate implementation of the recommended practices, as well as sustain those that are already present. These guides work for both large and small ambulatory physician practices as well as for large and small hospitals.

Chapter 9 helps organizations learn how to assess their patient identification-related practices. It begins with a report of an investigation into matching patient identifiers that lead to significant numbers of duplicate patient records. Processes related to patient identification are complex and vulnerable to breakdown. In the EHR-enabled healthcare environment, we rely upon technology to help support and manage these complex identification processes and thus EHRs should optimize how information related to patient identification is displayed and communicated. Technology configurations alone cannot ensure accurate patient identification. Staff must also be supported with adequate training and procedures. This self-assessment is intended to increase awareness of EHR system characteristics related to design, configuration, and implementation decisions related to patient identification. This assessment can help identify and evaluate where breakdowns related to patient identification may occur in your healthcare delivery system. It focuses on the processes related to creation of new patients in the EHR, patient registration, and retrieval of information on previously registered patients and other types of patient identification processes in the EHR with the goal being to mitigate problems that arise from duplicative records and patient mix-ups. Thoughtful use of this assessment by EHR users is intended to stimulate implementation of the recommended practices, as well as sustain those that are already present. The assessment works for ambulatory physician practices and other outpatient settings as well as for hospitals.

Chapter 10 includes a report of the development and field assessment of the SAFER guide that addresses computer-based provider order entry (CPOE) with clinical decision support. CPOE practices are complex and vulnerable to breakdown. In the EHR-enabled healthcare environment, we rely upon technology to support and manage these complex workflow processes. EHRs that incorporate standardized and automated features can improve the safety and effectiveness of how order entry information is communicated. This self-assessment guide is intended to increase awareness of characteristics that can improve the safety of EHRs and support the proactive evaluation of select risk areas. It helps organizations identify and evaluate where CPOE breakdowns may occur in your healthcare delivery system. It focuses on processes pertaining to electronic medication and laboratory test ordering. Thoughtful use of this assessment guide by EHR users is intended to stimulate implementation of the recommended practices, as well as sustain those that are already present. In addition, this guide discusses processes related to Clinical Decision Support (CDS), which are also complex and vulnerable to breakdown. It helps you identify and evaluate where CDS breakdowns may occur in your healthcare delivery system. The guide works for ambulatory physician practices and other outpatient settings as well as for hospitals.

Chapter 11 includes three articles that discuss various aspects of improving the follow-up of abnormal diagnostic test reporting. The first is a case report of the investigation of a software configuration error, among several other errors, that lead to failure to follow-up on numerous abnormal cancer screening tests. The second is an editorial regarding a systematic review of studies that focused on abnormal test result follow-up that concludes that alerts alone are not sufficient to solve this difficult problem. The third article describes ten strategies that can be used to improve the management of the abnormal test result alerts within the EHR-enabled work system. The chapter concludes with a self-assessment guide that is intended to increase awareness of characteristics that can improve the safety of EHRs and support the proactive evaluation of select risk areas. It helps organizations identify and evaluate where test result reporting and follow-up breakdowns may occur in their healthcare delivery system. It focuses on processes after tests have been performed, when providers are notified electronically of the results and are then responsible for reviewing the results and follow-up with patients, as appropriate. EHRs that incorporate standardized and automated features can improve the safety and effectiveness of how information from diagnostic reports is communicated. Thoughtful use of this assessment guide by EHR users is intended to stimulate implementation of the recommended practices, as well as sustain those that are already present. The guide works for ambulatory physician practices and other outpatient settings as well as for hospitals.

Chapter 12 focuses on the assessment of clinician-to-clinician electronic communication. Communication is a key aspect of nearly all patient care processes and has enormous potential to impact patient safety [22–26]. Communication breakdowns between clinicians are one of the most common causes of medical errors and patient harm. Communication processes have become increasingly integrated into EHRs [27]. These include sending and receiving referral and consult communication, communication about transitioning a patient from the inpatient to the outpatient setting, and communicating clinical messages with the EHR. Several attributes of EHR-based communication can result in a disconnect between the sender and the receiver of clinical information, including the sender's uncertainty about whether or when a message has been received, and a mismatch between single patient vs. multiple patient interactions. Messages may be incomplete, misdirected, or directed to an unavailable clinician, and may overload the recipient. The guide works for ambulatory physician practices and other outpatient settings as well as for hospitals.

Chapter 13 includes an evaluation of a novel tablet-based, handheld computing device designed to support primary care practice in rural India. Based on our 8-dimension conceptual model, we developed an assessment guide for the tablet system that was informed by literature review, interviews, and observations of health workers and supervisors. This article includes a SAFER-like guide that can be used to proactively assess similar handheld computing devices.

Chapter 14 discusses how organizations can increase their resilience, or their ability to recover from difficulties. Given the rapid adoption of EHRs by many organizations that are still early in their experiences with EHR safety, it is important to understand practices for maintaining resilience used by organizations with a track record of success in EHR use.

The chapter concludes with a SAFER guide that articulates general principles and practices relevant to any organization that provides health care, whether inpatient or ambulatory, large or small. The universal, minimum goal is to assure that the EHR does not negatively impact care. Thoughtful use of this assessment guide by EHR stakeholders is intended to introduce or sustain safety practices that are already present. The focus here is on activities, processes, and tasks, rather than on individual roles or titles. Some of the recommended activities are best conducted by teams or committees rather than one individual.

Chapter 15 offers our vision of, and the need for, an oversight infrastructure for EHR-related patient safety hazards. Specifically, we propose the creation of a national EHR oversight program to provide dedicated surveillance of EHR-related safety hazards and to promote learning from identified errors, close calls, and adverse events. The program calls for data gathering, investigation/analysis, and regulatory components. The final article describes our vision for the recently proposed federal health information technology safety center.

Taken together, the information provided in this book should help any organization, whether large or small, well on their way or just beginning their EHR journal to improve the safety and effectiveness of their EHR-enabled healthcare system.

REFERENCES

1. Blumenthal D, Glaser JP. Information technology comes to medicine. N Engl J Med. 2007 Jun 14;356(24):2527-34.
2. Blumenthal D. Launching HITECH. N Engl J Med. 2010 Feb 4;362(5):382-5. doi: 10.1056/NEJMp0912825.
3. Wright A, Henkin S, Feblowitz J, McCoy AB, Bates DW, Sittig DF Early results of the meaningful use program for electronic health records. N Engl J Med. 2013 Feb 21;368(8):779-80. doi: 10.1056/NEJMc1213481.
4. Propp DA. Successful introduction of an emergency department electronic health record. West J Emerg Med. 2012 Sep;13(4):358-61. doi: 10.5811/westjem.2012.1.11564.
5. Powsner SM, Wyatt JC, Wright P. Opportunities for and challenges of computerisation. Lancet. 1998 Nov 14;352(9140):1617-22.
6. Bates DW, Leape LL, Cullen DJ, Laird N, Petersen LA, Teich JM, Burdick E, Hickey M, Kleefield S, Shea B, Vander Vliet M, Seger DL. Effect of computerized physician order entry and a team intervention on prevention of serious medication errors. JAMA. 1998 Oct 21;280(15):1311-6.
7. Sittig DF, Singh H. Improving test result follow-up through electronic health records requires more than just an alert. J Gen Intern Med. 2012 Oct;27(10):1235-7.
8. Campbell EM, Sittig DF, Ash JS, Guappone KP, Dykstra RH. Types of unintended consequences related to computerized provider order entry. J Am Med Inform Assoc. 2006 Sep-Oct;13(5):547-56.
9. Myers RB, Jones SL, Sittig DF. Review of Reported Clinical Information System Adverse Events in US Food and Drug Administration Databases. Appl Clin Inform. 2011;2(1):63-74.
10. ECRI Institute PSO Deep Dive: Health Information Technology," ECRI Institute PSO, December 2012. Available at: https://www.ecri.org/EmailResources/PSRQ/ECRI_Institute_PSO_Deep%20Dive_HIT_TOC.pdf
11. Farley HL, Baumlin KM, Hamedani AG, et al. Quality and Safety Implications of Emergency Department Information Systems. Annals of Emergency Medicine (in press) June 21, 2013. Available at: http://www.annemergmed.com/article/S0196-0644(13)00506-4/abstract
12. Institute of Medicine. Health IT and Patient Safety: Building Safer Systems for Better Care. The National Academies Press, Washington DC. (2012).
13. Office of the National Coordinator for Health Information Technology. Health Information Technology Patient Safety Action & Surveillance Plan, July 1, 2013. Available at: http://www.healthit.gov/sites/default/files/safety_plan_master.pdf

14. Weingart SN, Seger AC, Feola N, Heffernan J, Schiff G, Isaac T. Electronic drug interaction alerts in ambulatory care: the value and acceptance of high-value alerts in US medical practices as assessed by an expert clinical panel. Drug Saf. 2011 Jul 1;34(7):587-93. doi: 10.2165/11589360-000000000-00000.
15. Carspecken CW, Sharek PJ, Longhurst C, Pageler NM. A clinical case of electronic health record drug alert fatigue: consequences for patient outcome. Pediatrics. 2013 Jun;131(6):e1970-3. doi: 10.1542/peds.2012-3252.
16. Shortliffe EH. Biomedical informatics in the education of physicians. JAMA. 2010 Sep 15;304(11):1227-8. doi: 10.1001/jama.2010.1262.
17. Singh H, Ash JS, Sittig DF. Safety Assurance Factors for Electronic Health Record Resilience (SAFER): study protocol. BMC Med Inform Decis Mak. 2013 Apr 12;13:46. doi: 10.1186/1472-6947-13-46.
18. Cliff R. A Systems Implementation Project Planning Guide. 2007. Available at: http://www.cliffconsulting.net/CCI%20Publications/System%20Guidelines%20w%20Matrix%20v4.pdf
19. Magrabi F, Ong MS, Runciman W, Coiera E. Using FDA reports to inform a classification for health information technology safety problems. J Am Med Inform Assoc. 2012 Jan-Feb; 19(1): 45-53. doi: 10.1136/amiajnl-2011-000369.
20. Sittig DF, Singh H. Defining health information technology-related errors: new developments since to err is human. Arch Intern Med. 2011 Jul 25; 171(14): 1281-1284. doi: 10.1001/archinternmed.2011.327.
21. Singh H, Mani S, Espadas D, Petersen N, Franklin V, Petersen LA. Prescription errors and outcomes related to inconsistent information transmitted through computerized order entry: a prospective study. Arch Intern Med. 2009 May 25; 169(10): 982-989. doi: 10.1001/archinternmed.2009.102.
22. Esquivel A, Sittig DF, Murphy DR, Singh H. Improving the effectiveness of electronic health record-based referral processes. BMC Med Inform Decis Mak. 2012;12:107.
23. Gandhi TK, Sittig DF, Franklin M, Sussman AJ, Fairchild DG, Bates DW. Communication breakdown in the outpatient referral process. J Gen Intern Med. 2000;15:626-631.
24. Saxena K, Lung BR, Becker JR. Improving patient safety by modifying provider ordering behavior using alerts (CDSS) in CPOE system. AMIA Annu Symp Proc. 2011;1207-1216.
25. McDonald CJ. Protocol-based computer reminders, the quality of care and the non-perfectability of man. N Engl J Med. 1976;295:1351-1355.
26. Murphy DR, Reis B, Kadiyala H, et al. Electronic health record-based messages to primary care providers: valuable information or just noise? Arch Intern Med. 2012;172:283-285.
27. Saleem JJ, Russ AL, Neddo A, Blades PT, Doebbeling BN, Foresman BH. Paper persistence, workarounds, and communication breakdowns in computerized consultation management. Int J Med Inform. 2011;80:466-479.

CHAPTER 1

THE CONTEXT OF EHR SAFETY AND THE NEED FOR RISK ASSESSMENT

DEFINING HEALTH INFORMATION TECHNOLOGY-RELATED ERRORS: NEW DEVELOPMENTS SINCE *TO ERR IS HUMAN*

Dean F. Sittig and Hardeep Singh

1.1.1 INTRODUCTION

Two Institute of Medicine (IOM) reports have recommended the use of information technologies to improve patient safety and reduce errors in health care [1,2]. Broadly speaking, health information technology (HIT) is the overarching term applied to various information and communication technologies to collect, transmit, display, or store patient data.

Despite HIT's promise in improving safety, recent literature has revealed potential safety hazards associated with its use, often referred to

---

Sittig DF and Singh H. Defining Health Information Technology-related Errors: New Developments Since To Err is Human. *Archives of Internal Medicine* **171**,14 (2011). *Copyright © 2011 American Medical Association. All rights reserved.*

as e-iatrogenesis [3,4]. For example, Koppel et al. described 22 types of errors facilitated by a commercially-available EHR system's computerized provider order entry (CPOE) application [5]. In response to similar emerging concerns, the Office of the National Coordinator for HIT recently sponsored an IOM committee to "review the available evidence and the experience from the field" on how HIT use affects patient safety. Given the national impact of HIT, this initiative is a major step forward in ensuring safety and well-being of our patients. However, the field currently lacks acceptable definitions of HIT-related errors and it's unclear how best to measure or analyze "HIT errors".

The goal of this manuscript is to advance the understanding of HIT-related errors and explain how adverse events, near misses, and patient harm can result from problems with HIT itself or from interactions between HIT, its users, and the work system. HIT errors almost always jeopardize patient outcomes and have high potential for harm [6]. This is because many of these errors are latent errors that occur at the "blunt end" of the healthcare system [7], with potential to affect large numbers of patients if not corrected. Furthermore, if important structural or process-related HIT problems are not addressed proactively, care of millions of patients may be affected due to impending widespread adoption and implementation of EHRs [8]. We thus focus this manuscript heavily on errors related to the use of EHR systems.

1.1.2 GENERAL CRITERIA FOR A HIT ERROR

We define the "HIT work system" as the combination of all the hardware and software required to implement the HIT, as well as the social environment in which it is implemented. We thus propose that HIT errors should be defined from the socio-technical viewpoint of end users (including patients, when applicable), rather than the purely technical view of manufacturers, developers, vendors, and personnel responsible for implementation. HIT-related error occurs anytime the HIT system is unavailable for use, malfunctions during use, is used by someone incorrectly, or when HIT interacts with another system component incorrectly, resulting in data being lost or incorrectly entered, displayed, or transmitted [9,10]. HIT errors may involve failures of either structures or processes and can occur in the design and development, implementation and use, or evaluation and optimization phases of the IT lifecycle [11]. This approach is consistent with the currently recommended systems and human factors approaches used to understand and reduce error [1].

The HIT system is considered to be *unavailable* for use if for any reason the user cannot enter, review, transmit or print data (e.g., patient's medication allergies or most recent laboratory test results). Reasons could include unavailable *computer hardware* (e.g., missing keyboard, or problems with the computer's monitor, the network routers that connect the computer to the data servers and printers, or the server where data is stored), *software* (e.g., problems with the operating system that manages either the computer applications such as the internet browser and EHR, or the interface between an EHR and the information system of an ancillary service such as radiology or lab), and *power sources* (e.g., a power outage that results in hospital-wide computer failure) [4].

The HIT system is considered to be *malfunctioning* (i.e., available, but not working correctly) whenever a user cannot accomplish the desired task despite using the HIT system as designed. In this situation, error results from any hardware or software defect (or "bug") that prohibits a user from entering or reviewing data, or any defect that causes the data to be entered, displayed, transmitted, or stored incorrectly. For example, the clinician enters the patient's weight in pounds and the weight-based dosing algorithm fails to convert it to kilograms before calculating the appropriate dose, resulting in a 2-fold overdose.

Finally, errors can occur even when hardware and software are functioning as designed. For instance, errors may result when users do not use the hardware or software as intended. For example, users might enter free-text comments (e.g., take 7.5mg Mon-Fri only) that contradict information contained in the structured section of the medication order (e.g., Warfarin tabs 10mg QD) [12]. Errors may also arise when two or more parts of the HIT system (e.g., CPOE application and the pharmacy's medication dispensing system) interact in an unpredicted manner, resulting in inaccurate, incomplete, or loss of data during entry, display, transmission or storage [13].

TABLE 1.1.1: Examples of the types of errors that can occur within each dimension of the socio-technical model[16] and corresponding suggested mitigating procedures.

Socio-technical model dimension	Examples of types of errors that could occur in each dimension	Examples of potential ways to reduce likelihood of these errors....
Hardware and Software - required to run the healthcare applications	Computer or network is not functioning	Provide redundant hardware for all essential patient care activities
	Input data truncated (ie, buffer overflow) – some entered data lost	Warn users when data entered exceeds amount that can be stored
Clinical Content – data, information, and knowledge entered, displayed, or transmitted	Allowable item can't be ordered(eg., no "amoxicillin" in the antibiotic pick-list) [5]	Conduct extensive pre-release testing on all system-system data and human-computer interfaces to insure that new features are working as planned and that existing features are working as before
	Incorrect default dose for given medication [5]	
Human Computer Interface - aspects of the system that users can see, touch, or hear	Data entry/review screen does not show the patient name, medical record number, birthdate, etc.	Encourage and provide methods for clinicians to report when patient-specific screens do not contain key patient demographics so that the software can be fixed
Human Computer Interface (cont.)	Wrong decision about KCl administration based on poor data presentation on the computer screen	Improve data displays and train users to routinely review and cross-validate all data values for appropriateness before making critical decisions.
	Two buttons with same label, but different functionality	Pre-release inspection of all screens for duplicate button names
People - the humans involved in the design, development, implementation, and use of HIT	Two patients with same name; data entered on wrong patient	Alert providers to potential duplicate patients and require re-confirmation of patient ID before saving data (eg, display patient photo before signing)
	Incorrect merge of two patient's data	Develop tools to compare key demographic data and calculate a probability estimate of similarity

TABLE 1.1.1: *Cont.*

Socio-technical model dimension	Examples of types of errors that could occur in each dimension	Examples of potential ways to reduce likelihood of these errors…
Workflow and Communication - the steps needed to ensure that each patient receives the care they need at the time they need it	RNs scan duplicate patient barcode taped to their clipboard rather than barcode on patient to save time	Improve user training, user interfaces, work processes, and organizational policies to reduce need for work-arounds
	Computer discontinues a medication order without notifying a human	Implement fail-safe communication (eg, re-send message to another hospital designee if no response from MD or RN) for all computer-generated actions,
	Critical abnormal test result alerts not followed up	Implement robust quality assurance systems to monitor critical alert follow-up rates ; use "dual notification" for alerts judiciously
Organizational Policies and Procedures - internal culture, structures, policies, and procedures that affect all aspects of HIT management and healthcare	Policy contradicts physical reality (eg, required Barcode med administration readers not available in all patient locations) [21]	
	Policy contradicts personnel capability (eg, 1 pharmacist to verify all orders entered via CPOE in large hospital)	
	Conduct pre- and post-implementation interviews with all affected users to better gauge workload	
	Conduct pre- and post-implementation inspections, interviews, and monitor feedback from users in all physical locations	

TABLE 1.1: *Cont.*

Socio-technical model dimension	Examples of types of errors that could occur in each dimension	Examples of potential ways to reduce likelihood of these errors…
	Incorrect policy allows "hard-stops" on clinical alerts causing delays in needed therapy	Disallow "hard-stops" on alerts; users should be able to override the computer in all but the most egregious cases (eg, ordering promethazine as IV push by peripheral vein)
	Billing requirements lead to inaccurate documentation in EHR (eg, inappropriate copy & paste)	Highlight all "pasted" material and include reference to source of material
External Rules, Regulations, and Pressures - external forces that facilitate or place constraints on the design, development, implementation, use, and evaluation of HIT in the clinical setting	Joint Commission required medication reconciliation processes causing rushed development of new medication reconciliation applications that were difficult to use and caused errors ; rescinded safety goal only to reinstate it 7/1/2011	Carefully consider potential adverse unintended consequences before making new rules or regulations; conduct interviews and observations of users to gauge effects of rules and regulations on patient safety, quality of care, and clinician work-life
System Measurement and Monitoring - of system availability, use, effectiveness, and unintended consequences of system use	Incomplete or inappropriate (eg, combining disparate data) data aggregation leads to erroneous reporting	Increase measurement and monitoring transparency by providing involved stakeholders with access to raw data, analytical methods, and reports
	Incorrect interpretation of results	

1.1.3 ORIGIN-SPECIFIC TYPOLOGY FOR AN HIT ERROR

Leveson [14] proposes that new technologies have fundamentally altered the nature of errors and asserts that these changes necessitate new models and methods for investigating technology-related errors. Thus, technological advances could potentially give rise to increasingly complex and multifaceted errors in healthcare. In view of the resultant expanding and evolving context of safe HIT implementation and use, we illustrate how a new socio-technical model for HIT evaluation and use can provide an origin-specific typology for HIT errors [15]. The model's 8 dimensions (Table 1.1.1) comprehensively account for the technology, its users and their respective workflow processes and how these two elements interface with the technology, the work system context including organizational and policy factors that affect HIT, and notably, the interactions between all of these factors [16]. Table 1.1.1 provides examples of specific EHR-related errors that can occur within each of the 8 dimensions of the socio-technical model, along with examples of potential ways that the likelihood of each error could be reduced. Thus, the model not only illustrates the complex relationships between active and latent errors but also lays a foundation for error analysis.

1.1.4 CONCLUSION

Rapid advances in HIT development, implementation, and regulation have complicated the landscape of HIT-related safety issues. Erroneous or missing data, or the decisions based on them, increase the risk of an adverse event and unnecessary costs. Because these errors can and frequently do occur post-implementation, simply increasing oversight of HIT vendors' development processes will not address all HIT–related errors. Comprehensive efforts to reduce HIT errors must start with clear definitions and an origin-focused understanding of HIT errors that addresses important socio-technical aspects of HIT use and implementation. To this end, we believe this commentary provides a much needed foundation for coordinating safety initiatives of HIT designers, developers, implementers, users, and policy-makers, who must continue to work together to achieve a high-reliability HIT work system for safe patient care.

REFERENCES

1. Institute of Medicine. To err is human: Building a safer health system. [Report by the Committee on Quality of HealthCare in America] Washington, DC: National Academy Press; 1999.
2. Institute of Medicine. Patient Safety: Achieving a new standard for care. [Report by the Committee on Data Standards for Patient Safety] Washington, DC: National Academy Press; 2004.
3. Weiner JP, Kfuri T, Chan K, Fowles JB. "e-Iatrogenesis:" The most critical unintended consequence of CPOE and other HIT. J Am Med Inform Assoc. 2007 Feb 28;
4. Myers RB, Jones SL, Sittig DF. Reported Clinical Information System Adverse Events in US Food and Drug Administration Databases. Applied Clinical Informatics, 2011; 2: 63–74. doi: 10.4338/ACI-2010-11-RA-0064.
5. Koppel R, Metlay JP, Cohen A, Abaluck B, Localio AR, Kimmel SE, Strom BL.Role of computerized physician order entry systems in facilitating medication errors. JAMA. 2005 Mar 9;293(10):1197-203.
6. Hofer TP, Kerr EA, Hayward RA. What is an error? Eff Clin Pract. 2000 Nov-Dec;3(6):261-9.
7. Reason J. Human error: models and management. BMJ. 2000 Mar 18;320(7237):768-70.
8. Stead W, Lin H, eds. Computational technology for effective health care: immediate steps and strategic directions. Washington, DC: National Academies Press, 2009.
9. Mangalmurti SS, Murtagh L, Mello MM. Medical malpractice liability in the age of electronic health records. N Engl J Med. 2010 Nov 18;363(21):2060-7.
10. Perrow C. "Normal Accidents: Living with High-Risk Technologies", Princeton University Press. Princeton, New Jersey, 1999.
11. Walker JM, Carayon P, Leveson N, Paulus RA, Tooker J, Chin H, Bothe A Jr, Stewart WF. EHR safety: the way forward to safe and effective systems. J Am Med Inform Assoc. 2008 May-Jun;15(3):272-7.
12. Singh H, Mani S, Espadas D, Petersen N, Franklin V, Petersen LA. Prescription errors and outcomes related to inconsistent information transmitted through computerized order entry: a prospective study. Arch Intern Med. 2009 May 25;169(10):982-9.
13. Kleiner B. Sociotechnical System Design in Health Care. In: Carayon P, editor. Handbook of Human Factors and Ergonomics in Health Care and Patient Safety. Mahwah, NJ: Lawrence Erlbaum; 2007.
14. Leveson N. A New Accident Model for Engineering Safer Systems. Safety Science, April 2004; 42(4): 237-270.
15. Sittig DF, Singh H. Eight rights of safe electronic health record use. JAMA. 2009 Sep 9;302(10):1111-3.

16. Sittig DF, Singh H. A new sociotechnical model for studying health information technology in complex adaptive healthcare systems. Qual Saf Health Care. 2010 Oct;19 Suppl 3:i68-74.
17. Kilbridge P. Computer crash--lessons from a system failure. N Engl J Med. 2003 Mar 6;348(10):881-2.
18. Horsky J, Kuperman GJ, Patel VL. Comprehensive analysis of a medication dosing error related to CPOE. J Am Med Inform Assoc. 2005 Jul-Aug;12(4):377-82.
19. Shojania KG. Patient Mix-Up. AHRQ WebM&M [serial online]. February 2003. Available at: http://www.webmm.ahrq.gov/case.aspx?caseID=1
20. AHIMA MPI Task Force. "Merging Master Patient Indexes." September 1997. Available at: http://www.cstp.umkc.edu/~leeyu/Mahi/medical-data6.pdf
21. Koppel R, Wetterneck T, Telles JL, Karsh BT. Workarounds to barcode medication administration systems: their occurrences, causes, and threats to patient safety. J Am Med Inform Assoc. 2008 Jul- Aug;15(4):408-23.
22. Kuperman GJ, Teich JM, Tanasijevic MJ, Ma'Luf N, Rittenberg E, Jha A, Fiskio J, Winkelman J, Bates DW.Improving response to critical laboratory results with automation: results of a randomized controlled trial. J Am Med Inform Assoc. 1999 Nov-Dec;6(6):512-22.
23. Singh H, Wilson L, Petersen LA, Sawhney MK, Reis B, Espadas D, Sittig DF. Improving follow-up of abnormal cancer screens using electronic health records: trust but verify test result communication. BMC Med Inform Decis Mak. 2009 Dec 9;9:49.
24. Singh H, Thomas EJ, Sittig DF, Wilson L, Espadas D, Khan MM, Petersen LA. Notification of abnormal lab test results in an electronic medical record: do any safety concerns remain? Am J Med. 2010 Mar;123(3):238-44.
25. Singh H, Thomas EJ, Mani S, Sittig D, Arora H, Espadas D, Khan MM, Petersen LA. Timely follow-up of abnormal diagnostic imaging test results in an outpatient setting: are electronic medical records achieving their potential? Arch Intern Med. 2009 Sep 28;169(17):1578-86.
26. Strom BL, Schinnar R, Aberra F, Bilker W, Hennessy S, Leonard CE, Pifer E. Unintended effects of a computerized physician order entry nearly hard-stop alert to prevent a drug interaction: a randomized controlled trial. Arch Intern Med. 2010 Sep 27;170(17):1578-83.
27. Grissinger M. Preventing serious tissue injury with intravenous promethazine (phenergan). Pharmacy & Therapeutics. 2009 Apr;34(4):175-6.
28. Medication reconciliation. 2005 National Patient Safety Goal #8 by the Joint Commission.
29. Poon EG, Blumenfeld B, Hamann C, Turchin A, Graydon-Baker E, McCarthy PC, Poikonen J, Mar P, Schnipper JL, Hallisey RK, Smith S, McCormack C, Paterno M, Coley CM, Karson A, Chueh HC, Van Putten C, Millar SG, Clapp M, Bhan I, Meyer GS, Gandhi TK, Broverman CA. Design and implementation of an application and associated services to support interdisciplinary medication reconciliation efforts at an integrated healthcare delivery network. J Am Med Inform Assoc. 2006 Nov-Dec;13(6):581-92.
30. APPROVED: Will Not Score Medication Reconciliation in 2009. Joint Commission. Available at: http://www.jcrinc.com/common/PDFs/fpdfs/pubs/pdfs/JCReqs/JCP-03-09-S1.pdf

31. Revised National Patient Safety Goal on medication reconciliation is approved. Joint Commission Online - December 8, 2010.

EIGHT RIGHTS OF SAFE ELECTRONIC HEALTH RECORD USE

Dean F. Sittig and Hardeep Singh

Computers can improve the safety, quality, and efficiency of health care [1]. The pressure on hospitals and physicians to adopt electronic health records (EHRs) has never been greater [2,3]. However, concerns about the immaturity and rigidity of currently available clinical application software, the inexperience of clinicians and information technologists in implementation and use of EHRs, and potentially harmful side effects of EHRs like provider order-entry, have raised questions regarding the safe use of EHRs [4,5,6].

President Obama has often referred to EHRs as a solution to reduce medical errors. To avoid these pitfalls and achieve the promise of EHRs, we propose eight "Rights of Safe EHR Use" grounded in Carayon's systems engineering initiative for patient safety model [7].

1.2.1 RIGHT HARDWARE/SOFTWARE

An EHR must be capable of supporting required clinical activities. For instance, it should calculate the medication dose based on the patient's weight, transmit the order to the appropriate ancillary department, and notify the nurse that an order has been placed. A medication error could eas-

---

Sittig DF and Singh H. Eight Rights of Safe Electronic Health Record Use. Journal of the American Medical Association *302,10 (2009). Copyright © (2009) American Medical Association. All rights reserved.*

ily follow a breakdown in any of these functions. Furthermore, if hardware or software is inadequately sized, configured, or maintained, the EHR will function poorly. Anything that slows or disrupts the clinician's workflow has the potential to negatively affect patient safety.

Local software oversight committees are one way to ensure that hardware and software are functioning safely [8]. Another solution may be "cloud computing," reliable computing services that are accessible from remote locations via the Internet; potentially reducing hardware procurement, configuration, and maintenance burdens for healthcare organizations. Before clinicians can rely on EHRs in the "cloud", internet speed, reliability, and access must be improved.

1.2.2 RIGHT CONTENT

Right content includes standard medical vocabularies used to encode clinical findings and the clinical knowledge used to create specialty-specific features (e.g., post-transplant orders) and functions (e.g., health maintenance reminders). Content must be evidence-based, carefully constructed, monitored, complete, and error-free.

The federal government has taken a significant positive step towards advancing a controlled vocabulary with its strong support of SNOMED-CT; the most comprehensive, multilingual clinical healthcare terminology in the world. Through its membership in The International Health Terminology Standards Development Organization, SNOMED-CT is free. Adoption of a standard vocabulary is prerequisite to implementing advanced clinical decision support (CDS). In an effort to increase access to a standards-based set of validated, evidence-based CDS, an open-access clinical knowledge base of interventions should be developed that primarily focuses on helping clinicians achieve the quality and safety targets for "meaningful" EHR use. These interventions could be downloaded and utilized directly, or perhaps accessed over the internet as a service, by any EHR.

1.2.3 RIGHT HUMAN-COMPUTER USER-INTERFACE

The right user-interface allows clinicians to quickly learn and utilize a complex EHR safely and efficiently. The interface should present all the relevant patient data in a format allowing clinicians to rapidly perceive problems, formulate responses, and document their actions. A key design consideration is the trade-off between clinicians' desire to "see everything on one screen" and limited screen space. Clinicians miss crucial information in applications that overload information on one screen, leading to subsequent errors. On the other hand, systems that offer users too many nested menu options, or multiple, step-wise pathways can be difficult to learn and time consuming to use. The physical aspects of the interface (e.g., the keyboard, mouse, or touch screen) may also interfere with the data-entry process and make input or selection of information error prone.

A particularly difficult problem facing busy clinicians is the requirement to navigate different EHR interfaces safely and efficiently at different practice sites. Although a complex undertaking, the federal government along with the EHR vendors, should develop common user interface standards for healthcare applications.

1.2.4 RIGHT PEOPLE

As emphasized in Carayon's model of patient safety, trained and knowledgeable people are essential to safe EHR use. Clinicians require not only basic computing skills but also knowledge of how to integrate the EHR into their workflows, which may necessitate one-on-one training sessions; and how to function when the EHR is unavailable.

We must have adequately trained EHR software designers, developers, trainers, and implementation/maintenance staff. System developers should posses extensive software engineering skills, be able to design effective user interfaces, utilize existing standardized clinical vocabularies, and have a sound understanding of the practice of clinical medicine. EHR trainers, implementers, and maintenance staff should have clinical experience, understanding of EHR capabilities and limitations, and excellent project management skills. Close interaction among informatics experts, clinical application coordinators, and end users is essential for safe design and use.

In an attempt to create the "right people," the American Medical Informatics Association (AMIA) has created the "10x10 Training Programs" [10] and identified the knowledge and skills necessary for clinical informatics subspecialty fellowship programs [11]. Similar programs need to be bolstered nationwide.

1.2.5 RIGHT WORKFLOW / COMMUNICATION

Any disruption in workflow or information transfer is fertile ground for error. Prior to system implementation, a careful workflow analysis that accounts for EHR use could lead to identification of potential breakdown points. For example, vulnerabilities in hand-offs could be exposed in such an analysis [12], and communication tasks deemed critical could be required to have a traceable electronic receipt acknowledgement.

Errors also perpetuate if CDS interventions (i.e., alerts and reminders) are not well-focused or judiciously delivered at the point in the workflow that best supports the clinician's decision making or data entry. Delivering CDS interventions streamlined with clinicians' electronically-enabled workflow through a standard set of EHR functions (e.g., pop-up alerts, pick lists, or order sets) can lead to safer care.

1.2.6 RIGHT ORGANIZATIONAL CHARACTERISTICS

Organizational factors including a, culture of innovation, exploration, and continual improvement just as in other safety models, are key to safe EHR use. Organizations should adopt and actively encourage methods for users to report errors, or barriers to care, resulting from EHR use even if the findings are used for local or internal improvement. Organizations must also carefully review their existing policies and procedures before EHR implementation. For instance, EHRs can improve transmission of critical information through electronic notifications, but may do more harm than good if there are no standard operating procedures regarding information follow-up [13]. We believe the Veterans Affairs health system exhibits many model organizational features, including a fair amount of central control, standardized procedures for collecting error data and implementing upgrades, and a recent emphasis on studying innovations from field-users.

1.2.7 RIGHT STATE AND FEDERAL RULES AND REGULATIONS

State and federal regulations act as barriers or facilitators for achieving safe use of EHRs.

The American Recovery and Reinvestment Act (ARRA) stipulates that clinicians and healthcare organizations can receive incentive payments for "meaningful use" of EHRs. Depending on the definition and timeline for "meaningful use", this legislation could result in a rush to implement systems that have the potential to decrease patient safety.

Furthermore, ARRA includes language designed to protect patients' privacy that will require significant modifications to existing EHRs. For example, one provision requires organizations to provide a list of data disclosures to third parties for patients. Identifying and reporting such disclosures will be difficult and expensive given current technical constraints.

Regulations to safeguard patient privacy are clearly important but may also have the greatest unintended consequence on national EHR implementation. Policies must address safety and effectiveness of national health information exchange, which may call for reopening the unique national patient identifier debate. Currently used probabilistic patient matching algorithms, used to link patient information from disparate healthcare organizations, are prone to error, and many matches are never made. We recommend that state and federal governments create a regulatory environment compatible with widespread EHR use and interoperability. This will enable systems to continue evolving while maintaining appropriate safety and privacy oversight.

1.2.8 RIGHT MONITORING

The creation of the Certification Commission for Health Information Technology is a significant step towards accelerating EHR adoption, but an equally detailed post-implementation usability inspection process is also needed. Several recent reports have described serious errors related to the use or misuse of EHR systems, many of which were the result of faulty system design, configuration or implementation processes [14]. Organizations must continually evaluate the usability and performance of EHRs after implementation and reliably measure benefits, and potential iatrogenic effects of EHRs Furthermore, the federal government should mandate the development and use of a vendor-independent EHR hazard reporting database [1] and a national EHR implementation accreditation test. An EHR accreditation test would help ensure that EHRs are functioning as designed and are safe to use. The LeapFrog clinical decision support functionality test is an example for how such a test could be constructed.

SUMMARY

EHR developers have encountered many roadblocks on their journey to achieving safe and effective EHRs for all. If we are to succeed in the next 10 years we must have a coordinated multi-disciplinary research and development effort, much like the formation of NASA following President Kennedy's promise of a moon landing. This effort must bring the best scientists, engineers, and clinicians together to address the myriad problems described in this and other publications. Our efforts must move beyond the lone informatics researcher in an isolated laboratory if we are to truly understand and address the complex interaction of organizational, technical, and cognitive factors that affect the safety of EHRs. Without this understanding, any solutions are sure to be far from optimal. But without high-quality, well-designed and carefully implemented EHRs, we may never achieve highly reliable, safe health care.

REFERENCES

1. Chaudhry B, Wang J, Wu S, et al. Systematic review: impact of health information technology on quality, efficiency, and costs of medical care. Ann Intern Med. 2006 May 16;144(10):742-52.

2. Committee on Quality Health Care in America. Using information technology. Crossing the quality chasm: A new health system for the 21st century. Washington, DC: Institute of Medicine; 2001.
3. Han YY, Carcillo JA, Venkataraman ST, et al. Unexpected increased mortality after implementation of a commercially sold computerized physician order entry system. Pediatrics. 2005 Dec;116(6):1506-12.
4. Koppel R, Metlay JP, Cohen A, et al. Role of computerized physician order entry systems in facilitating medication errors. JAMA. 2005 Mar 9;293(10):1197-203.
5. The Joint Commission. Safely implementing health information and converging technologies. Issue 42, December 11, 2008. Accessed April 2009. Available at: http://www.jointcommission.org/SentinelEvents/SentinelEventAlert/sea_42.htm
6. Carayon P, Schoofs Hundt A, Karsh BT, Gurses AP, Alvarado CJ, Smith M, Flatley Brennan P. Work system design for patient safety: the SEIPS model. Qual Saf Health Care. 2006 Dec;15 Suppl 1:i50-8.
7. Miller RA, Gardner RM. Recommendations for responsible monitoring and regulation of clinical software systems. American Medical Informatics Association, Computer-based Patient Record Institute, Medical Library Association, Association of Academic Health Science Libraries, American Health Information Management Association, American Nurses Association. J Am Med Inform Assoc. 1997 Nov-Dec;4(6):442-57.
8. Microsoft Corporation. Microsoft Health Common User Interface home page. Accessed April 2009. Available at: http://www.codeplex.com/mscui.
9. American Medical Informatics Association. AMIA 10x10 Goal. Accessed April 2009. Available at: http://www.amia.org/10x10.
10. Safran C, Shabot MM, Munger BS, et al. Program requirements for fellowship education in the subspecialty of clinical informatics. J Am Med Inform Assoc. 2009 Mar-Apr;16(2):158-66.
11. Singh H, Naik A, Rao R, Petersen L. Reducing Diagnostic Errors Through Effective Communication: Harnessing the Power of Information Technology. Journal of General Internal Medicine. 2008;23:489-94.
12. Singh H, Arora HS, Vij MS, Rao R, Khan M, Petersen LA. Communication outcomes of critical imaging results in a computerized notification system. J Am Med Inform Assoc. 2007;14:459-66
13. Sittig DF, Ash JS, Jiang Z, Osheroff JA, Shabot MM. Lessons from "unexpected increased mortality after implementation of a commercially sold computerized physician order entry system". Pediatrics. 2006 Aug;118(2):797-801.
14. Stead WW, Lin HS (eds.) Computational Technology for Effective Health Care: Immediate Steps and Strategic Directions. The National Academies Press, Washington, DC, 2009.

ELECTRONIC HEALTH RECORD-RELATED SAFETY CONCERNS: A CROSS-SECTIONAL SURVEY

Shailaja Menon, Hardeep Singh, Ashley N.D. Meyer, Elisabeth Belmont, and Dean F. Sittig

1.3.1 INTRODUCTION

The Health Information Technology for Economic and Clinical Health (HITECH) Act has encouraged the adoption of health information technology (HIT) [1] through incentive payments to physicians and healthcare organizations for meaningful use of electronic health records (EHRs). [2] As a result, recently there has been a significant increase in EHR implementation. [3] Nearly three-quarters of office-based physicians now use some form of an EHR system. Since 2009, physicians' capability to prescribe electronically has more than doubled. [4] To date, physicians, hospitals, and other healthcare providers have received over $12.6 billion in incentive payments under the provisions of the HITECH Act. [5]

The widespread adoption of EHRs is expected to transform healthcare through benefits such as complete availability of patient records and clinical decision support. [6] Despite the benefits of EHRs, there is a growing concern regarding risks associated with use of these technologies. [7–11] Because HIT is implemented in highly complex healthcare settings, new and unanticipated sources of errors are beginning to emerge. [9,10,12,13] For example, partial use of EHRs can result in loss of critical information or documentation between the twin worlds of paper and electronic records. The introduction of EHRs could also alter preexisting workflows and introduce new types of cognitive challenges and unsafe workarounds. [14] For instance, several types of errors have been associated with incorrect entry of information into the EHR and inadequate provider training. [15] Finally, even long after implementation, there are potential risks related to system-wide EHR downtimes that could result in widespread adverse effects on clinical care. [16]

Electronic Health Record–Related Safety Concerns: A Cross-Sectional Survey, Menon S, Singh H, Meyer AND, Belmont E, and Sittig DF, Journal of Healthcare Risk Management **34**,1. *Copyright 2014 American Society for Healthcare Risk Management of the American Hospital Association.*

Although there is emerging evidence of these safety concerns, comprehensive data on EHR-related safety events are lacking, partly because of limited disclosure of HIT-related medical errors. [7] The Pennsylvania Patient Safety Authority [17] recently identified EHR-related errors and problems through an analysis of HIT-related incident reports. The 2012 Institute of Medicine report on HIT and patient safety identified the lack of risk reporting and hazard data on HIT as a major barrier in building safer systems. [7] Given the increasing number of EHR implementations, as well as the proliferation of EHR vendors with different clinical information systems, additional data are needed to identify the extent of EHR-related safety concerns.

EHR-related safety concerns might not always be visible to users, or users can be unaware of the origin of the problem. Conversely, risk managers and healthcare system lawyers have access to quality and safety data from multiple sources and are often privy to additional safety data from sources unavailable to HIT personnel and clinicians (e.g., malpractice claims). In order to gain new knowledge and learn about their experiences, we conducted a cross-sectional survey of risk managers and health lawyers to obtain exploratory data about EHR-related serious safety events. Our study objectives were to identify (1) the most frequent types of EHR-related serious safety events reported by these respondents, (2) possible factors they believed to be associated with EHR-related serious safety events, and (3) patterns of measurement related to tracking and reporting of EHR-related safety concerns within their institutions.

1.3.2 METHODS

1.3.2.1 PARTICIPANTS

Members of the American Health Lawyers Association (AHLA) and the American Society for Healthcare Risk Management (ASHRM) participated in the survey. The membership of these 2 associations includes individuals who represent large and small hospital systems and long-term care facilities. Members include patient safety professionals, such as risk managers and attorneys practicing healthcare law. The risk managers are responsible for promoting risk management policies and programs through education and communication among senior management and governing bodies, medical staff members, and employees at all levels of the organizations. The health lawyers represent and counsel hospital systems, physicians, managed care organizations, and other healthcare entities on health-related legal issues. All registered AHLA and ASHRM members were invited to participate in the survey through an e-mail invitation that was distributed by the organizations using their mailing lists. The one-time invitation informed potential participants about the purpose of the study and assured confidential and voluntary participation. An independent survey firm managed survey administration and data collection, all of which was conducted using a secure Web-based platform.

1.3.2.2 SURVEY DEVELOPMENT

We performed a literature search to identify previously developed surveys about EHR implementations and their impact on healthcare practices. We did not find any surveys that specifically addressed the frequency and nature of EHR-related serious safety events. Therefore, we developed a new survey to address the study questions. The survey focused on 5 content areas:
1. *Degree of EHR implementation at the respondents' healthcare organization* (ie, for a lawyer, where the respondent was hired for legal representation). We asked respondents to indicate the extent of EHR implementation defined as the percentage of patient health records that were maintained in electronic form. [18] The response categories were none, 1%–10%, 11%–25%, 26%–50%, 51%–75%, 76%–99%, and 100%.
2. *Frequency of EHR-related serious safety events.* Participants rated the frequency of 11 types of EHR-related serious safety events, such as hardware and software malfunctioning, issues related to data display, incorrect patient identification, subversion of clinical decision support protocols, and issues related to data aggregation. [19] Frequencies were reported on a 5-point Likert scale with the following categories: frequently, occasionally, seldom, never, and

don't know. A separate item asked respondents to indicate their concern about the potential occurrence of EHR-related serious safety events over the next 5 years, rated on a 5-point scale as very concerned, moderately concerned, somewhat concerned, slightly concerned, or not at all concerned.
3. *Factors affecting EHR-related serious safety events.* Respondents chose from a list of 7 EHR characteristics (eg, EHR workflow process, type of users, degree of integration of new EHR) that might have affected the type or frequency of EHR-related serious safety events they had witnessed in the past. [14,20]
4. *Best practices to avoid EHR-related serious safety events.* Participants rated 12 good clinical practices (eg, prompt vendor and organization-level response to EHR-related system errors, EHR downtime training, oversight and accountability structure) that can be used to avoid occurrences of serious safety events related to use of or transition to EHRs. Respondents rated each practice as very important, important, moderately important, somewhat important, or not important. [16]
5. *Tracking of EHR-related safety measurements.* [21] Respondents were asked to indicate whether any of 12 EHR-related safety measures (eg, EHR-related serious safety events, EHR system response time, open or incomplete patient orders, EHR system uptime rate) were tracked and reported at their facility. Separately, respondents were asked to indicate which tracked measures were routinely shared with the governing boards of their healthcare organizations.

Most survey items were closed-ended. For each closed-ended question, we used expert opinion and an extensive literature review to generate a list of responses.

1.3.2.3 ANALYSIS

We used IBM SPSS Statistics software to analyze the survey data. We used descriptive statistics to summarize frequencies of degree of EHR implementation, types of EHR-related serious safety events, factors affecting EHR-related serious safety events, and tracking of EHR-related safety measurements. We also investigated whether EHR-related safety measures that were tracked were successively shared with the governing body of healthcare organizations represented by the respondents. Additionally, we compared frequency of EHR-related serious safety events experienced in the past 5 years and concerns expressed about future EHR use and potential for serious safety events. Because we were interested in highlighting most common types of EHR-related serious safety events experienced in the past, we combined frequently and occasionally response categories. Similarly, we were interested in highlighting the presence of relatively greater concern, and thus combined very concerned and moderately concerned response categories to represent respondents who had expressed more concern about future EHR use and potential for serious safety events.

1.3.3 RESULTS

The online survey was open to 15,400 AHLA and ASHRM members between August and September 2012. We were unable to get a more accurate denominator for respondents (ie, the number of members eligible to answer the survey) because many AHLA and ASHRM members' institutions either do not have an EHR or the members do not directly work on clinical issues related to the EHR. We estimated that about one-third of members were affiliated with institutions with EHRs, based on the most recent national EHR adoption rates available. [22] Based on input from senior members, we further assumed that only one-half of those remaining were working closely enough with an EHR to be able to respond to the survey. Thus, we estimated that approximately 2500 members were eligible to participate. Three hundred sixty-nine respondents completed the survey, and hence our estimated response rate was about 15%. Most respondents were risk managers (53%), followed by an equal proportion of patient safety officers and attorneys exclusively practicing healthcare law (14%). Other participants included attorneys who practiced law within and outside healthcare (about 10%), compliance officers (9%), and vice presidents of quality (4%). Two-thirds of respondents (66%) worked for hospitals or healthcare systems. Other respondents represented physician practice groups (18%), long-term care facilities (5%), and health plans (5%). As shown in Figure 1.3.1, healthcare organizations represented in the survey had variable degrees of EHR implementation, with about half having at least 76% of their medical records maintained in electronic form and 2% having no electronic records.

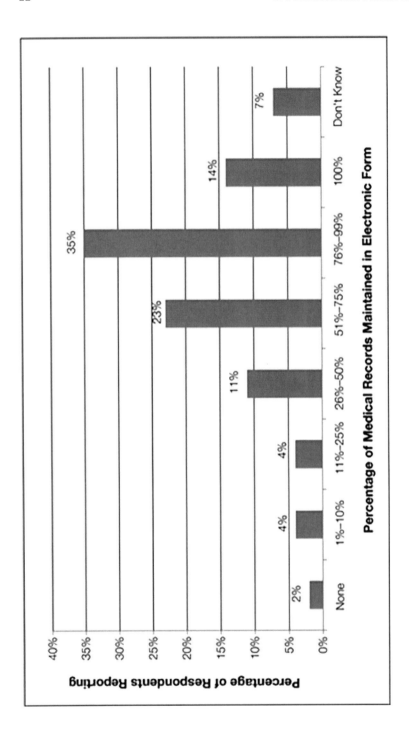

FIGURE 1.3.1: Percentage of Medical Records Maintained in Electronic Form

TABLE 1.3.1: Type and Frequency of Safety Events in the Past 5 Years

Survey question: For each of the following types of serious safety events, please indicate how frequently the healthcare organization for which you are employed or provide legal representation has experienced those events in the past 5 years—frequently, occasionally, seldom, never, don't know

	Frequently N	(%)	Occasionally N	(%)	Sum of Frequently and Occasionally N	(%)
Type of safety event						
Some aspect of data display in the hardware is incomplete, missing, or misleading	55	(15.4)	130	(36.5)	185	(52.0)
Open or incomplete patient orders	40	(11.3)	140	(39.7)	180	(51.0)
Procedures and policies that are ineffective given equipment and/or staffing realities	48	(13.5)	115	(32.3)	163	(45.8)
Failure to follow up abnormal test results due to computer or user input error	26	(7.3)	133	(37.4)	159	(44.7)
Confusing one patient with another because of similar names, incorrect input or other errors	20	(5.7)	130	(36.9)	150	(42.6)
Reliance upon inaccurate or incomplete patient-generated health data (eg, personal health records)	31	(8.7)	105	(29.6)	136	(38.3)
Intentionally or accidentally subverting clinical decision support protocols that issue an alert based on the entry of a certain clinical finding, result, or adverse drug interaction	29	(8.1)	93	(26.1)	122	(34.3)
Automatic discontinuation of a prescription	14	(4.0)	87	(24.8)	101	(28.8)
Data aggregation leading to erroneous data reporting and/or incorrect interpretation of data	19	(5.4)	75	(21.1)	94	(26.5)
Prolonged downtime of EHR systems resulting in unavailability of patient information	11	(3.1)	59	(16.6)	70	(19.7)
Errors resulting from implementing accrediting body, regulatory, or legal mandates	10	(2.8)	50	(14.1)	60	(16.9)

1.3.3.1 FREQUENCY OF SERIOUS SAFETY EVENTS IN THE PAST 5 YEARS

More than half (53%) of respondents surveyed admitted to having at least one EHR-related serious safety event in the previous 5 years; 10% of all respondents experienced more than 20 such events in the same time frame. About half (47%) reported that they had not experienced or were unaware of any EHR-related serious safety events in their organization in the past 5 years.

The 2 most common types of EHR-related safety concerns identified by the respondents related to data display and open or incomplete patient orders (Table 1.3.1). These were followed closely by failure to follow up on abnormal test results and wrong patient identification. Errors due to unavailability of patient data during downtime and errors resulting from implementing accrediting body, regulatory, or legal mandates were perceived as less common.

When asked about the variables that have affected the type and frequency of EHR-related serious safety events in the past, the 3 most frequently reported variables included EHR workflow processes, user familiarity with and training on the EHR, and degree of integration of the new EHR system (Figure 1.3.2). Vendor-specific variables, such as EHR vendor reliability and contractual protection such as acceptance testing or uptime guarantees, were less often endorsed as contributing to EHR-related serious safety events.

A majority of respondents indicated that serious EHR-related adverse events were tracked in their respective institutions; other EHR-related measures were tracked less frequently and with considerable variability (Table 1.3.2). For instance, a number of potentially hazardous EHR-related safety measures such as "open or incomplete patient orders," "incorrect reporting of laboratory and other diagnostic test results," and "alert override and adjustment rate" were reported as being used by less than half of the respondents. Change in mortality rate following EHR system implementation was the least tracked measure. Even when EHR-related measures were tracked, they were not automatically reported to the leadership. Compared to overall tracking rates, rates of reporting these measures to the institutional or system governing boards were consistently lower, sometimes markedly, for all of the measures we assessed.

TABLE 1.3.2: Tracking and Reporting of EHR-Related Safety Measures

Survey question 1: What measures does the healthcare organization for which you are employed or provide legal representation track relating to its EHR system(s)? (Check all that apply)

Survey question 2: For which of the following measures is tracking information shared with the governing board of healthcare organization for which you are employed or provide legal representation? (Check all that apply)

	Question 1: Tracked N	(%)	Question 2: Shared N	(%)
EHR-Related Measure†				
All serious EHR-related adverse events	229	(62.1)	173	(46.9)
Open or incomplete patient orders after a set period	182	(49.3)	45	(12.2)
Laboratory and other diagnostic test results incorrectly reported	159	(43.1)	50	(13.6)
Alert override and adjustment rate	150	(40.7)	43	(11.7)
Results of network penetration to assess the confidentiality, integrity, and availability of e-Protected Health Information (PHI)	149	(40.4)	67	(18.2)
EHR system uptime rate	134	(36.3)	40	(10.8)
Adherence to the Joint Commission Sentinel Event Alert #42—Safely Implementing Health Information and Converging Technologies	129	(35.0)	63	(17.1)
Adherence to clinical decision support protocols	105	(28.5)	32	(8.7)
EHR system response time	101	(27.4)	25	(6.8)
Clinical user satisfaction survey	98	(26.6)	46	(12.5)
Serious EHR fix rate	93	(25.2)	32	(5.7)
Change in mortality rate following EHR systems implemented	48	(13.0)	24	(6.5)
None of the above	51	(13.8)	0	0.0

†Questions 1 and 2: Respondents could choose all measures that are tracked and shared. The total for each measure represents number of respondents who chose that measure.

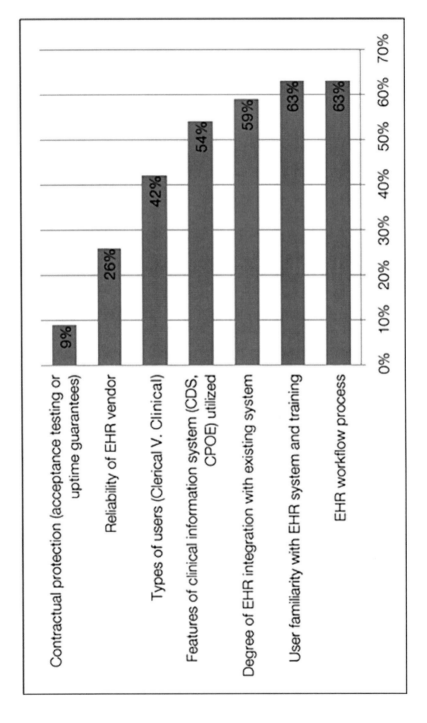

FIGURE 1.3.2: Percentage Distribution of Variables Affecting EHR-Related Serious Safety Events

TABLE 1.3.3: Concerns About Future EHR Use and Potential for Serious Safety Events and Frequency of Safety Events Experienced in the Past 5 Years

Survey question 1: Concerns about future EHR use and potential for serious safety events (very/moderately concerned)

Survey question 2: Frequency of safety events in the past 5 years (frequent/occasional)

	Question 1: Future Concerns		Question 2: Frequency of Past Concerns	
	N	(%)	N	(%)
Type of Serious Safety Events				
Failure to follow up on abnormal test results due to computer or user input error	291	(59.3)	159	(43.1)
Some aspect of data display is incomplete, inaccurate, or misleading	205	(55.6)	185	(50.1)
Reliance upon inaccurate or incomplete patient-generated health data (eg, personal health record)	196	(53.1)	136	(36.9)
Open or incomplete patient orders	189	(51.2)	180	(48.8)
Intentionally or accidentally subverting clinical decision support protocols that issue an alert based upon the entry of a certain clinical finding, result or adverse drug interaction	184	(49.9)	122	(33.1)
Confusing one patient with another because of similar names, incorrect input, or other error	176	(47.7)	150	(40.7)
Procedures and policies that are ineffective given equipment and/or staffing realities	174	(47.2)	163	(44.2)
Prolonged downtime of EHR systems resulting in the unavailability of patient information	145	(39.3)	70	(19.0)
Automatic discontinuation of prescription	132	(35.8)	101	(27.4)
Data aggregation leading to erroneous data reporting and/or incorrect interpretation of data	120	(32.5)	94	(25.5)
Errors resulting from implementing accrediting body, regulatory, or legal mandates	9	(26.3)	60	(16.3)

When asked how concerned they were about future EHR use and potential for serious safety events, more than half of respondents indicated they were very or moderately concerned about the following 3 serious safety events: (1) failure to follow up on abnormal test results due to computer or user input error; (2) some aspect of EHR data display that is incomplete, inaccurate, or misleading; and (3) reliance on inaccurate or incomplete patient-generated health data (Table 1.3.3). To understand how serious safety events experienced in the past might affect the respondent's perceptions about potential problems in the future, we looked at the frequency of EHR-related serious safety events reported in the past 5 years. As shown in Table 1.3.3, concerns about future EHR-related serious safety events were not entirely consistent with past experiences with serious safety events. For instance, although 37% of respondents reported inaccurate patient-generated health data as a common safety event in the past, over half of respondents expressed concern about this safety risk, perhaps due to an anticipated increase in patients' involvement in managing their health records. Similarly, though only 19% of respondents reported prior frequent events related to unavailability of patient information due to prolonged downtime, a much higher number of respondents (39%) were concerned about this issue arising in the future.

1.3.4 DISCUSSION

We conducted a Web-based survey of members of the AHLA and ASHRM to elicit information about factors associated with EHR-related serious safety events. More than half of respondents reported that their facilities had experienced at least 1 EHR-related serious safety event in the previous 5 years. Issues related to data display, open or incomplete patient orders, and failure to follow up on abnormal test results were the 3 most common types of EHR-related serious safety events. Although a majority of respondents stated that all EHR-related serious safety events were tracked at their facilities, fewer reported regular monitoring of EHR safety measures that could have flagged hazardous conditions. Only a few measures were reported to the leadership/governing boards of the healthcare organizations.

A growing body of literature suggests that EHRs and other forms of HIT can introduce new types of errors. [7–11] Although these errors can have serious implications for patient safety, [23] few reports about the nature and magnitude of these errors have been published. This is largely due to the fact that EHR-related errors are not clearly defined [13] and are rarely reported. [7] Whereas some prior studies have used reported events to classify errors, [17] there is little empirical data on the frequency and types of EHR-related errors in real practice. [17] Our survey offers additional insights to understand the risks posed by using EHRs.

Respondents viewed EHR workflow processes, user familiarity with EHR system and training, and degree of integration of new EHRs as the most significant factors affecting EHR-related serious safety events. These findings lend support to the argument that EHR implementation invariably alters existing workflows and introduces new types of risks, and that organizations must work closely with their EHR vendor and frontline clinicians to create new EHR-enabled clinical workflows that are both efficient for clinicians and safe for patients. [14] Specific features and configurations of new clinical information systems (clinical decision support, computerized provider order entry [CPOE]), along with their degree of integration with existing legacy systems, also contribute to serious safety events. In addition, we found that respondents considered user training and familiarity with EHR systems to be important variables linked to EHR safety events. In the current regulatory environment that encourages rapid implementation of EHR systems to meet time-sensitive criteria for monetary incentives, these findings serve as a cautionary note.

Our findings regarding types of EHR-related errors support the Pennsylvania Patient Safety Authority's report that found inaccurate data display as one of the most frequently reported safety events. For example, this report [17] found that "wrong data"–related events (data are missing, not updated, not entered, or incorrectly entered) were involved in a majority of EHR-related error reports. Clinical data entered into the EHR are among the most important components of the patient record, and the ability of EHR systems to share these data within and among healthcare organizations magnifies the risks associated with inaccurate data. Our study also found that more than half of respondents indicated that open or incomplete patient orders were the second most frequent type of serious safety event. A patient order is considered incomplete when important components such as date and time of order, drug name, drug dose, drug route, schedule, and duration are not entered. Incomplete patient orders can lead to serious medication errors and resulting harm. In addition to CPOE risks identified by Koppel et al, [24] about 81% of HIT events reported in the Pennsylvania Patient Safety Study involved medication errors and many of these involved medication orders. Additionally, risks related to follow-up of abnormal test results in EHRs (the third most common EHR-related serious safety event) have been identified in other studies as well. [25,26]

Safety risks associated with EHR use can be mitigated with use of a comprehensive monitoring mechanism. [27,28] For instance, tracking of EHR-related safety measures can provide information about potentially hazardous practices within an organization. To change these practices, this information must be shared with the organization's leadership. However, data about EHR safety measures is rarely available, and EHR-related serious safety events are underreported. [28] Measurement in this area is clearly underdeveloped; only some institutions appear to be monitoring EHR-related safety measures (Table 1.3.2). Furthermore, much of the data about safety measures were not consistently shared with the leadership. To enable EHR safety-related improvement, sharing data with organizational leadership is important; what cannot be measured cannot be improved.

This study has several limitations. A low response rate was an obvious limitation. While the estimated 15% response rate is significantly lower than what we expected, the relatively large number of total responses (369) represents one of the largest samples to date of organizations reporting EHR-related serious safety events. The knowledge of EHR-related safety concerns is still evolving, and it is possible that by providing a list of potential EHR-related patient events created by survey developers, we biased the respondents. However, some of the findings, such as data-related errors and errors related to follow-up of test results, have also been found in other studies.

1.3.5 CONCLUSION

Although EHR-related patient safety concerns are difficult to detect and measure, some risk managers and health lawyers appear to be witnessing serious EHR-related safety concerns in their respective organizations and could provide useful data on areas of improvement. Data display, open or incomplete patient orders, and failure to follow up on abnormal test results were identified as common types of EHR-related serious safety events. Most respondents did not use EHR safety measures comprehensively, and of the safety data that were being measured, relatively little was shared with their leadership. Because EHR-related serious safety events are underreported and understudied, organizations should consider implementing robust measures within their institution for mitigating risks from EHR-related safety concerns.

REFERENCES

1. Blumenthal D, Tavenner M. The "meaningful use" regulation for electronic health records. N Engl J Med. 2010;363:501–504.
2. Wright A, Henkin S, Feblowitz J, McCoy AB, Bates DW, Sittig DF. Early results of the meaningful use program for electronic health records. N Engl J Med. 2013;368:779–780.
3. King J, Patel V, Furukawa MF. Adoption of Health Record Technology to Meet Meaningful Use Objectives: 2009–2010. ONC Data Brief, No.7. 2012. Washington, DC: Office of the National Coordinator for Health Information Technology.
4. Charles D, Furukawa MF, Hufstader M. Electronic Health Record System and Intent to Attest to Meaningful Use among Non-Federal Acute Care Hospitals in the United States: 2008–2011. ONC Data Brief, No. 1. 2012.Washington, DC: Office of National Coordinator for Health Information Technology.
5. EHR incentive programs: monthly payment and registration summary report. Available at: www cms gov/Regulations-and-Guidance/Legislation/EHRIncentivePrograms/DataAndReports html [serial online], 2013. Accessed July 1, 2013.
6. Blumenthal D, Glaser JP. Information technology comes to medicine. N Engl J Med. 2007;356:2527–2534.
7. Health IT and Patient Safety: Building Safer Care. Washington, DC: Institute of Medicine; 2011.
8. Harrington L, Kennerly D, Johnson C. Safety issues related to the electronic medical record (EMR): synthesis of the literature from the last decade, 2000–2009. J Healthc Manag. 2011;56:31–43.
9. Magrabi F, Ong MS, Runciman W, Coiera E. Using FDA reports to inform a classification for health information technology safety problems. J Am Med Inform Assoc. 2012;19:45–53.
10. Myers RB, Jones SL, Sittig DF. Review of reported clinical information system adverse events in US Food and Drug Administration databases. Appl Clin Inform. 2011;2:63–74.
11. Warm D, Edwards P. Classifying health information technology patient safety related incidents—an approach used in Wales. Appl Clin Inform. 2012;3:248–257.
12. Campbell EM, Sittig DF, Ash JS, Guappone KP, Dykstra RH. Types of unintended consequences related to computerized provider order entry. J Am Med Inform Assoc. 2006;13:547–556.

13. Sittig DF, Singh H. Defining health information technology-related errors: new developments since to err is human. Arch Intern Med. 2011;171:1281–1284.
14. Campbell EM, Guappone KP, Sittig DF, Dykstra RH, Ash JS. Computerized provider order entry adoption: implications for clinical workflow. J Gen Intern Med. 2009;24:21–26.
15. Weir CR, Hurdle JF, Felgar MA, Hoffman JM, Roth B, Nebeker JR. Direct text entry in electronic progress notes. An evaluation of input errors. Methods Inf Med. 2003;42:61–67.
16. Sittig DF, Singh H. Electronic health records and national patient-safety goals. N Engl J Med. 2012;367:1854–1860.
17. Sparnon E, Marella WM. The role of electronic health records in patient safety events. Pa Patient Saf Advis. 2012;9(4):113–121.
18. Sittig DF, Guappone K, Campbell EM, Dykstra RH, Ash JS. A survey of U.S.A. acute care hospitals' computer-based provider order entry system infusion levels. Stud Health Technol Inform. 2007;129:252–256.
19. Sittig DF, Ash JS. Clinical Information Systems: Overcoming Adverse Consequences. Sudbury, MA: Jones and Bartlett; 2011.
20. Ammenwerth E, Schnell-Inderst P, Machan C, Siebert U. The effect of electronic prescribing on medication errors and adverse drug events: a systematic review. J Am Med Inform Assoc. 2008;15:585–600.
21. Sittig DF, Guappone K, Dykstra R, Ash JS. Recommendations for monitoring and evaluation of in-patient computer-based provider order entry systems: results of a Delphi survey. AMIA Annu Symp Proc. 2007 Oct 11;671–675.
22. Hsiao CJ, Hing E. Use and characteristics of electronic health record system among office-based physician practices: United States, 2001–2012. NCHS Data Brief No. 111, December 2012; 1–8.
23. Weiner JP, Kfuri T, Chan K, Fowles JB. "e-Iatrogenesis": the most critical unintended consequence of CPOE and other HIT. J Am Med Inform Assoc. 2007;14:387–388.
24. Koppel R, Metlay JP, Cohen A, et al. Role of computerized physician order entry systems in facilitating medication errors. JAMA. 2005;293:1197–1203.
25. Singh H, Thomas EJ, Sittig DF, et al. Notification of abnormal lab test results in an electronic medical record: do any safety concerns remain? Am J Med. 2010;123:238–244.
26. Singh H, Thomas EJ, Mani S, et al. Timely follow-up of abnormal diagnostic imaging test results in an outpatient setting: are electronic medical records achieving their potential? Arch Intern Med. 2009;169:1578–1586.
27. Singh H, Classen DC, Sittig DF. Creating an oversight infrastructure for electronic health record-related patient safety hazards. J Patient Saf. 2011;7:169–174.
28. Sittig DF, Classen DC. Safe electronic health record use requires a comprehensive monitoring and evaluation framework. JAMA. 2010;303:450–451.

There are several supplemental files that are not available in this version of the article. To view this additional information, please use the citation provided.

CHAPTER 2

ANALYSIS OF EHR SAFETY

REVIEW OF REPORTED CLINICAL INFORMATION SYSTEM ADVERSE EVENTS IN US FOOD AND DRUG ADMINISTRATION DATABASES

Risa B. Myers, Stephen L. Jones, and Dean F. Sittig

2.1.1 BACKGROUND

A popular anti-virus program update led to cancelled surgeries (1). Routine maintenance on the Australian Medicare patient verification system caused an estimated 1,300–1,800 pathology report results to be assigned to the wrong family member (2). A drug formulary update altered the default and alternate dosage amounts for certain medications (3). These are just a few examples of the risks and consequences of updating Clinical Information Systems (CISs) reported recently in the mainstream press.

Much discussion has ensued over who bears responsibility for these types of adverse events involving clinical information systems, which can have catastrophic consequences, and what level of reporting and oversight is appropriate (4-6). Koppel and Sittig, in separate articles, have called for

---

Myers RB, Jones SL, Sittig DF. Review of Reportel Clinical Information System Adverse Events in US Food and Drug Administration Databases. Appl Clin Inf *2011; 2: 63-74.*

increased reporting of near misses and errors to increase patient safety, as well as a review of the ongoing permissibility of "Hold Harmless" clauses included in many vendor licensing agreements (4,5). In a recent position paper, the American Medical Informatics Association (AMIA) issued a similar statement labeling "Hold Harmless" clauses "unethical" under circumstances where software defects or errors are integral to the adverse event (7).

The call for increased vendor accountability and centralized reporting is not unique to clinical information systems. The Brennan Center for Justice, a non-partisan institute focused on democratic and judicial topics, recently released a report calling for the creation of a federal clearinghouse and oversight agency for voting machine failures (8). This situation has many similarities to CISs: high cost of entry; relatively new technology; strong vendor control; new certification organization; minimal required reporting. The Brennan Center has called for the creation of a centralized, publically available database with mandatory reporting as well as the empowerment of a federal agency to investigate and to enforce correction of alleged issues. The Center expects these recommendations to result in higher quality systems and increased public confidence.

There are three sources of information regarding adverse events related to CISs. These include U.S. Food and Drug Administration (FDA) device databases, academic research on CISs and anecdotal reports in both the mainstream and academic press.

Foremost are the FDA databases. Since 1984, the FDA has collected voluntary reports of significant adverse events associated with medical devices. The core FDA requirement pertaining to manufacturers requires reporting within 30 days of awareness of a problem with a device. Key criteria for inclusion are devices that "(1) May have caused or contributed to a death or serious injury; or (2) Has malfunctioned and this device or a similar device that you market would be likely to cause or contribute to a death or serious injury, if the malfunction were to recur." Reports must be submitted to the FDA either through a paper report or electronically with prior approval (9). The FDA currently considers clinical information systems to be medical devices, but to date they have refrained from enforcing their regulatory requirements (10).

The data sources for the aforementioned research and for the current study are three databases supported by the FDA: Medical Device Reporting (MDR) (11), Manufacturer and User Facility Device Experience (MAUDE) (3) and Medical Product Safety Network (MedSun) (12,13). All three systems provide mechanisms for submitting and reporting adverse events resulting from the use of medical devices. All have on-line search capabilities, and all de-identify the reporting source. Event descriptions in the systems ranged from 60–4,000 characters in length. Basic metadata, such as the date the report was received by the FDA, is associated with each report.

MDR is the oldest of the systems, covering mandatory reporting from 1984–1996 and voluntary reporting thru June 1993 (11). MAUDE contains voluntary reports starting in June 1993, facility reports starting in 1991, distributor reports from 1993 and manufacturer reports since August 1996 (3).

The third database, MedSun, is leveraged by an organization focused on medical device safety. The approximately 350 member organizations, including hospitals, nursing homes and other healthcare organizations receive training and regular communications regarding medical device safety (13).

The second source of information is the academic research on effects, both intentional and unintentional, of CIS deployment (14,15), rights of users and consumers (16), and on deployment lessons learned (17).

Finally, there are reports of adverse events associated with CIS deployment. A well-known example is that of Children's Hospital of Pittsburgh, where the rollout of a Computerized Provider Order Entry system (CPOE) resulted in a significant increase in patient mortality (18). In this particular case, the findings from the study have been largely attributed to the roll-out of the CPOE and the impact on hospital workflow, as a second hospital deploying the same software a year later saw no significant change in mortality (18,19). However, these publications further highlight the potentially severe side effects that can result from CIS implementation, independent of system malfunctions.

These in-depth studies of CIS usage have focused on a small number of sites or a particular hospital network. This study, instead, looks across the board at adverse events reported from both manufacturers and user facilities around the world.

A related area of development is the use of automated systems in feeding Spontaneous Reporting Systems (SRSs) for adverse drug events

(ADEs). The ADE Spontaneous Triggered Event Reporting (ASTER) system was implemented to automatically submit ADE reports to the FDA with minimal additional physician involvement. This pilot was successful in that 217 reports were submitted during a five month period from 26 physicians who frequently discontinued medications (on average 1422/year) due to ADEs but had not submitted any reports in the previous year. This type of semi-automated reporting can increase medical knowledge and physician involvement in ADE reporting (20). This input was welcomed and well received by the FDA, which provided advice during system design (21).

2.1.2 OBJECTIVES

This paper examines historical FDA data in order to categorize reports and to gain understanding of the documented issues.

2.1.3 METHODS

Based on the current national focus on Meaningful Use in Electronic Health Records, we focused solely on reports on general clinical information systems, excluding isolated, clinical domains such as Blood Bank systems or Picture Archiving and Communication Systems (PACS). The clinical information systems we studied included: Laboratory Information Systems, Perioperative Systems, and electronic health records (EHRs). It was challenging to identify these systems, because, unlike PACS and Blood Bank, there are no product codes in the FDA databases for these types of systems. Another problem was deciding where to draw the line–is a charting system or image viewer considered a clinical information system? We decided to assemble a list of manufacturers and product names from the Certification Commission for Health Information Technology (CCHIT) list (22) and the list of vendors evaluated by KLAS, a third party reviewer of healthcare systems (23) and use these vendors as our starting point and basic inclusion criteria.

Analysis of EHR Safety

Unique lists of manufacturer names were extracted from downloads of the MAUDE and MDR databases. These lists were manually searched for logical variations of the vendor names (e.g. "MEDITECH" and "MEDI-TECH" and "CERNER" and "CERNER CORP."). These vendor names were used to find the corresponding generic product names (e.g. "INFORMATION", "S/W" and "SOFTWARE") and product codes. The databases were then searched for these terms, using wildcards to permit as many matches as possible and the lists scrutinized for relevant reports. The focus of this study was CIS systems, so other types of systems, such as Blood Bank software, patient monitoring and treatment planning systems were excluded. Furthermore, the list of reports is limited to commercial products. We recognize that many hospitals and physician practices utilize homegrown Clinical Information Systems. However, the FDA databases only include reports on commercial systems, as in-house systems developed solely for internal use are exempt from the reporting requirement (10). Consequently, reports on in-house systems are not included in this analysis. The MedSun database does not support downloads, so all of the search terms (including common misspellings) were entered into the search engine manually.

Once a set of records was found, the generic product description terms for these records were used to conduct additional searches in all of the databases. This process was iterative, and "finds" in one database were used as search terms in the other databases. In addition, in the MAUDE database, the device type for discovered reports was used as a search term. This iterative process was repeated until no new terms or records were discovered. Next, the reports were abstracted and maintained off-line.

Twenty-eight duplicate reports were eliminated, including multiple reports from the same user facility to a particular vendor and multiple reports on the same issue from different user facilities. In one case there were ten (total) reports of a single error. Another case had five reports. All of the other reports were duplicates with only a single additional report. Virtually identical text was considered evidence of duplicate entries. In addition, the date the report was received by the FDA, date of event (when reported) and problem description, including manufacturer comments, were all used to identify and eliminate duplicate reports.

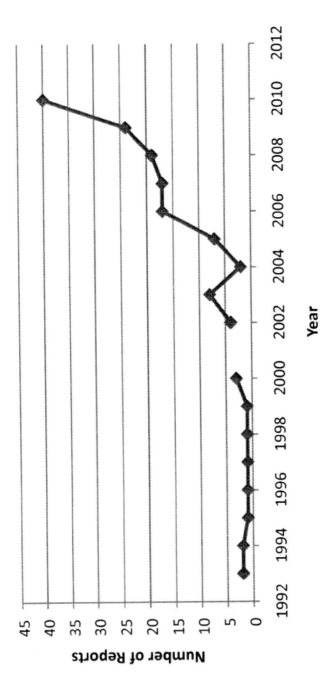

FIGURE 2.1.1: CIS adverse event reports by year. 2010 total is extrapolated from current rate of recorded reports through March 31, 2010.

TABLE 2.1.1: Identified causes in reported CIS errors

Cause	Explanation
Functionality	Particular system feature was assumed by users, but was not present, or the system behaved in an unexpected manner. This type of error includes drug or allergy rules that were not triggered as expected or in progress (versus final) notes that are available for sign out, incorrect delivery of messages within the system or updated orders not being discontinued under certain circumstances.
Incorrect calculation	Incorrect values derived from available data, or missing data or values assigned to the wrong patient. Includes errors in calculations such as date of delivery or incorrect drug dose calculation as well as interchanges of data between patients.
Incorrect content	"Rule" based logic incomplete or incorrect. Includes drug-allergy or drug-drug alerts, incorrect test reference ranges, system allowing "absurd" combinations of drugs or doses that are not possible with existing pill sizes, etc.
Insufficient detail	Insufficient detail was available to determine cause of the issue. These reports blamed the system for the adverse event, but did not provide specifics.
Integration	Pertaining to data exchange between products, which may, or may not, belong to the same vendor.
Large data volume	Errors that occurred when large values or numbers of items were present (often buffer overflows).
Other malfunction	Referencing an error other than one of those listed above. This category includes software "bugs", such as reuse of unique identifiers.
Support	Statements describing issues with, or lack of, vendor support.
Transition of care	Errors associated with CISs that involved patients moving between levels of care.
Upgrade	Related to process errors or side effects from system upgrades
User behind on patches	Vendor response indicated that the specified problem had been fixed and a software patch released.
User interface	Problem due to poor display of information or difficult to use system.

If another occurrence of an error was found at a later time, the second report was attributed to user error if the vendor had issued a patch and notified users. Similarly, the report was tagged as a support issue when the vendor failed to carry the fix forward to future releases.

One of the authors (RBM) initially evaluated all candidate reports and assigned cause categories. We used a grounded theory approach to establish the categories of errors contained in the three databases rather than any of the existing error classification systems since we were striving to understand people's perceptions, intentions and actions in reporting these errors regardless of their source, the time, or place from which they originated. Furthermore, the inconsistent reporting of available data did not allow for a consistent application of any formal error classification system. The classifications and categories were then reviewed, discussed, and revised by all of the authors until full agreement was reached. The final set of categories is described in Table 2.1.1.

2.1.4 RESULTS

A total of 120 unique reports were discovered. These reports identified 32 different manufacturers over a period of 18 years. Three vendors were identified in over half of all reported events. Seventy-four percent of the reports were filed by healthcare professionals. The break down of the report sources is show in Table 2.1.2. Note that more than one category could be indicated on the report and half of the records did not contain a reporting source, so the total does not match the number of records.

TABLE 2.1.2: Reports by source

Source	Count	Percentage
Health professional	47	39%
Company Representation	15	13%
User Facility	13	11%
Foreign	7	6%
Other	2	2%

Analysis of EHR Safety

TABLE 2.1.3: Reports by discovered cause

Cause	Count	Example of reported problem
User interface	63 (52.5%)	"The sound of the beep is the same whether it is a 'correct patient' scan or it is an incorrect patient."
Integration	21 (17.5%)	"When an update is made to the frequency field on an existing prescription, the frequency schedule ID is not simultaneously updated on new orders sent to the pharmacy via [application]."
Calculation	18 (15.0%)	"An additive value of the metric and english converted to metric by sys."
Functionality	16 (13.3%)	"The final document then displayed in the sign out option for final signature, however, the temporary document was actually signed out and available for printing as the final signed out report."
Incorrect Content	13 (10.8%)	"Patient had a known allergy to Tylenol which had been entered into the system last year. We cannot show that the pharmacist entering the medication or the nurse documenting the medication got an alert to say the patient was allergic to the medication, as they should have."
Support	12 (10.0%)	"This is not the first time that a safety issue with the software has been reported to the software vendor without any further communication to all end users with the warning of the issue at hand."
Upgrade	11 (9.2%)	"There was an error in the procedure used to push the new program to the workstations that resulted in the demo driver being activated."
Other malfunction	7 (5.8%)	"In the message center inbox, a user can make changes to a new pending message and save the changes without saving the message to a patient chart. If the user then performs the same task on a second pending message, the system replaces the entire text of the second message with the entire text of the first message."
Large data volume	5 (4.2%)	"If the date range is too extensive [i.e., in the request to display results] and the volume of cases to scan is over 10,000, some reports are not printed. There is no audit trail to trace the unprinted documents."
User behind on patches	5 (4.2%)	"Product was already corrected in initial report of problem under [release code]. The client validated the correction in live in 2002." [Report filed with FDA in 2006.]
Insufficient Detail	2 (1.7%)	"Hospital wide breakdown of system of electric charts and electric order gadgets resulted in confusion, neglect, failed communications and delayed treatments in the days immediately following the surgery…"

TABLE 2.1.3: *Cont.*

Cause	Count	Example of reported problem
Transition of care	2 (1.7%)	"Examples include orders to transfer patient from ICU to a non-ICU bed. Patient is moved to another bed but recipient care team does not receive communication and was not aware patient was under their charge."

The inferred causes, report counts and representative examples are tabulated in Table 2.1.3.

The manifestations of these adverse events included: missing or incorrect data, data displayed for the wrong patient, chaos during system downtime and hung systems (system unavailable for use). In addition, while many reports note that the problem was detected before harm could be done; adverse patient outcomes were reported including delays in diagnosis or treatments, unnecessary or emergency procedures and/or treatments, incorrect medication administration, patient injury or disability, and death. Given the brevity of the reports it is difficult to classify the severity of each problem. Approximately 80% of the records provided classification as to whether or not there was an associated adverse event and the type of outcome (disability, death, etc.).

2.1.5 DISCUSSION

2.1.5.1 VOLUME OF RECORDS

In the past few years, the number of reports attributed to CISs, including Electronic Health Records (EHRs) and Computerized Provider Order Entry (CPOE) has grown. If the number of reports recorded to date in 2010 continues at its current rate, the annual rate will be 2/3 higher than 2009, as shown in Figure 2.1.1.

The volume of reports is low, but so is current market penetration of Clinical Information Systems. According to a 2008 survey, fewer than 2% of U.S. hospitals had a comprehensive EHR system, and only 8–12% had a basic system in one or more departments (24). EHRs have been touted as part of the solution for improving the quality of medical care as well as having the potential to reduce costs in the long term. Additionally, with the financial incentives for providers and hospital systems to adopt an EHR and financial penalties for failure to adopt an EHR and use it in a "meaningful" way (25) we can expect these numbers to increase. There is already evidence to support the hypothesis that EHR adoption in the US is increasing (26). A reasonable consequence of increased deployment is a corresponding increase of these systems' involvement in adverse patient events, regardless of the entity mainly responsible for the adverse event.

In his testimony to the Health Information Technology (HIT) Policy Committee Adoption/Certification Workgroup on February 25, 2010, Jeffrey Shuren, Director of the FDA's Center for Devices and Radiological Health reported that over the past 10 years 260 voluntary reports have been filed with the FDA (27). Shuren's count included reports from Blood Bank and Radiology Systems, which were excluded from this study. We identified and excluded 164 Blood Bank reports and over 200 other reports that were classified the same as our reports but did not meet our inclusion criteria of being on the list of certified and recommended vendors and being a general CIS. Our results align well with those reported by Shuren, with key categories of concern being errors of commission, which include wrong patient errors; omission, including loss of patient data; analysis or calculation errors, and incompatibility or interface errors. We categorized these problems as Functionality, Calculation, Incorrect Content and Integration. These types of errors comprised over half of the errors we saw.

2.1.5.2 KEY LESSONS IDENTIFIED

A number of lessons can be inferred from these results. First, that very few reports are being filed. Second, that there are known dangerous areas in the design, implementation, use, and support of CISs, and finally, that there is clearly a need for three way communication between manufacturers, organizations, and users.

The number of unique reports found was very small—a mere 120 from the over 1.4 million reports in the combined databases. Based on our considerable experience with CISs, it is safe to say that people are not reporting all adverse events related to CISs. Causes for the low number of reports likely include the current low market penetration of CISs and the lack of knowledge on the part of users of what type of incidents should be reported (6). In addition, the difficulty of assigning responsibility for an adverse event in a complex, integrated environment can be a contributing factor. The aforementioned "Hold Harmless" clause in many vendor end user licensing agreements documents may also be a deterrent to reporting. Additionally, perhaps a general lack of awareness of the availability and anonymity of the FDA reporting system among end users contributes to the low numbers of reports as well. Lastly, reporting of errors and adverse events involving CIS and EHR's is voluntary and as discussed above, there are substantial barriers to knowing when and where to report an event.

The number of manufacturers represented in the list of reports is even smaller, a mere 32. It is implausible that no other vendors have ever had a software "bug" that resulted in an adverse event in a clinical setting over the past 18 years. Furthermore, over half of all reported events were associated with only 3 vendors. In addition to the reasons listed above regarding difficulties in reporting, it may be that these three vendors have a much larger market share, that these products have a large number of issues, or that these manufacturers have both a) submitted many of the events they have identified themselves, and b) encouraged, or at least not discouraged, their clients to submit reports.

There is limited awareness of these databases. Even when known, it requires significant effort on the part of an individual to submit a report. When there is no requirement or expectation of value for the reporter (i.e. "what's in it for me?"), the likelihood of a report being submitted is low. In addition, it is extremely difficult to extract reports from these databases; therefore the value of submitting a report is further diminished since access to the data is virtually non-existent. A potential solution to this problem might be to add a "File Report with FDA" feature to all certified CIS systems to help facilitate the reporting process akin to that piloted by AS-TER for adverse drug events (28). This enhancement would link users directly to the FDA web site. Such a function could auto-populate many of the necessary fields (e.g. date, vendor ID, system type, screen print, etc.) leaving only the details of the problem for the user to fill-in thus further reducing the reporting burden for end users. Copies of these reports could also be sent to vendors to facilitate their ability to improve their own software quality assurance processes.

Finally, there are no clear-cut guidelines on what types, or severity of events, should be reported. Should user errors be included? Are there threshold events akin to the Joint Commission "Never Events" that can be identified for Clinical Information Systems (29)? These could include unplanned system downtime over a predefined number of hours or situations where a patient is harmed and a computer is involved in some manner, such as when a patient is given a wrong medication or dose, or medications or treatments are administered at the wrong time (6). Many of the reports from the FDA databases specifically stated that the error was detected before harm could be done. Should these "near misses" continue to be reported, as with the Federal Aviation Authority that requires reporting of all accidents and near misses (30)?

Another lesson learned from this analysis is that there are key events that are frequently associated with these CIS-related problems. These areas must be addressed collaboratively with manufacturers, organizations, and users. While it is difficult to address user interface issues overall, particular attention must be paid to system upgrades and data integration points, for example, when new hardware, software or clinical content is added to an existing system. More transparent systems should be designed to allow the user to more easily detect the results of these errors such as wrong patient errors and missing data.

It is important to note that adverse events are not solely due to system malfunctions. A few reports described chaotic environments when systems were unavailable. These reports illustrate the need for 1) trained clinical staff, 2) appropriate sanity checks and testing before implementing new applications or versions in the production environment, 3) checks and balances in clinical workflow, 4) increased software testing during the development processes, 5) appropriate system backup plans, and 6) safe and effective downtime procedures in the event of system failure (14).

These reported events likely greatly underestimate the incidence of all adverse events attributable to CISs. The small sample size, the voluntary nature of the reporting system, the relative lack of vendor self-reporting and the barriers to reporting discussed above all contribute to the biased nature of the available data. Further sources of bias include: the possibility that some vendors may selectively report events based on errors that could easily be fixed or had no negative impact on patient outcomes and the restricted list of vendors used as a starting point. Finally, the data sources used are among the only publically available longitudinal collections of data for researchers to analyze.

2.1.5.3 RECOMMENDATIONS TO IMPROVE THE SYSTEM

2.1.5.3.1 CLINICAL INFORMATION SYSTEM VENDORS

Clearly, there is a need for three-way communication between manufacturers, healthcare organizations, and users. Vendors must:

1. Listen and solicit input from users;
2. Track and escalate identified errors in a manner visible to their customers and potential customers (7);
3. Act on the submitted reports and
4. Target communication to all their clients regarding a particularly problematic feature or function in a way that reduces overall noise (i.e., separate from routine system enhancement notes).
5. Provide a prepopulated Adverse Event form in a format acceptable to the FDA system(s), similar to the one produced by ASTER for Adverse Drug Events (20).

2.1.5.3.2 HEALTHCARE ORGANIZATIONS AND USERS

Similarly, users and their organizations must

1. Be encouraged to report errors to their local EHR oversight committee (31) or if none exists to one of the FDA data bases;
2. Request enhancements and fixes from their vendors;
3. Be aware of vendor reports regarding their products; and
4. Stay up to date on system patches and updates.

2.1.5.3.3 U.S. FOOD AND DRUG ADMINISTRATION

Greater value for evaluating and reviewing Clinical Information Systems could be achieved by providing more consistency in data reporting. Recently, the Agency for Healthcare Research and Quality (AHRQ) released version 1.1 (beta) of their "Common Format–Device or Medical/Surgical Supply" in an attempt to provide this consistency (32). In addition, the iHealth Alliance which is "comprised of industry leaders from medical societies, liability carriers, patient advocacy groups and others dedicated to protecting the interest of patients and providers" has created a website (www.ehrevent.org) to provide a safe and secure means of reporting EHR-related safety events. Briefly, this site provides a structured list of manufacturers and a predefined list of causes submitters can use to characterize the reports. Unfortunately, to date, the data collected via this website is not available in the public domain so it cannot be analyzed or monitored.

Finally, the Institute of Medicine has recently convened a committee to study "Patient Safety and Health Information Technology" (33). This committee has been asked to review the evidence and experience from the field on how the use of clinical information systems affects the safety of patient care. Their report is due at the end of 2011.

2.1.6 CONCLUSIONS

The FDA databases offer a centralized location for reporting on and finding information about significant adverse events related to Clinical Information Systems. However, there is a need for increased awareness of the reporting requirements and enhanced user participation.

The discovered causes provide insight into common problems encountered with CISs and should be leveraged by manufacturers to improve their products. Users and facilities should be aware of potential issues, in particular with regard to workflow impact, integration between systems and upgrades. Everyone would benefit from forming partnerships to explore and resolve these issues.

This study demonstrates the complexity of deploying and maintaining Clinical Information Systems. These systems have the potential to add significant value in healthcare delivery, but require vigilance on the part of manufacturers, healthcare information technology facilities and providers. Steps that are not expected to be complex often are. Working together, users, facilities, CIS vendors and the FDA can build safe and effective clinical information systems that improve the efficiency and quality of healthcare in the United States.

2.1.6.1 IMPLICATIONS OF RESULTS FOR PRACTITIONERS AND CONSUMERS

In order to improve quality of care, end users must be aware of and able to report issues with Clinical Information Systems. A federally administered centralized repository with mandatory reporting could potentially increase awareness of issues and, consequently, improve system quality.

REFERENCES

1. NPR Staff. Anti-Virus Program Update Wreaks Havoc With PCs : NPR [Internet]. 2010 Apr 21 [cited 2010 Apr 25];Available from: http://www.npr.org/templates/story/story.php?storyId=126168997&sc=17&f=1001
2. Dearne K. Medicare glitch affects records | The Australian [Internet]. 2010 Apr 20 [cited 2010 Apr 25];Available from: http://www.theaustralian.com.au/australian-it/medicare-glitch-affects-records/story-e6frgakx-1225855706275
3. MAUDE - Manufacturer and User Facility Device Experience [Internet]. [cited 2010 Apr 15];Available from: http://www.accessdata.fda.gov/scripts/cdrh/cfdocs/cfMAUDE/search.CFM
4. Koppel R, Kreda D. Health care information technology vendors' "hold harmless" clause: implications for patients and clinicians. JAMA. 2009 Mar 25;301(12):1276-1278.
5. Sittig DF, Classen DC. Safe electronic health record use requires a comprehensive monitoring and evaluation framework. JAMA. 2010 Feb 3;303(5):450-451.
6. Koppel R. Monitoring and evaluating the use of electronic health records. JAMA. 2010 May 19;303(19):1918; author reply 1918-1919.
7. Goodman KW, Berner ES, Dente MA, Kaplan B, Koppel R, Rucker D, et al. Challenges in ethics, safety, best practices, and oversight regarding HIT vendors, their

Analysis of EHR Safety 49

customers, and patients: a report of an AMIA special task force. J Am Med Inform Assoc. 2011 Jan 1;18(1):77-81.
8. Norden L. Voting system failures: a database solution [Internet]. New York N.Y.: Brennan Center for Justice; 2010 [cited 2010 Sep 20]. Available from: http://www.brennancenter.org/content/resource/voting_system_failures_a_database_solution/
9. CFR - Code of Federal Regulations Title 21 [Internet]. [cited 2011 Jan 8];Available from: http://www.accessdata.fda.gov/scripts/cdrh/cfdocs/cfcfr/CFRSearch.cfm?CFRPart=803&showFR=1&subpartNode=21:8.0.1.1.3.5
10. FDA POLICY FOR THE REGULATION OF COMPUTER PRODUCTS, 11/13/89 (Draft) [Internet]. [cited 2011 Jan 28];Available from: http://www.janosko.com/documents/FDA%20Policy%20Computer%20Products/FDAPolicyComputers1989.htm
11. MDR Database Search [Internet]. [cited 2010 Apr 20];Available from: http://www.accessdata.fda.gov/scripts/cdrh/cfdocs/cfmdr/search.cfm?searchoptions=1
12. Medsun Reports [Internet]. [cited 2010 Apr 15];Available from: http://www.accessdata.fda.gov/scripts/cdrh/cfdocs/medsun/searchReport.cfm
13. MedSun: Medical Product Safety Network [Internet]. [cited 2010 Apr 15];Available from: http://www.fda.gov/MedicalDevices/Safety/MedSunMedicalProductSafetyNetwork/default.htm
14. Campbell EM, Sittig DF, Guappone KP, Dykstra RH, Ash JS. Overdependence on technology: an unintended adverse consequence of computerized provider order entry. AMIA Annu Symp Proc. 2007;:94-98.
15. Campbell EM, Sittig DF, Ash JS, Guappone KP, Dykstra RH. Types of unintended consequences related to computerized provider order entry. J Am Med Inform Assoc. 2006 Oct;13(5):547-556.
16. Sittig DF, Singh H. Eight rights of safe electronic health record use. JAMA. 2009 Sep 9;302(10):1111-1113.
17. DeVore SD, Figlioli K. Lessons premier hospitals learned about implementing electronic health records. Health Aff (Millwood). 2010 Apr;29(4):664-667.
18. Han YY, Carcillo JA, Venkataraman ST, Clark RSB, Watson RS, Nguyen TC, et al. Unexpected increased mortality after implementation of a commercially sold computerized physician order entry system. Pediatrics. 2005 Dec;116(6):1506-1512.
19. Del Beccaro MA, Jeffries HE, Eisenberg MA, Harry ED. Computerized provider order entry implementation: no association with increased mortality rates in an intensive care unit. Pediatrics. 2006 Jul;118(1):290-295.
20. Linder JA, Haas JS, Iyer A, Labuzetta MA, Ibara M, Celeste M, et al. Secondary use of electronic health record data: spontaneous triggered adverse drug event reporting. Pharmacoepidemiol Drug Saf. 2010 Dec;19(12):1211-1215.
21. Dal Pan GJ. Commentary on "Secondary use of electronic health record data: spontaneous triggered adverse drug event reporting" by Linder et al. Pharmacoepidemiol Drug Saf. 2010 Dec;19(12):1216-1217.
22. CCHIT [Internet]. [cited 2010 Apr 15];Available from: http://www.cchit.org/
23. Ambulatory EMR - Segment Profile - KLAS Helps Healthcare Providers by Measuring Vendor Performance [Internet]. [cited 2010 Apr 15];Available from: http://www.klasresearch.com/Research/Segments/Default.aspx?id=3&evProductID=336

09&ReturnURL=%2fResearch%2fSegments%2fDefault.aspx%3fid%3d3%26evPr oductID%3d33609
24. Jha AK, DesRoches CM, Campbell EG, Donelan K, Rao SR, Ferris TG, et al. Use of electronic health records in U.S. hospitals. N. Engl. J. Med. 2009 Apr 16;360(16):1628-1638.
25. H.R. 1 [111th]: American Recovery and Reinvestment Act of 2009 (GovTrack.us) [Internet]. [cited 2011 Jan 14];Available from: http://www.govtrack.us/congress/bill.xpd?bill=h111-1
26. Hsiao C, Beatty P, Hing E, Woodwell D, Rechtsteiner E, Sisk J. Products - Health E Stats - EMR and EHR Use by Office-based Physicians [Internet]. [cited 2011 Jan 14];Available from: http://www.cdc.gov/nchs/data/hestat/emr_ehr/emr_ehr.htm
27. Shuren J. Testimony of Jeffrey Shuren, Director of FDA's Center for Devices and Radiological Health [Internet]. 2010. Available from: http://healthit.hhs.gov/portal/server.pt/gateway/PTARGS_0_10741_910717_0_0_18/3Shuren_Testimony022510.pdf
28. ASTER Study [Internet]. [cited 2010 Aug 10];Available from: http://www.aster-study.com/
29. Issue 42: Safely implementing health information and converging technologies | Joint Commission [Internet]. [cited 2010 Aug 7];Available from: http://www.jointcommission.org/SentinelEvents/SentinelEventAlert/sea_42.htm
30. Federal Aviation Administration. Aeronautical Information Manual–Official Guide to Basic Flight Information and ATC Procedures [Internet]. 2010 Feb 11;Available from: http://www.faa.gov/air_traffic/publications/atpubs/aim/
31. Miller RA, Gardner RM. Recommendations for responsible monitoring and regulation of clinical software systems. American Medical Informatics Association, Computer-based Patient Record Institute, Medical Library Association, Association of Academic Health Science Libraries, American Health Information Management Association, American Nurses Association. J Am Med Inform Assoc. 1997 Dec;4(6):442-457.
32. PSO Privacy Protection Center - Device or Medical/Surgical Supply, including HIT Device (Beta) [Internet]. [cited 2011 Jan 18];Available from: https://www.psoppc.org/web/patientsafety/device-or-medical/surgical-supply-including-hit-device-beta
33. Patient Safety and Health Information Technology - Institute of Medicine [Internet]. [cited 2011 Jan 18];Available from: http://www.iom.edu/Activities/Quality/PatientSafetyHIT.aspx

EXPLORING THE SOCIOTECHNICAL INTERSECTION OF PATIENT SAFETY AND ELECTRONIC HEALTH RECORD IMPLEMENTATION

Derek W. Meeks, Amirhossein Takian, Dean F. Sittig, Hardeep Singh, and Nick Barber

2.2.1 BACKGROUND

The USA federal government, through stimulus spending and the Affordable Care Act, is encouraging widespread implementation of health information technology (HIT) to improve healthcare quality and patient safety. [1] These efforts are founded on expectations of increased coordination of care, improved follow-up, and increased efficiency throughout the continuum of care. [2] However, research suggests that technology may lead to new uncertainties and risks for patient safety through disrupting established work patterns, creating new risks in practice, and encouraging workarounds. [3–10] In particular, the increasing adoption of electronic health records (EHR) has revealed potential safety implications related to EHR design, implementation, and use. [11–15] These risks are not related solely to the technological features of the EHR but may involve EHR users and their workflows, aspects of the organizations in which they function, and the rules and regulations that govern or oversee their activities. Furthermore, patient safety risks associated with EHR may vary along the EHR adoption and implementation timeline. Given the complexity and multifaceted nature of EHR-related safety risks, a comprehensive model is needed to understand and anticipate these risks in a sociotechnical context.

Sittig and Singh [16,17] developed an eight-dimensional sociotechnical model to study the safety and effectiveness of HIT at all levels of design, development, implementation, use, and evaluation. Four earlier sociotechnical models informed the development of the eight-dimensional

---

Reproduced from Journal of the American Medical Informatics Association, *Meeks DW, Takian A, Sittig DF, Singh H, and Barber N., 21(e1), pp. e28-34, copyright 2014 with permission from BMJ Publishing Group Ltd.*

model: the model of Henriksen et al, [1,8] the framework for analyzing risk and safety of Vincent et al, [1,9] the systems engineering initiative of patient safety of Carayon et al, [20] and the interactive sociotechnical analysis of Harrison et al. [2,1] The model's dimensions represent interdependent domains of an EHR-enabled healthcare system: hardware and software; clinical content; human–computer interface; people; workflow and communication; internal organization policies, procedures, and culture; external rules, regulations, and pressures; system measurement and monitoring (Figure 2.2.1). [16,17] For example, failure to follow up a critical laboratory result could be attributable to a software error that prevented transmission of the laboratory result to the correct provider (hardware and software), faulty display of information in the provider's EHR window (human–computer interface), or inadequate coordination of roles within the clinical care team (workflow and communication). [22] Efforts to improve EHR-related patient safety rely on identification of underlying risks as well as an appreciation of contributing areas of vulnerability (eg, people, organization policies and procedures, or system measurement). [23]

The sociotechnical intersection of patient safety and EHR is complex. First, this intersection conceptualizes the healthcare system as an evolving, complex adaptive system in which safety risks often emerge from users' interactions with the EHR that lead to new clinical workflow processes. These new workflow processes involve different environmental (eg, human interaction with physical devices and their workspace), [24] cultural (eg, role changes of clinicians in the EHR-enabled workflow), [25] or even sociopolitical (eg, clinical power structure) factors. [26] Second, these safety risks are multifactorial and rarely involve a single contributing factor. Third, improving patient safety within an EHR-enabled healthcare system requires a journey in which the sociotechnical infrastructure and functionalities evolve over time. The sociotechnical model does not itself convey how it fits into the continuum of HIT safety that includes safe transition from paper to fully integrated EHR. Therefore, to understand the intersection of EHR and patient safety, Sittig and Singh [27] further proposed a three-phase model to account for the variation in the stages of implementation, levels of complexity, and related patient safety concerns within an EHR-enabled healthcare system. The first phase is concerned with safety events that are unique and specific to technology (ie, unsafe technology), which often emerge early in the process of implementation. The second phase addresses unsafe or inappropriate use of technology as well as unsafe changes in the overall workflow that emerge due to technology use. The third phase addresses use of technology proactively to identify and monitor potential safety concerns before harm occurs to the patient. While the boundaries between the phases may not always be distinct, the three-phase model could be useful for goal setting and identification of threats to patient safety. [27]

In light of emerging and often novel risks associated with EHR, comprehensive models such as those described above are needed to assess the variety of safety threats and near misses. Such efforts will advance the understanding of EHR-related safety events to allow for the planning of safer systems and processes. Previously, we conducted a longitudinal, sociotechnical evaluation of the implementation and adoption of EHR in English National Health Service (NHS) hospitals. [28,29] As part of that study, we conducted interviews that yielded a large volume of open-ended comments, some of which reflected concerns about patient safety. That study demonstrated the importance of considering the sociotechnical context of EHR implementation, although the UK investigators did not apply a formal framework to assess patient safety until now. [30] Our aim was to explore and illustrate the application of the eight-dimensional sociotechnical and three-phase EHR safety models to organize and interpret EHR-related patient safety concerns elicited during evaluation. Rather than conduct hypothesis testing, our goal was to highlight the 'real-world' usefulness of practical sociotechnical approaches to ensuring safe and effective EHR implementation and future use.

2.2.2 MATERIALS AND METHODS

2.2.2.1 SETTING AND DESIGN

In 2002, the UK Department of Health decided to implement three centrally procured national EHR applications, both made to order and commercially available, in the English NHS hospitals. Implementation was supported by a small number of centrally contracted local service providers, each responsible for delivering standard software systems to local hospitals, ensuring system integration, interoperability, and national connectivity within a geographical region. This was part of an overall US$19.6 (£12.7) billion strategic initiative to transform the NHS's HIT infrastructure into an integrated set of electronic systems connected to national databases and a messaging service (the 'NHS spine'). [30] The data presented here were extracted from a 30-month (September 2008 to March 2011) prospective, longitudinal, and real-time case study-based evaluation during EHR implementation and adoption in 12 hospitals (nine acute and three mental health). [31] The original research proposal was approved as a service evaluation by a NHS ethics committee.

2.2.2.2 DATA COLLECTION

The methods of data collection have been described elsewhere. [28–30] Interviews were conducted at all stages of EHR implementation and adoption from initial awareness and planning to sustained use. In order to explore the implementation processes across hospitals, interviewers sought to determine the organizational activities undertaken and their consequences for professional roles, workflows, and clinical practices. Participating hospitals were purposefully selected according to their projected implementation timelines and included a range of hospital types (ie, teaching, non-teaching, acute care, and mental health) to allow comparisons.

The original investigators conducted semistructured interviews with a broad range of stakeholders: managers, implementation team members, information technology (IT) staff, junior and senior physicians, nurses, allied health professionals, administrative staff, external implementation-related stakeholders, and software developers. The six interviewers did not explicitly ask interviewees questions regarding patient safety. Interviews were audio-recorded and transcribed verbatim. Data were anonymized by redacting information that identified the individual participant or site.

2.2.2.3 DATA ANALYSIS

One author (AT) asked the original UK investigators to review transcripts for content related to patient safety. Out of 480 interviews conducted in the evaluation, AT confirmed 49 interviews in which patient safety content was

Analysis of EHR Safety

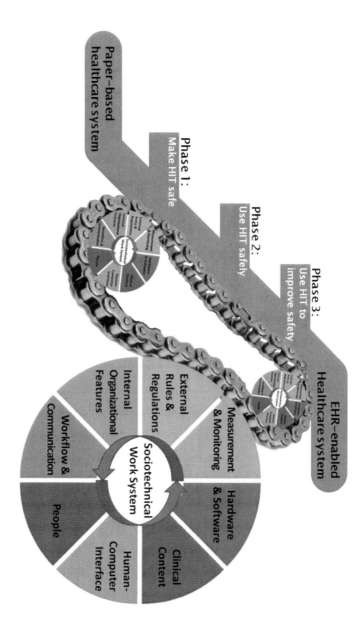

FIGURE 2.2.1: Diagram illustrating the interaction between the eight-dimension sociotechnical and three-phase electronic health record (EHR) safety models. The goal is for organizations to move from a paper-based medical record system 'up the escalator' to become an EHR-enabled healthcare system. Within each phase of the three-phase model, all eight dimensions of the sociotechnical model come into play. HIT, health information technology.

present. The data were then analyzed using a framework analysis approach, a qualitative research method that has pre-set aims but accommodates new themes from the data. [32] Framework analysis has five stages: familiarization; thematic analysis; indexing (coding); charting; and mapping and interpretation. We began by reviewing and summarizing relevant quotes regarding EHR-related patient safety concerns. Using the eight dimensions of the sociotechnical model as the framework, three reviewers (DWM, DFS, and HS) indexed the data. While acknowledging the interrelatedness of the models, for clarity we coded the dimension and phase most directly implicated in the safety concern. The data were then arranged according to the three-phase model (charting). This analysis was performed iteratively until consensus was obtained among the reviewers. Interrater reliability was not assessed as the aim of the study was to explore themes of patient safety and EHR implementation (mapping and interpretation), not rigorous classification with the two models. ATLAS.ti 6 by ATLAS.ti Scientific Software Development (http://www.atlasti.com) was used for data management.

2.2.3 RESULTS

The interviewees' roles in EHR implementation and the number of hospital represented are shown in Table 2.2.1. The sociotechnical domains were not mutually exclusive, but were seen to interact in the data; however, they are presented within the domain judged to be most involved with the safety concern. Some dimensions of the sociotechnical model are better represented than others in the dataset, as demonstrated by the mappings of phases and dimensions in Table 2.2.2. Similarly, most data were mapped to phases one and two of the three-phase model. Table 2.2.3 provides a high-level summary of the safety concerns present in the data. This table reveals that certain dimensions have heterogeneity while others have more homogeneous concerns expressed. For instance, in hardware and software concerns regarding EHR availability were prominent in phase one; data sharing and system–system interface issues were also seen. Conversely, in clinical content, most concerns were regarding phase two, in which users experienced difficulties (perceived or actual) with order entry through the EHR. We present the data according to the three-phase model to illustrate safety risks that emerged as most relevant to each phase of implementation.

TABLE 2.2.1: Interviewee role and hospital representation

Interviewee role	No of interviewees	No of hospitals represented
Senior manager	7	6
EHR implementation/IT team	9	6
Healthcare practitioners	16	6
Clinical managers	6	5
Administrators	3	5
Strategic health authorities	3	N/A
Local IT service providers	2	N/A
EHR software developers	3	N/A
Total	49	N/A

EHR, electronic health record; IT, information technology.

TABLE 2.2.2: Types of safety concerns categorized by sociotechnical dimensions and phases of EHR implementation and use

	Phase 1	Phase 2	Phase 3
Hardware and software	11	2	0
Clinical content	3	7	0
Human–computer interface	4	4	0
People	1	4	0
Workflow and communication	1	6	0
Internal organization policies, procedures, and culture	3	0	0
External rules, regulations, and pressures	2	0	0
System measurement and monitoring	0	0	1

EHR, electronic health record.

TABLE 2.2.3: Summaries of interview data demonstrating safety concerns by phase and dimension

Sociotechnical dimension	Phase of use	Summary of safety concern
Hardware and software	Phase one	Problems with EHR availability (login or network access) (n=4)
		Lack of basic EHR functionality (n=4)
		Problems related to data maintenance, sharing, or security (n=3)
	Phase two	Problems with accessing appropriate clinical information
		Problem with system–system interfaces
Clinical content	Phase one	Undeveloped or non-standardized clinical content in the EHR (n=3)
	Phase two	Parallel use of paper and EHR
		Problems or difficulties with use of order entry (n=6)
Human–computer interface	Phase one	User interface too burdensome or error prone for data entry (n=4)
	Phase two	User interface does not support clinical workflow (n=3)
		Risk of copy and paste functionality
People	Phase one	Data security concerns
	Phase two	Users sharing EHR access (n=3)
		Poor training leads to improper use
Workflow and communication	Phase one	Errors related to appointment scheduling applications
	Phase two	EHR not integrated into clinical workflow
		EHR causes delays in work (n=3)
		Laboratory result routing unreliable (n=2)
Internal organizational policies, procedures, and culture	Phase one	Multiple medical record numbers per patient increase risk of wrong selection
		Data confidentiality risks
		Local IT budget must support ongoing IT infrastructure requirements
External rules, regulations, and pressures	Phase one	National IT budgeting important for safe EHR use after implementation
		Complexity of software and business models of vendors may affect future use
System measurement and monitoring	Phase three	Challenges and benefits of EHR-based quality reporting

EHR, electronic health record; IT, information technology.

2.2.3.1 PHASE ONE

In accordance with the model, phase one EHR safety concerns were unique and specific to technology. Within the framework of the sociotechnical model, specific comments were frequently mapped to the domains of hardware and software, clinical content, and human–computer interface. An example of a phase one safety concern regarding hardware and software was the acknowledgment of an insufficient data center and back-up procedures.

> *The danger with [hospitals] doing their own thing is that instead of having a proper data centre meeting certain standards you get it sort of in a shed out the back sort of thing and it's not 24/7, it's not resilient, it doesn't have a fail over site that it can go to, it doesn't have a fail over within, guaranteed two hours service level and it's up to what they can negotiate with the supplier, so cost effectively it's not as cost effective and from a resilience and safety point of view it's not as good. I think the safety is probably one of the key things that doing it centrally and nationally is a lot more secure.*
> IT Manager, Site H
> Sociotechnical model: hardware and software

A recurring safety concern, also related to hardware and software, was implementation of an EHR without necessary software features to support a clinical workflow that demanded those features.

> *If you think someone's at risk of suicide and you kind of tick the box there and put some text in, you expect that will bounce through to the care plan module so they could then put a response to it and it stops things getting lost and what have you. It doesn't do anything like that. When you identify needs it doesn't bounce it through to the care planning functionality so that it's already there so that you know what you've got to address, and if you forget to transfer the fact that this person is at risk of stabbing someone, then the system doesn't offer any safeguards to drag it through.*
> Healthcare provider, Site G
> Sociotechnical model: hardware and software

In contrast to the absence of a feature, some users identified a design or implementation they perceived to be error prone. For instance, users described EHR hardware and software issues or human–computer interface problems that contributed to patient safety concerns.

> *We've had a couple of instances in Radiology where we've not been able to cancel requests and patients have been scanned twice, so they've had a double exposure of radiation.*
> Director, Site E
> Sociotechnical model: hardware and software

> *...[It's] terribly easy to make a mistake, because you can bring up several Maria Smiths and if you are not careful and you don't look at the date of birth, because they are just a list and they are right on top of each other, you could pick the wrong one.*
> Receptionist, Site E
> Sociotechnical model: human–computer interface

2.2.3.2 PHASE TWO

In this phase patient safety is compromised through unsafe use of technology or unsafe changes in workflow. The most common dimensions in this phase were workflow and communication, people, human–computer interface, and clinical content. The prevailing theme from the data was the risk introduced when EHR was placed within a clinical context that did not facilitate safe use. For instance, a phase two concern was the improper integration of computers into clinical encounters in which EHR use cannot occur simultaneously with delivery of care (ie, in procedural or sterile areas). Another example was the barrier associated with the requirement to sign into the EHR, which resulted in password sharing and generic password use.

> *...you go to your colleague and you say, log me in and then you use other people's cards. They had to have this generic access in*

A&E (emergency department) because actually this was a crazy situation. It broke all the rules for information and governance and data protection.
Manager, Site E
Sociotechnical model: people

Certain EHR features, such as copy and paste, were recognized as safety risks due to inappropriate use. In the example below, pathology specimens were mislabeled and the EHR was understood, in this instance, to increase risk of patient harm.

The ability to copy and paste in fields is dangerous. Incorrect details are being pasted into incorrect patient fields (i.e., prostate as specimen details in female patient request or missed miscarriage in clinical details for male patient).
Healthcare provider, Site D
Sociotechnical model: human–computer interface

Some workflow and communication problems were specific to certain practice areas for which use of the EHR, as implemented, was thought to be particularly ill suited. For instance, EHR users in the mental health hospitals felt the effort needed to document in the EHR was not only potentially unsafe, but impeded the ability to see patients in a timely manner.

The psychiatric assessments are quite lengthy and there are quite a lot of notes that go with it. Doctors are not going to be able to do it while they are with the patient, because of issues like risk... So it's going to increase the time spent and you are then delayed seeing the next patient which is I think the big anxiety.
Doctor, Site M
Sociotechnical model: workflow and communication

Finally, as clinical workflow and communication was noted to become error prone when the medical record was in transition from paper to electronic form, clinical content also arose as an area of potential risk.

We have to print out now anyway and put into the paper notes because not everyone is on [software X]... But I can also see the fact that when everyone is on it you won't have to do it.
Healthcare provider, Site H
Sociotechnical model: clinical content

2.2.3.3 PHASE THREE

This phase addresses EHR use to monitor and identify safety concerns before patients are harmed. This ultimate use of technology was reflected in only one interview. The participant noted the difficulty in reporting quality measures before EHR implementation and the potential advantages of an EHR-enabled healthcare system.

If everybody is using the same system, they have the same functionality available to them. There is only a limited amount of ways that you can record information from reporting and performance indicator and assessment sort of point of view. We often have difficulty meeting certain targets, because we don't have a way of reporting it. It's a real struggle. But, at least if everybody has the same struggle then you are comparable to everybody else and there aren't these gaps. You are more easily able to make a comparison across organizations. I think that's an advantage.
Manager, Site M
Sociotechnical model: system measurement and monitoring

2.2.4 DISCUSSION

IT and EHR could potentially have large quality and safety benefits. However, there is increasing acknowledgement that the use of EHR could introduce unintended risks, and simultaneous efforts are needed to establish safe EHR design and implementation. [14] As with other patient safety issues, a piecemeal, reactive approach to identifying and correcting EHR-related safety issues is unlikely to be efficient or effective. Systematic analysis of EHR-related safety concerns must be performed within a context that accounts for the evolving sociotechnical infrastructure and functionality that defines the journey to a safe EHR-enabled healthcare system. In this analysis from the evaluation of the NHS's implementation of EHR, we attempted to demonstrate the 'real-world' usefulness of analyzing spontaneously reported safety concerns through two operational models related to HIT: an eight-dimension sociotechnical model and a three-phase EHR safety model. A sociotechnical approach may allow developers, IT managers, administrators, clinicians, and others to understand risks in the development, implementation, and use of EHR and HIT while accounting for complex interactions of technology within the healthcare system. Further application of these models may be helpful as government bodies make HIT safety a greater priority within clinical environments. [33]

The three-phase model was useful to understand the context of safety risks given that our sites were still early in their EHR implementation journey, and therefore both phases one and two were sufficiently represented. Unfortunately, we were unable to identify many activities within phase three of the model. Furthermore, the eight-dimension model was found to have face validity to understand and classify EHR-related safety concerns within the technical, social, or clinical context in which they occur. Applications of such models could be useful to inform or prioritize implementation efforts. For example, we found, as anticipated, that phase one safety concerns arose most commonly in the hardware and software domains of the sociotechnical model. Therefore, organizations should ensure that proper hardware requirements are in place before EHR implementation (eg, adequate number of workstations, appropriate data center). Phase two concerns were frequently mapped to clinical content and workflow and communication. Phase two priorities could therefore involve understanding and changing the clinical workflow or the EHR configuration to facilitate safe care. Organizational and leadership factors are commonly recognized as important for success, [34] but we suggest that understanding the local culture, workflow, and potential impact on productivity is equally necessary. [31,35] Our combined model also suggests that as an organization evolves, both patient safety improvement activities and patient safety hazards also evolve from concerns about safe functionality and ensuring safe and appropriate use, to using the EHR itself to provide ongoing surveillance and monitoring of patient safety. Further exploration of this evolution could inform sociotechnical approaches to improving safety in future large-scale EHR implementations.

The strengths of this qualitative analysis include the large scale of the EHR implementation and evaluation involving simultaneous interviews. Other qualitative investigations have analyzed EHR implementations, but primarily focused on barriers to implementation, system-wide challenges, or overall benefits and concerns rather than patient safety. [35–39] Our high-level approach differs from that of other classification systems, notably that of Magrabi and colleagues, [40,41] which includes both technical and human elements. [42] For instance, the human elements it encompasses are generally related to the direct use of the computer, and to actions closely linked in time to the error at hand. By contrast, the model used in this paper encompasses a broader range of sociotechnical factors (eg, workflow and organizational factors) that are more temporally dissociated. Each approach might have its own advantages and limitations depending on what type of data is available for analysis, the depth and breadth of available data, and the rationale of why the analysis was undertaken.

We also build on previous work demonstrating the use of sociotechnical models. For instance, in our previous work, we found this sociotechnical model applicable in specific clinical contexts (eg, test results and referral communication), [43–48] but until this analysis, a formal model to study patient safety issues with EHR implementation was lacking (including within the previous body of work done by the UK investigators). Our sociotechnical model was adapted by the Institute of Medicine in their report on HIT safety albeit without the detailed technology dimensions that we believe are essential to appreciate the nuances involved with EHR use. [14]

To our knowledge, there are few if any practical models that are specific to HIT that provide guidance in this area. The combination of the sociotechnical model with the three-phase model allows us to view EHR safety from a systems engineering perspective. Through this lens, interaction of the two models is considered from four fundamental perspectives of complex systems: scale (quantitative size); function (the reason for existence); structure (the interconnection of system elements); and temporality (scales of time). [49] In our combined model (Figure 2.2.1), the phases differ in their 'sociotechnical' scale, function, structure, and temporality. Within each phase, the eight-dimensional sociotechnical model can be used to understand unique safety issues. For instance, a phase one software problem may encompass a single function such as inappropriate matching of blood products due to a software coding or content error. While in phase three, errors in blood typing would be identified in real time through an organization-wide monitoring program that alerts clinicians whenever the blood type of a patient has 'changed'. In other words, in phase one, we view the sociotechnical scale of the problem to be much more isolated and contained, while in the latter phases, the scale increases significantly: including users and the physical environment in phase two and, potentially, the entire organization in phase three.

Another example is the different skills and roles of people involved in phase one who are responsible for configuring the hardware (eg, moving database servers to a physically secure location) and software (eg, setting up encryption keys on the periodic back-up systems) to ensure patient confidentiality. While in phase three, people ensuring patient safety would probably include informaticians developing surveillance and monitoring capabilities to identify potential breaches of patient confidentiality or health information management and human resource professionals to investigate these potential breaches and enforce policies to protect health information. [50,51]

The limitations of this study include the interview protocol's lack of specificity to patient safety issues and the inability to assess impact on patient safety. The interviewers broadly focused on EHR implementation and did not intentionally seek detailed responses about patient safety. While safety concerns arose in several interviews, the interviews did not necessarily elicit the full range of potential EHR-related safety concerns. Although the concerns of those involved during implementation appeared appropriate, no additional effort was made to validate these concerns. As this was a secondary analysis of previously collected data, interview data regarding safety potentially could have been overlooked during the initial review by the original UK investigators because the data collection did not anticipate this use. The case study design may have reduced the generalizability of the findings, but despite different EHR software, cultures, and methods of healthcare delivery, we believe the usefulness of our analysis is the potential ability of the two models to identify EHR-related safety concerns and priorities to address them.

2.2.5 CONCLUSION

Examining the intersection of HIT and patient safety with practical conceptual models can advance the EHR-enabled healthcare system towards the goal of improving patient safety. 'Safe technology' and 'safe use of technology' are necessary for efforts to improve and monitor patient safety; for example, phase three of the EHR-enabled healthcare system. We demonstrated how the combined use of two models has face validity to facilitate understanding of the sociotechnical aspects of safe EHR implementation and the complex interactions of technology within the evolving healthcare system. Our sociotechnical approach, along with other existing frameworks, may be beneficial to help stakeholders understand, synthesize, and anticipate risks within the continuum of HIT safety that includes safe transition from paper to integrated EHR.

REFERENCES

1. Blumenthal D. Stimulating the adoption of health information technology. N Engl J Med 2009;360:1477–9.
2. Schiff GD, Bates DW. Can electronic clinical documentation help prevent diagnostic errors? N Engl J Med 2010;362:1066–9.
3. Ash JS, Berg M, Coiera E. Some unintended consequences of information technology in health care: the nature of patient care information system-related errors. J Am Med Inform Assoc 2004;11:104–12.
4. Balka E, Doyle-Waters M, Lecznarowicz D, et al. Technology, governance and patient safety: systems issues in technology and patient safety. Int J Med Inform 2007;76(Suppl. 1):S35–47.
5. Bates W, Cohen M, Leape L, et al. Reducing the frequency of errors in medicine using information technology. J Am Med Inform Assoc 2001;8:299–308.
6. Coleman RW. Translation and interpretation: the hidden processes and problems revealed by computerized physician order entry systems. J Crit Care 2004;19:279–82.
7. Hundt AS, Adams JA, Schmid JA, et al. Conducting an efficient proactive risk assessment prior to CPOE implementation in an intensive care unit. Int J Med Inform 2013;82:25–38.

8. Koppel R. Role of computerized physician order entry systems in facilitating medication errors. JAMA 2005;293:1197–203.
9. Patterson ES, Cook RI, Render ML. Improving patient safety by identifying side effects from introducing bar coding in medication administration. J Am Med Inform Assoc 2002;9:540–53.
10. Pirnejad H, Niazkhani Z, van der SH, et al. Impact of a computerized physician order entry system on nurse–physician collaboration in the medication process. Int J Med Inform 2008;77:735–44.
11. Singh H, Mani S, Espadas D, et al. Prescription errors and outcomes related to inconsistent information transmitted through computerized order entry: a prospective study. Arch Intern Med 2009;169:982–9.
12. Myers RB, Jones SL, Sittig DF. Review of reported clinical information system adverse events in US Food and Drug Administration databases. Appl Clin Inform 2011;2:63–74.
13. Weiner JP, Kfuri T, Chan K, et al. "e-Iatrogenesis": the most critical unintended consequence of CPOE and other HIT. J Am Med Inform Assoc 2007;14:387–8.
14. IOM (Institute of Medicine). Health IT and patient safety: building safer systems for safer care. Washington, DC: The National Academies Press, 2012.
15. Harrington L, Kennerly D, Johnson C. Safety issues related to the electronic medical record (EMR): synthesis of the literature from the last decade, 2000–2009. J Healthc Manag 2011;56:31–43.
16. Sittig DF, Singh H. Defining health information technology-related errors: new developments since to err is human. Arch Intern Med 2011;171:1281–4.
17. Sittig DF, Singh H. A new sociotechnical model for studying health information technology in complex adaptive healthcare systems. Qual Saf Health Care 2010;19(Suppl. 3):i68–74.
18. Henriksen K, Kaye R, Morisseau D. Industrial ergonomic factors in the radiation oncology therapy environment. Advances in Industrial Ergonomics and Safety V. Taylor and Francis, 1993:325.
19. Vincent C, Taylor-Adams S, Stanhope N. Framework for analysing risk and safety in clinical medicine. BMJ 1998;316:11547.
20. Carayon P, Schoofs Hundt A, Karsh BT, et al. Work system design for patient safety: the SEIPS model. Qual Saf Health Care 2006;15(Suppl. 1):i50–8.
21. Harrison MI, Koppel R, Bar-Lev S. Unintended consequences of information technologies in health care—an interactive sociotechnical analysis. J Am Med Inform Assoc 2007;14:542–9.
22. Singh H, Wilson L, Petersen LA, et al. Improving follow-up of abnormal cancer screens using electronic health records: trust but verify test result communication. BMC Med Inform Decis Mak 2009;9:49.
23. Singh H, Thomas EJ, Sittig DF, et al. Notification of abnormal lab test results in an electronic medical record: do any safety concerns remain? Am J Med 2010;123:238–44.
24. Koppel R, Wetterneck T, Telles JL, et al. Workarounds to barcode medication administration systems: their occurrences, causes, and threats to patient safety. J Am Med Inform Assoc 2008;15:408–23.
25. Campbell EM, Guappone KP, Sittig DF, et al. Computerized provider order entry adoption: implications for clinical workflow. J Gen Intern Med 2009;24:21–6.

26. Ash JS, Sittig DF, Campbell E, et al. An unintended consequence of CPOE implementation: shifts in power, control, and autonomy. American Medical Informatics Association, 2006:11.
27. Sittig DF, Singh H. Electronic health records and national patient-safety goals. N Engl J Med 2012;367:1854–60.
28. Robertson A, Cresswell K, Takian A, et al. Implementation and adoption of nationwide electronic health records in secondary care in England: qualitative analysis of interim results from a prospective national evaluation. BMJ 2010;341:c4564.
29. Sheikh A, Cornford T, Barber N, et al. Implementation and adoption of nationwide electronic health records in secondary care in England: final qualitative results from prospective national evaluation in "early adopter" hospitals. BMJ 2011;343:d6054.
30. Takian A, Petrakaki D, Cornford T, et al. Building a house on shifting sand: methodological considerations when evaluating the implementation and adoption of national electronic health record systems. BMC Health Serv Res 2012;12:105. [
31. Takian A, Sheikh A, Barber N. We are bitter, but we are better off: case study of the implementation of an electronic health record system into a mental health hospital in England. BMC Health Serv Res 2012;12:484.
32. Pope C, Ziebland S, Mays N. Qualitative research in health care. Analysing qualitative data. BMJ 2000;320:114–16.
33. Health Information Technology Safety Action & Surveillance Plan. The Office of the National Coordinator of Health Information Technology 2013 July 2 [cited 2013 Jul 11]. http://www.healthit.gov/sites/default/files/safety_plan_master.pdf
34. Takian A. Envisioning electronic health record systems as change management: the experience of an English hospital joining the National Programme for Information Technology. Stud Health Technol Inform 2012;180:901–5.
35. Scott JT, Rundall TG, Vogt TM, et al. Kaiser Permanente's experience of implementing an electronic medical record: a qualitative study. BMJ 2005;331:1313–16.
36. Greiver M, Barnsley J, Glazier RH, et al. Implementation of electronic medical records: theory-informed qualitative study. Can Fam Physician 2011;57:e390–7.
37. Zwaanswijk M, Verheij RA, Wiesman FJ, et al. Benefits and problems of electronic information exchange as perceived by health care professionals: an interview study. BMC Health Serv Res 2011;11:256.
38. Yoon-Flannery K, Zandieh SO, Kuperman GJ, et al. A qualitative analysis of an electronic health record (EHR) implementation in an academic ambulatory setting. Inform Prim Care 2008;16:277–84.
39. Spetz J, Burgess JF, Phibbs CS. What determines successful implementation of inpatient information technology systems? Am J Manag Care 2012;18:157–62.
40. Magrabi F, Ong MS, Runciman W, et al. An analysis of computer-related patient safety incidents to inform the development of a classification. J Am Med Inform Assoc 2010;17:663–70.
41. Magrabi F, Ong MS, Runciman W, et al. Using FDA reports to inform a classification for health information technology safety problems. J Am Med Inform Assoc 2012;19:45–53.
42. Sparnon E, Marella WM. The role of the electronic health record in patient safety events. Harrisburg, PA: Pennsylvania Patient Safety Authority, 2012.

43. Singh H, Esquivel A, Sittig DF, et al. Follow-up actions on electronic referral communication in a multispecialty outpatient setting. J Gen Intern Med 2011;26:64–9.
44. Singh H, Spitzmueller C, Petersen NJ, et al. Primary care practitioners' views on test result management in EHR-enabled health systems: a national survey. J Am Med Inform Assoc 2013;20:727–35.
45. Hysong SJ, Sawhney MK, Wilson L, et al. Understanding the management of electronic test result notifications in the outpatient setting. BMC Med Inform Decis Mak 2011;11:22.
46. Singh H, Thomas EJ, Mani S, et al. Timely follow-up of abnormal diagnostic imaging test results in an outpatient setting: are electronic medical records achieving their potential? Arch Intern Med 2009;169:1578–86.
47. Singh H, Spitzmueller C, Petersen NJ, et al. Information overload and missed test results in electronic health record based settings. JAMA Intern Med 2013;173:7024.
48. Sittig DF, Singh H. Improving test result follow-up through electronic health records requires more than just an alert. J Gen Intern Med 2012;27:12357.
49. De Weck OL, Roos D, Magee CL. Engineering systems: meeting human needs in a complex technological world. MIT Press, 2011.
50. Boxwala AA, Kim J, Grillo JM, et al. Using statistical and machine learning to help institutions detect suspicious access to electronic health records. J Am Med Inform Assoc 2011;18:498–505.
51. Powell C. Akron General fires employees for patient privacy violations in hospital shooting case. Akron Beacon J 2012.

AN ANALYSIS OF ELECTRONIC HEALTH RECORD-RELATED SAFETY CONCERNS

Derek W. Meeks, Michael W. Smith, Lesley Taylor, Dean F. Sittig, Jeanie Scott, and Hardeep Singh

2.3.1 BACKGROUND AND SIGNIFICANCE

Investments in health information technology (HIT) can enhance the safety and efficiency of patient care and enable knowledge discovery.[1]

Reproduced from Journal of the American Medical Informatics Association, *An Analysis of Electronic Health Record-Related Patient Safety Concerns.* Meeks DW, Smith MW, Taylor L, Sittig DF, Scott JM, Singh H., J Am Med Inform Assoc. 2014 Nov;21(6):1053-9. with permission from BMJ Publishing Group Ltd.

However, emerging evidence suggests that HIT may cause new patient safety concerns and other unintended consequences due to usability issues, disruptions of clinical processes, and unsafe workarounds to circumvent technology-related constraints.[2-14] In particular, rapid adoption of electronic health records (EHRs) has revealed potential safety concerns related to EHR design, implementation, and use.[13,15-18] Patient safety concerns are broadly defined as adverse events that reached the patient, near misses that did not reach the patient, or unsafe conditions which increase the likelihood of a safety event.[19,20] Detecting and preventing EHR-related safety concerns is challenging because concerns are often multifaceted, involving not only potentially unsafe technological features of the EHR but also EHR user behaviors, organizational characteristics, and rules and regulations that guide EHR-related activities. Thus, comprehensive and newer "sociotechnical" approaches that account for these elements are required to address the complexities of EHR-related patient safety.[21-24]

Despite a clear need to define and understand EHR-related safety concerns,[25] data that describe the nature and magnitude of these concerns are scarce. A few studies have attempted to quantify and classify HIT-related safety concerns by mining patient safety incident reporting databases.[16,26-28] In addition, conceptual frameworks or models have been developed to incorporate the breadth of technical and nontechnical factors into the analysis of HIT safety and effectiveness.[22,24,29-31] For instance, we previously developed a sociotechnical model that proposes eight interdependent dimensions that are essential to understand EHR-related safety (Table 2.3.1).[21,32] The model accounts for the complexities of technology, its users, the involved workflow, and the larger external or organizational policies and context in assessment of EHR-related safety concerns.[33,34]

We conducted a qualitative "sociotechnical analysis" of completed EHR-related safety investigations from voluntary reports within a large, integrated healthcare system. Using Sittig and Singh's sociotechnical model as a guiding framework, our aim was to describe common EHR-related safety concerns and understand the nature and context of these safety concerns in order to build a foundation for future work in this area.

2.3.2 METHODS

2.3.2.1 DESIGN AND SETTING

We performed a retrospective analysis of completed investigation reports about EHR-related safety concerns from healthcare facilities within the Department of Veterans Affairs (VA). The VA operates the largest integrated healthcare system in the United States with over 1700 sites of care (e.g., hospitals, clinics, community living centers, domiciliaries, readjustment counseling centers).[35] A comprehensive EHR, nationally mandated in 1999, is used at all its facilities to provide care to approximately 8.3 million Veterans.[36] The VA is considered a leader in the design, development, and use of EHRs to address healthcare quality.[37-39] The established HIT infrastructure is comprised of internally developed and commercially procured systems that provide a range of applications (e.g. laboratory, pharmacy, radiology, patient record, scheduling, registration, billing). VA facilities have the ability to customize the available administrative, financial, and clinical applications to match local processes and practice conditions while the core functionality is centrally updated and distributed. In conjunction with other patient safety initiatives such as sentinel event monitoring, root cause analysis, and proactive risk assessment, the VA created an Informatics Patient Safety (IPS) Office in 2005 to establish a mechanism for non-punitive, voluntary reporting of EHR-related safety concerns.

The IPS reporting system, which includes only health IT-related reports, is the foundation for a rigorous approach that includes not only event investigation and analysis, but also feedback to reporters and developers of solutions to mitigate future risks to patients. Clinical or administrative EHR users along with EHR developers can report EHR-related patient safety concerns through an intranet website or by using the national VA information technology helpdesk system. The most common process for clinical users to report a safety concern is by notification of local IT staff. The local IT staff investigate the safety concern, determine if national support is needed, and report the incident to the helpdesk. At the national level, if the event is patient safety related, an initial IPS report is populated with the initial reporter's contact information, the description of the event, the applications in use at the time of the event, any harm or potential for harm, and any known corrective actions. IPS analysts with healthcare, safety, and informatics training and human factors specialists investigate reports; at an average, it takes about 30 days per incident. The goals of analysis are understanding user actions that immediately preceded the safety concern, identifying the underlying root causes, and, if possible, safely replicating the event with "test" patients in the "live" EHR system. At a minimum, an account of the incident is elicited from the person who detects it, and this account is further reviewed by an IPS patient safety and informatics specialist with expertise in human factors. Most incidents are then subjected to attempts to replicate the incident in the EHR, reviews of logs, discussions with technical specialists, or other efforts to determine the exact nature of the incident. The reports are analyzed and scored according to potential severity, frequency, and detectability. The score prioritizes the need for solution development: a solution could be considered depending on resources but is not mandatory (low), a solution required an action plan such as training or request for software modification (intermediate), a solution required an immediate action (high), such as a software patch. After analysis, the IPS makes recommendations to software developers, individual medical facilities, or other relevant stakeholders within the VA healthcare system to mitigate the risk of error or harm.[40] Investigation-related information is maintained in a database and tracked until the investigation is "closed." The final, closed investigation for each report contains a narrative as provided by the initial reporter, the technical narrative by IPS and information technology staff that includes details of the investigation, and any solution that might have been identified.

2.3.2.2 DATA COLLECTION

We searched the IPS database for closed investigations that contained full analyses and narratives that provided meaningful information, excluding duplicate entries. We also excluded safety concerns related to erroneous editing or merging of patient records resulting in co-mingled or overlaid records. Although these are known safety concerns,[41] they were excluded because they are not routinely analyzed by IPS and are handled primarily by a separate office in the VA. We extracted 100 consecutive records that met our search criteria. Previous exploratory studies in patient safety have been able to shed powerful light on contributory factors with a similar sample size and, given the rich nature of the qualitative data, we believed this number was both valuable and feasible.[42]

2.3.2.3 DATA ANALYSIS

We analyzed narrative data in the completed investigation reports using a framework analysis method, which allows emerging themes to be incorporated into a previously established framework.[43,44] Framework analysis consists of five stages: familiarization, thematic analysis, indexing, charting, and mapping and interpretation. First, two authors (D.W.M. and M.W.S.) independently reviewed and summarized the investigation reports to become familiar with the data, but at this secondary stage of analysis, we made no further effort to replicate the investigation, determine additional causes, or offer additional solutions. Thematic analysis was guided primarily by the application of the eight-dimension sociotechnical model. A coding scheme was created so that each concern could be described and indexed according to one or more sociotechnical dimensions that underlay or contributed to the safety concern. Additionally, we categorized incidents by "phases" of safe EHR implementation and use: incident related to inherently unsafe technology or technology failures ("phase 1"), incidents related to unsafe or inappropriate use of technology ("phase 2"), and incidents related to lack of monitoring of potential safety concerns before harm occurs ("phase 3").[45]

Our coding scheme allowed a safety concern to be classified in multiple dimensions from the sociotechnical model, but in only one of the EHR safety phases. When more than one sociotechnical dimension was involved in a safety incident we noted this interaction by counting co-occurring dimensions. The two coding authors (i.e., a physician with informatics training and a human factors engineer) independently indexed each safety concern after analyzing the results of the IPS investigation. Discrepancies in coding were resolved by consensus. The emergent safety concerns were generated via collaborative, iterative analysis of the whole set of coding results. This included re-reading and re-arranging the data (charting) with members of our multidisciplinary project team whose areas of expertise included clinical medicine, patient safety, informatics, human factors, and information technology. Finally, emergent and recurring safety concerns were identified and described (mapping and interpretation) according to their sociotechnical origins and EHR safety phase.

We used the software package Atlas.ti version 6.2 to facilitate coding of the investigation narratives and Microsoft Excel to arrange and structure the data.

2.3.3 RESULTS

We extracted 100 consecutive, unique, closed investigations between August 2009 and May 2013 from 344 reported incidents. The selected incidents were reported from 55 unique VA facilities. The priority scores for solution development were 48 low priority, 38 intermediate priority, and 14 high priority incidents. Table 2.3.1 summarizes our analysis of the safety concerns along the sociotechnical model's dimensions and EHR safety phases. Approximately three-fourths of safety concerns were categorized as phase 1 (i.e., concerns related to unsafe technology). Sociotechnical dimensions of phase 1 concerns most commonly involved hardware and software, workflow and communication, and clinical content. One-quarter were classified as phase 2 (i.e., unsafe EHR use) and most commonly involved the dimensions of people, clinical content, workflow and communication, and human-computer interface. Only one safety concern involving phase 3 (i.e., failure to use the EHR to monitor patient safety) was represented in our analysis. Incidents frequently reflected occurrence of more than one sociotechnical dimension: 40 incidents were classified with two sociotechnical dimensions, 23 incidents had three, and 7 involved four dimensions.

During charting, mapping, and interpretation of the interactions of social and technical components of EHR use, several distinct (although not mutually exclusive) safety concerns emerged. We classified these concerns into four types: unmet display needs in the EHR, safety concerns with software modifications or upgrades, concerns related to data transmis-

Analysis of EHR Safety

sion at system-system interfaces, and concern of "hidden dependencies" in distributed systems (i.e., when one EHR component unexpectedly or unknowingly is affected by the state or condition of another). Table 2.3.2 provides definitions and examples of these four types of concerns, which accounted for 94% of the incidents analyzed. All four types of safety concerns affected or had the potential to affect multiple patients although we did not further analyze outcomes data except as noted below.

TABLE 2.3.1 EHR-related Safety Concerns Categorized by Sociotechnical Dimensions and Phases of EHR Implementation and Use

Sociotechnical Dimension	Phase 1 Unsafe technology or technology failures (n=74)	Phase 2 Unsafe or inappropriate *use* of technology (n=25)	Phase 3 Lack of monitoring of safety concerns (n=1)	Total
Hardware and software: The computing infrastructure used to power, support, and operate clinical applications and devices	67	9	0	76
Clinical content: The text, numeric data, and images that constitute the "language" of clinical applications	22	15	1	38
Human-computer interface: All aspects of technology that users can see, touch, or hear as they interact with it	16	12	1	29
People: Everyone who interacts in some way with technology, including developers, users, IT personnel, and informaticians	5	15	0	20
Workflow and Communication: Processes to ensure that patient care is carried out effectively	24	11	0	35

TABLE 2.3.1 *(Continued)*

Sociotechnical Dimension	Phase 1 Unsafe technology or technology failures (n=74)	Phase 2 Unsafe or inappropriate *use* of technology (n=25)	Phase 3 Lack of monitoring of safety concerns (n=1)	Total
Internal Organizational Features: Policies, procedures, work-environment and culture	4	2	0	6
External Rules and Regulations: Federal or state rules that facilitate or constrain preceding dimensions	1	1	0	2
System Measurement and Monitoring: Processes to evaluate both intended and unintended consequences of health IT implementation and use	1	0	0	1

TABLE 2.3.2 EHR-related Safety Concerns with Definitions and Examples

Category of Concern	Definition	Examples
Unmet display needs (n=36)	Information needs and content display mismatch	User required to review multiple screens to determine status of orders or review active medications User working on two patients with two instances of EHR orders medication for wrong patient User interface wording and function inconsistent throughout EHR Order entry dialog allows conflicting information to be entered
Software modifications (n=24)	Concerns due to upgrades, modifications, or configuration	Software designed at remote facility conflicts with local software use Despite testing, a new feature allows unauthorized users to sign orders Corrupted files or databases prevent entry of diagnoses, orders Corrupted files or databases prevent retrieval of complete patient information

Analysis of EHR Safety   77

TABLE 2.3.2 *(Continued)*

Category of Concern	Definition	Examples
Hidden dependencies in distributed system (n=17)	One component of the EHR is unexpectedly or unknowingly affected by the state or condition of another component	Transition of patients between wards or units not reflected in EHR, resulting in missed medications or orders Bulk ordering of blood products results in prolonged delay due to matching algorithm Template completion depends on remote data and user is unaware that network delays have caused failure User assigns surrogate signer for patient alerts, but alerts not forwarded due to logical error not seen by user
System-system interface (n=17)	Concerns due to failure of interface between EHR systems or components	Failure of patient context manager Remote internal server failure prevents relevant patient data to be retrieved Radiology studies canceled in EHR remain active in Picture Archiving and Communication System (PACS) workflow Interface flaw causing duplicate patient record creation from external source

2.3.3.1 CONCERNS RELATED TO UNMET DATA DISPLAY NEEDS IN THE EHR

Unmet display needs was the most common type of concern observed (36 incidents). This category represented a pattern of hazards in which human-EHR interaction processes did not adequately support the tasks of the end-users. These events reflected a poor fit between information needs and the task at hand, the nature of the content being presented (e.g., patient specific information requiring action, such as drug-allergy warnings or information required for successful order entry), and the way the information was displayed. As a result of these conditions, the displayed information available to the end-user failed to reduce uncertainty or led to increased potential for patient harm.

As an example, one incident described a situation in which a patient was administered a dose of a diuretic that exceeded the prescribed amount.

This error occurred due to a number of interacting sociotechnical factors. First, a pharmacist made a data entry error while approving the order for a larger-than-usual amount of diuretic. Although a dose error warning appeared upon order entry, this particular warning was known to have a high false positive rate. Due to diminished user confidence in the warning's reliability, the warning was overridden. The override released the incorrect dose for administration by nursing staff. The nurse, unaware of the discrepancy between the prescribed amount and the amount approved by the pharmacist, administered the larger dose. This event highlights complex interactions between the hardware and software, human-computer interface, people, and workflow and communication dimensions, which served to either prevent or obscure the users' receipt of appropriate information. Across the 36 concerns within this concern type, the contributory dimensions were hardware and software (22 incidents), human-computer interface (22 incidents), workflow and communication (10 incidents), clinical content (9 incidents), people (9 incidents), organizational policies and procedures (2 incidents), and system measurement and monitoring (1 incident). Most (22 of 36) of these concerns were classified as phase one issues, followed by phase two, and 1 phase three.

2.3.3.2 CONCERNS RELATED TO BOTH INTENDED AND UNINTENDED SOFTWARE MODIFICATIONS

The second most frequent concern type involved upgrades to the EHR or one of its components, or improperly configured software (24 incidents). One configuration error included a disease management package that, after local implementation, was found to have erroneously escalated user privileges to place and sign orders. Another concern involved "legacy" software (i.e., an older system that has not evolved despite newer technologies[46]) that needed an upgrade or maintenance, but support staff did not have sufficient knowledge of these systems. For example, one incident described an inadvertent change to a configuration file during an update to the EHR that prevented the EHR from communicating with the printing system used to label laboratory specimens. Since these printers were installed and configured prior to recruitment of the current staff, the configuration error was not immediately recognized. The main contributing sociotechnical dimensions of this concern type were hardware and software (21 incidents), clinical content (10 incidents), and workflow and communication (5 incidents). This concern type was most often associated with phase one EHR safety (21 incidents). Three concerns were classified as phase two, and none were phase three.

2.3.3.3 CONCERNS RELATED TO SYSTEM-SYSTEM INTERFACES

We analyzed 17 cases where the primary safety concern involved, system-system interfaces, the means by which information is transferred from one EHR component to another. Patient safety concerns in this category often involved maintaining a unique patient's context, a process designed to keep various individual EHR components centered on a single patient as the user traverses the EHR components.[47] For example, if patient context is not maintained between the user's EHR screen and the radiology viewing screen, a different patient's data will be shown in the two EHR components and the user may incorrectly assume the data is associated with the original patient. Patient context-related concerns were caused by network failures, conflicts created by non-EHR software, and EHR upgrades that were not compliant with context protocols.

Another example of a system-system interface concern occurred when a patient who was allergic to angiotensin converting enzyme (ACE) inhibitors presented to an emergency department with elevated blood pressure. The patient was prescribed an ACE inhibitor and subsequently required treatment for allergic reactions and angioedema. Although the patient's medication allergy list at a remote facility included ACE inhibitors, a network problem prevented remote allergy checking. As highlighted in this example, the system-system interface concern involved interactions from multiple sociotechnical dimensions: hardware and software (17 incidents), workflow (6 incidents), and content (5 incidents). All incidents of this concern category were coded as phase one EHR use.

2.3.3.4 CONCERN OF HIDDEN DEPENDENCIES IN DISTRIBUTED SYSTEMS

Concerns may develop not only because the EHR fails to support a particular task, but also because other processes within the EHR system conflict with the safe execution of that task. The concern of hidden dependencies or "cascading" effects[48] occurs if one component of the EHR system is unexpectedly or unknowingly affected by the state or condition of another component. While safety concerns involving hidden dependencies and system-system interfaces are not mutually exclusive, system-system interfaces are usually known and therefore potential points of failure and possible safety concerns may be more readily identified. For example, one safety concern involved medications that were ordered for a patient who was admitted to the hospital, but temporarily placed in an outpatient unit. Once the patient was transferred to the regular inpatient unit, certain medications were automatically removed from the active medication list because they were previously ordered on an "outpatient" status. This "hidden dependency" (i.e., between the patient's physical location and medication order status) can be potentially harmful to the patient because there was no clear expectation that medications would need to be re-ordered. Another example of a hidden dependency was a blood product compatibility matching algorithm that was not equipped to handle an incoming bulk order, which exponentially delayed the processing of blood products. This delay resulted in a disruption of the blood bank workflow by preventing further entry of blood product orders through the EHR and delaying release of blood products to the requesting clinical services.

The concerns of hidden dependencies primarily involved the dimensions of hardware and software (14 of 17 incidents), workflow (14 incidents), clinical content (9 incidents), and people (5 incidents). Incidents in this category were also noted to be largely dependent on multiple interactions between these dimensions, and only one incident was coded with a single dimension. These incidents also spanned both phase one (n=11) and phase two (n=6) of EHR safety.

2.3.4 DISCUSSION

We analyzed 100 unique, consecutive investigations of EHR-related safety concerns reported to and investigated by the VA's Informatics Patient Safety Office. Although the reports documented a variety of EHR-related safety incidents, four broad types of safety concerns were prominent. These were unmet data display needs within the EHR, problems with software modifications or upgrades, concerns related to system-system interfaces, and hidden dependencies within the EHR. Safety concerns typically emerged from complex interactions of multiple sociotechnical aspects of the EHR system. Although it is challenging to detect these concerns, let alone prevent them, our findings may be useful in guiding proactive efforts to monitor and improve safety as more institutions adopt EHRs.[49,50]

A novel feature and strength of our study is the use of an information-rich data source. Previous studies have largely used isolated event reports without benefit of an independent human factors investigation to analyze or replicate the event in the EHR.[16,26-28] Conversely, we analyzed the contents of both initial incident reports as well as the findings of the detailed safety investigations and analysis conducted by the VA's IPS office. Our data sources included detailed narratives that explained the circumstances in which safety concerns arose, the actions of users and EHR systems at the time of the concerns, and, when possible, the final determination of causes or preventive strategies. This level of detail enabled a more robust analysis in terms of understanding the larger sociotechnical context in which an event occurred.

Our sociotechnical analysis of completed IPS investigations provides additional opportunities for safety improvement. Other large reports of HIT-related safety concerns have focused on incident reports.[26,51] However, our study involved incident reports that had received further detailed investigation by informaticians and human factors experts. While studies using self-reported data, including this one, are limited by the possibility of reporters' recall bias or knowledge, [52] our methods may allow for a more complete representation of an incident and the underlying safety concern. Additional strengths of our study include the nationwide distribution of our sample of EHR-related safety events and the relatively sophisticated implementation and use of the EHR across the VA healthcare system.[37] As an early adopter of EHRs, the VA has evolved into a "learning system" that dedicates resources to investigating safety concerns and making EHR-related safety improvements decades after first launch.[40]

TABLE 2.3.3 EHR-related Safety Concerns and Suggested Mitigating Procedures

Category of Concern	Mitigating Procedures
Unmet display needs	• Testing information display in context of "real-world" tasks
	• Validating display with all expected information and reasonable unexpected information
	• Ensuring essential information is complete and clearly visible on the screen
	• System messages and labels are unambiguously worded
Software modifications	• Availability and testing of appropriate hardware and software occurs at the unit level and as-installed before go-live
	• Testing changes with full range of clinical content
	• Exploring impact of changes on workflows
System-system interface	• Understanding, documenting, and testing content and workflow requirements on both sides of interface.
	• Ensuring communication is complete (disallow partial transmission of information)
	• Developing workflows that incorporate back-up methods to transmit information
Hidden dependencies in distributed system	• Documenting ideal actions of EHR or components
	• Documenting assumptions or making dependencies explicit in software, workflows
	• Establishing monitoring and measurement practices with system-wide scope

Our findings underscore the importance of continuing the process of detecting and addressing safety concerns long after EHR implementation and "go-live" has occurred. Having a mature EHR system clearly does not eliminate EHR-related safety concerns and a majority of reported incidents were phase 1 or unsafe technology. However, few healthcare systems have robust reporting and analytic infrastructure similar to the VA's IPS. In light of increasing use of EHRs, activities to achieve a resilient EHR-enabled healthcare system should include a reporting and analysis infrastructure for EHR-related safety concerns as well as proactive risk assessments to identify safety concerns.[49]

Although we cannot make specific claims about the prevalence of various EHR-related concerns, it is notable that the vast majority of incidents could be classified into one of four types of concern. The categories that emerged from our analysis appear to represent common and significant safety concerns that need to be addressed with current and future EHR implementations. Some safety concerns had relatively straightforward origins, such as simultaneous use of multiple instances of an EHR application by a single user, leading to order entry on the wrong patient. Other problems had more complex origins, such as user misinterpretation of information presented through the EHR's user interface. Our study suggests that technology-based solutions alone will only partially mitigate concerns and that interventions to improve EHR-related safety should encompass the people, organizations, systems, and policies that influence how EHRs are used. We list several general mitigating procedures that could be used to address these concerns in Table 2.3.3.

This study has several limitations. All incidents were related to use of the same EHR within a single, albeit very large, healthcare system. Although the sample size is smaller than that of some other studies, the case descriptions were rich (i.e., 2-4 single-spaced pages), spanned a period of 3 years, and represented a continuum of care from home-based primary care to large, urban medical centers. Nevertheless, our findings may not represent all types of EHR-related safety concerns and might not be generalizable to other institutions with different organizational characteristics, HIT infrastructure, or patient safety reporting mechanisms. The data used for our analysis were composed of safety concerns that ranged from unsafe conditions to patient harm. Although the analysis of unsafe conditions or near misses is useful to illustrate concerns in EHR-enabled care, we acknowledge that their circumstances or implications may be different from adverse events that result in patient harm. All four emergent safety concerns affected or had the potential to affect multiple patients, but we did not analyze additional data on patient outcomes as a result of these concerns. In general, less than 10% of medical errors are captured through reporting and such data does not allow us to calculate prevalence rates.[52-54] Despite capturing a low percentage of errors, we were able to gain insight about non-technical aspects of EHR-related safety concerns that may not be routinely considered in technology-focused investigations.

In conclusion, our study demonstrates the potential utility of analyzing patient safety concerns using a sociotechnical approach to account for the complexities of using health information technology. We found that even within a well-established HIT infrastructure, many significant EHR-related safety concerns related to both unsafe technology and unsafe use of technology remain. The predominant concerns we identified can help to focus future safety assessment activities and, if confirmed in other studies, can be used to prioritize ongoing interventions or further research. Safety concerns we identified had complex sociotechnical origins and would need multifaceted strategies for improvement. Thus, institutions with long-standing EHRs as well as those currently implementing EHRs should consider building a robust infrastructure to monitor and learn from EHR-related safety concerns.

REFERENCES

1. Institute of Medicine (IOM). Crossing the quality chasm a new health system for the 21st century. National Academy Press; 2001.
2. Sittig DF, Singh H. Legal, ethical, and financial dilemmas in electronic health record adoption and use. Pediatrics 2011 Apr;127(4):e1042-e1047.
3. Ash JS, Berg M, Coiera E. Some Unintended Consequences of Information Technology in Health Care: The Nature of Patient Care Information System-related Errors. Journal of the American Medical Informatics Association 2004 Mar 1;11(2):104-12.
4. Balka E, Doyle-Waters M, Lecznarowicz D, FitzGerald JM. Technology, governance and patient safety: Systems issues in technology and patient safety. International Journal of Medical Informatics 2007 Jun;76, Supplement 1(0):S35-S47.
5. Bates W, Cohen M, Leape L, Marc Overhage J, Michael Shabot M, Sheridan T. Reducing the Frequency of Errors in Medicine Using Information Technology. Journal of the American Medical Informatics Association 2001 Jul 1;8(4):299-308.
6. Coleman RW. Translation and interpretation: the hidden processes and problems revealed by computerized physician order entry systems. J Crit Care 2004 Dec;19(4):279-82.

Analysis of EHR Safety

7. Hundt AS, Adams JA, Schmid JA et al. Conducting an efficient proactive risk assessment prior to CPOE implementation in an intensive care unit. International Journal of Medical Informatics 2013 Jan;82(1):25-38.
8. Koppel R. Role of computerized physician order entry systems in facilitating medication errors. JAMA: The Journal of the American Medical Association 2005 Mar 9;293(10):1197-203.
9. Patterson ES, Cook RI, Render ML. Improving Patient Safety by Identifying Side Effects from Introducing Bar Coding in Medication Administration. Journal of the American Medical Informatics Association 2002 Sep 1;9(5):540-53.
10. Pirnejad H, Niazkhani Z, van der SH, Berg M, Bal R. Impact of a computerized physician order entry system on nurse-physician collaboration in the medication process. Int J Med Inform 2008 Nov;77(11):735-44.
11. Fairbanks RJ, Wears RL. Hazards With Medical Devices: The Role of Design. Annals of Emergency Medicine 2008 Nov;52(5):519-21.
12. Karsh BT, Weinger MB, Abbott PA, Wears RL. Health information technology: fallacies and sober realities. J Am Med Inform Assoc 2010 Nov;17(6):617-23.
13. Institute of Medicine (IOM). Health IT and Patient Safety: Building Safer Systems for Safer Care. Washington, DC: The National Academies Press; 2012.
14. Middleton B, Bloomrosen M, Dente MA et al. Enhancing patient safety and quality of care by improving the usability of electronic health record systems: recommendations from AMIA. Journal of the American Medical Informatics Association 2013 Jun 1;20(e1):e2-e8.
15. Singh H, Mani S, Espadas D, Petersen N, Franklin V, Petersen LA. Prescription errors and outcomes related to inconsistent information transmitted through computerized order entry: a prospective study. Arch Intern Med 2009 May 25;169(10):982-9.
16. Myers RB, Jones SL, Sittig DF. Review of Reported Clinical Information System Adverse Events in US Food and Drug Administration Databases. Appl Clin Inform 2011;2(1):63-74.
17. Weiner JP, Kfuri T, Chan K, Fowles JB. "e-Iatrogenesis": the most critical unintended consequence of CPOE and other HIT. J Am Med Inform Assoc 2007 May;14(3):387-8.
18. Harrington L, Kennerly D, Johnson C. Safety issues related to the electronic medical record (EMR): synthesis of the literature from the last decade, 2000-2009. J Healthc Manag 2011 Jan;56(1):31-43.
19. Veterans Health Administration. VHA National Patient Safety Improvement Handbook. Washington, DC; 2011 Mar 4. Report No.: VHA Handbook 1050.1.
20. Clancy CM. Common Formats Allow Uniform Collection and Reporting of Patient Safety Data by Patient Safety Organizations. American Journal of Medical Quality 2010 Jan 1;25(1):73-5.
21. Sittig DF, Singh H. A new sociotechnical model for studying health information technology in complex adaptive healthcare systems. Qual Saf Health Care 2010 Oct;19 Suppl 3:i68-i74.
22. Harrison MI, Koppel R, Bar-Lev S. Unintended Consequences of Information Technologies in Health Care -- An Interactive Sociotechnical Analysis. Journal of the American Medical Informatics Association 2007 Sep 1;14(5):542-9.

23. Harrison MI, Henriksen K, Hughes RG. Improving the health care work environment: a sociotechnical systems approach. Jt Comm J Qual Patient Saf 2007 Nov;33(11 Suppl):3-6, 1.
24. Carayon P, Schoofs Hundt A, Karsh BT et al. Work system design for patient safety: the SEIPS model. Quality and Safety in Health Care 2006 Dec 1;15(suppl 1):i50-i58.
25. Health Information Technology Safety Action & Surveillance Plan. The Office for the National Coordinator of Health Information Technology 2013 July 2 [cited 2013 Jul 11];Available from: URL: http://www.healthit.gov/sites/default/files/safety_plan_master.pdf
26. Magrabi F, Ong MS, Runciman W, Coiera E. Using FDA reports to inform a classification for health information technology safety problems. J Am Med Inform Assoc 2012 Jan;19(1):45-53.
27. Magrabi F, Ong MS, Runciman W, Coiera E. An analysis of computer-related patient safety incidents to inform the development of a classification. J Am Med Inform Assoc 2010 Nov;17(6):663-70.
28. Sparnon E, Marella WM. The role of the electronic health record in patient safety events. Harrisburg, Pa.: Pennsylvania Patient Safety Authority; 2012 Dec.
29. Henriksen K, Kaye R, Morisseau D. Industrial ergonomic factors in the radiation oncology therapy environment. Advances in Industrial Ergonomics and Safety V. Taylor and Francis; 1993. p. 325.
30. Charles V, Sally T, Nicola S. Framework for analysing risk and safety in clinical medicine. BMJ 1998 Apr 11;316.
31. Meeks DW, Takian A, Sittig DF, Singh H, Barber N. Exploring the Sociotechnical Intersection of Patient Safety and Electronic Health Record Implementation. Journal of the American Medical Informatics Association 2013 Sep 19.
32. Sittig DF, Singh H. Defining health information technology-related errors: new developments since to err is human. Arch Intern Med 2011 Jul 25;171(14):1281-4.
33. Singh H, Wilson L, Petersen LA et al. Improving follow-up of abnormal cancer screens using electronic health records: trust but verify test result communication. BMC Med Inform Decis Mak 2009;9:49.
34. Singh H, Thomas EJ, Sittig DF et al. Notification of abnormal lab test results in an electronic medical record: do any safety concerns remain? Am J Med 2010 Mar;123(3):238-44.
35. U.S.Department of Veterans Affairs. About VHA. 2014 [cited 2014 Jan 15];Available from: URL: http://www.va.gov/health/aboutVHA.asp
36. Brown SH, Lincoln MJ, Groen PJ, Kolodner RM. VistA−U.S. Department of Veterans Affairs national-scale HIS. International Journal of Medical Informatics 2003 Mar;69(2−3):135-56.
37. Spetz J, Burgess JF, Phibbs CS. What determines successful implementation of inpatient information technology systems? Am J Manag Care 2012 Mar;18(3):157-62.
38. Perlin JB, Kolodner RM, Roswell RH. The Veterans Health Administration: quality, value, accountability, and information as transforming strategies for patient-centered care. Am J Manag Care 2004 Nov;10(11 Pt 2):828-36.
39. Bonner LM, Simons CE, Parker LE, Yano EM, Kirchner JE. 'To take care of the patients': Qualitative analysis of Veterans Health Administration personnel experiences with a clinical informatics system. Implement Sci 2010;5:63.

40. Taylor L, Wood SD, Chapman R. Analysis and Mitigation of Reported Informatics Patient Safety Adverse Events at the Veterans Health Administration.: Human Factors and Ergonomics Society; 2012.
41. McCoy AB, Wright A, Kahn MG, Shapiro JS, Bernstam EV, Sittig DF. Matching identifiers in electronic health records: implications for duplicate records and patient safety. BMJ Quality & Safety 2013 Mar 1;22(3):219-24.
42. Graber ML, Franklin N, Gordon R. Diagnostic error in internal medicine. Arch Intern Med 2005 Jul 11;165(13):1493-9.
43. Green J, Thorogood N. Qualitative methods for health research. Introducing qualitative methods.London: SAGE, 2004. p. xv, 262.
44. Pope C, Ziebland S, Mays N. Qualitative research in health care. Analysing qualitative data. BMJ 2000 Jan 8;320(7227):114-6.
45. Sittig DF, Singh H. Electronic Health Records and National Patient-Safety Goals. N Engl J Med 2012 Nov 7;367(19):1854-60.
46. Brodie ML, Stonebraker M. Migrating legacy systems: gateways, interfaces & the incremental approach. Morgan Kaufmann Publishers Inc.; 1995.
47. Sittig DF, Teich JM, Yungton JA, Chueh HC. Preserving context in a multi-tasking clinical environment: a pilot implementation. Proc AMIA Annu Fall Symp 1997;784-8.
48. Patterson ES, Roth EM, Woods DD. Facets of complexity in situated work. Macrocognition Metrics and Scenarios: Design and Evaluation for Real-World Teams Ashgate Publishing ISBN 2010;978-0.
49. Singh H, Ash J, Sittig D. Safety Assurance Factors for Electronic Health Record Resilience (SAFER): study protocol. BMC Medical Informatics and Decision Making 2013;13(1):46.
50. Wright A, Henkin S, Feblowitz J, McCoy AB, Bates DW, Sittig DF. Early results of the meaningful use program for electronic health records. N Engl J Med 2013 Feb 21;368(8):779-80.
51. Cheung KC, van der Veen W, Bouvy ML, Wensing M, van den Bemt PMLA, de Smet PAGM. Classification of medication incidents associated with information technology. Journal of the American Medical Informatics Association 2013 Sep 24.
52. Holden RJ, Karsh BT. A Review of Medical Error Reporting System Design Considerations and a Proposed Cross-Level Systems Research Framework. Human Factors: The Journal of the Human Factors and Ergonomics Society 2007 Apr 1;49(2):257-76.
53. Sari AB-A, Trevor AS, Alison C, Alastair T. Sensitivity of routine system for reporting patient safety incidents in an NHS hospital: retrospective patient case note review. BMJ 2007 Jan 11;334.
54. Brubacher JR, Hunte GS, Hamilton L, Taylor A. Barriers to and incentives for safety event reporting in emergency departments. Healthcare quarterly (Toronto, Ont) 2011;14(3):57-65.

CHAPTER 3

USER CONTEXT OF SAFE AND EFFECTIVE EHR USE

RIGHTS AND RESPONSIBILITIES OF ELECTRONIC HEALTH RECORD USERS

Dean F. Sittig and Hardeep Singh

3.1.1 INTRODUCTION

Over the last 10 years the governments of Australia [1], Belgium [2], Canada [3], Denmark [4], United Kingdom [5], and most recently the United States [6], have all made long-term, multi-billion dollar investments in health information technologies (HIT) including electronic health records (EHRs). Although the definition of an EHR might vary across countries, most EHRs include systems that are widely accessible across a health-

---

Reprinted with permission from the Canadian Medical Association. Sittig DF and Singh H. Rights and Responsibilities of Electronic Health Records. Canadian Medical Association Journal 2012 Sep 18;184(13):1479-83.

care network and provide a computer-based user interface that replaces the paper chart. The primary goal of these HIT initiatives is to transform the collection, display, transmission, and storage of patient data with the aim of improving health, while a secondary goal is to use these data to improve the healthcare delivery system. The rationale for these investments stems from numerous quality and safety concerns related to paper-based systems, which include legibility problems, access limited to a single provider at a single location, difficulties with aggregating information from multiple records, and problems maintaining record confidentiality and accurate backup copies [7]. Comprehensive, well-implemented EHRs with advanced clinical decision support interventions have potential to reduce medication errors [8] and increase the quality, efficiency, and reliability of information transfer [9,10].

Despite progress [11], EHR adoption has resulted in larger than expected challenges in day-to-day clinical processes [12]. For example, processing electronic information [13,14,15] can reduce clinician productivity and increase workload, while other disruptions can result in safety concerns due to loss of attention and situation awareness [16]. Thus, clinicians may perceive that the costs of EHRs (eg, time, monetary or required changes in workflow) outweigh direct benefits to themselves, whereas patients and payers appear to benefit more readily [17]. Clinicians require assurances that EHRs will deliver the features and functions they need and that the regulatory environment will support them.

Based on recent literature and our experiences in clinical informatics-focused health services research, we identified ten emergent topics that, if addressed, could overcome some of these challenges. Topics were grounded in our recently developed 8-dimension socio-technical model of safe and effective EHR use [18]. These topics were circulated among several colleagues, including practicing clinicians, informaticists, and computer scientists, who offered their feedback. This was followed by presentations at four international scientific meetings with multidisciplinary audiences who gave additional feedback. All this input was taken into consideration and used to iteratively refine the rights and responsibilities with a goal of making them as universally acceptable and applicable as possible.

3.1.2 CONTEXTUALIZING EMERGENT TOPICS AS CLINICIAN "RIGHTS" AND RESPONSIBILITIES

Some degree of workflow disruption is inevitable with EHR implementation, which requires modification of long-standing work processes derived from paper-based systems. In addition, EHR use often results in loss of clinician autonomy due to increased external oversight (ie, clinician profiling) and control (ie, orderable medications limited to formulary) facilitated by its features and functions. Concomitantly, practicing clinicians are often at a relative disadvantage when negotiating EHR-related issues with other stakeholders (eg, healthcare administrators, HIT vendors, governments, insurance companies or other payers, policy makers). To preserve a balance and to encourage debate between clinicians and other stakeholders involved, we discuss these topics as what front-line practicing clinicians would want as "professional rights", i.e. not merely desirable but "must-have" EHR features, functions, and user privileges that are important to provide the highest quality, safest, and most cost-effective care. Nevertheless, each "right" is accompanied by a corresponding clinician responsibility, without which the ultimate goal of improving health care quality might not be achieved [19]. We acknowledge that contextualizing these topics as clinician rights has significant implications for other stakeholders, but these issues must be addressed to move the field forward. Although these "rights" are clearly not of the same magnitude or universal importance as the World Health Organization's human rights-based approach to health [20] or the Hippocratic Oath [21], they can reduce the potential impact of unintended adverse consequences on patient care and clinicians' livelihoods. These "rights" could be a foundation upon which HIT designers, developers, implementers, policy makers, and most importantly, users can co-create a new age of computer-assisted healthcare [22].

3.1.3 TEN PROFESSIONAL "RIGHTS" AND RESPONSIBILITIES

- *Universal EHR access.* Extended EHR outages pose a significant risk to patient care. Therefore, clinicians have the right to have an EHR they can access via a secure, organizationally approved, network-attached device 24 hours per day, 7 days per week, 365 days per year. Although no device or system can be 100% reliable, EHR vendors, institutions, and physicians must work together to design, develop, implement, and use fail-safe equipment and downtime processes to ensure that patient care continues in the event of an outage. Clinicians have the responsibility to protect their passwords, log off the system when done, and access only records of patients under their care or within their administrative purview.

- *No "missing" data.* Clinicians have the right to see all clinical data that were captured in the normal course of care for each of their patients [23]. Amid concerns about patient privacy, some argue that patients or clinicians should be able to "hide" specific data (eg, records of psychiatric or substance abuse treatment) [24] or even to "opt-out" of having their data available to other clinicians [25,26]. This withheld data increases clinician liability unnecessarily. Clinicians have the responsibility to ensure that having all patient data on their desktops does not replace the time-honored tradition of observing, listening to, and examining patients [27].
- *Succinct patient summaries.* Current EHRs contain a wealth of clinical data. As more community-wide health information exchanges come online, the amount of data available for review will grow exponentially, increasing the likelihood that relevant information will be overlooked. Clinicians thus have the right to EHRs that provide succinct summaries of their patients' medical problems, medications, laboratory test results, vital signs, and progress notes [23]. Some EHRs currently have "summary" views that arrange data by type (eg, all laboratory results together) and time (eg, most recent data first) on different screens. However, future innovations in this area are needed. For example, problem-oriented summaries that integrate data from different sources on one screen could potentially facilitate better information processing and exert a lower cognitive load [7]. Clinicians conversely have the responsibility to maintain accurate, up-to-date problem lists using a controlled clinical terminology (eg, SNOMED-CT) and link them with corresponding diagnostic and treatment elements through the EHR to prevent "incomplete care" [28].
- *Overriding computer-generated interventions.* Clinicians receive a large number of computer-generated alerts, many of which are considered unnecessary [29]. These alerts can cause cognitive overload and fatigue. Even more troublesome a few cannot be overridden because of local institutional configuration decisions requiring "hard stops" (i.e., the computer prohibits completion of the task) [30]. Clinicians should have the right to override, but not permanently disable, any computer-generated clinical intervention. In the event of an exceptionally hazardous scenario or when the organization's clinical leadership decides that a particular order should never occur, clinicians should be required to obtain an overriding co-signature from a higher ranking or more experienced clinician before completing the task. Disallowing overrides through hard-stops implies that computers have access to more accurate data and greater medical knowledge and expertise

than clinicians. In reality, computers are often not able to interpret or convey the clinical context for many reasons: unavailable or inaccurate data; errors in logical processing (eg, software bugs); situation-specific clinical exceptions (eg, user request for blood transfusion denied by a computer-generated intervention that did not capture active bleeding since last hemoglobin result); and user-interface limitations (eg, limited screen space available to show most recent laboratory results near medication order). Clinicians have the responsibility to justify overrides and be accountable for decisions by agreeing to have their actions reviewed. Additionally, they must participate on clinical decision support (CDS) oversight committees and work with other stakeholders to review, redesign, test, re-implement, or remove CDS interventions judged ineffective [31].
- *CDS Rationale.* Advanced CDS interventions are necessary if EHRs are to generate expected improvements in healthcare quality, safety, and effectiveness. Nevertheless, clinicians have the "right" to request and receive a clear, evidence-based rationale at the point of care for all computer-generated clinical interventions (eg, alerts or reminders). Physicians have the responsibility to carefully consider computer-generated clinical interventions; either blindly following or ignoring CDS interventions can lead to errors [32].
- *Reliable performance measurement.* EHR-based performance measurement is inevitable. Current data collection and measurement methods are not fail-safe and often measure what is easy to measure [33]. To correct discrepancies, clinicians have the right to review all EHR-based processes used to generate reports that inform policy decisions or performance measurement [34]. All computer-based measurements should have unambiguous exclusion criteria and allow clinicians to identify patients to whom the measure does not apply (eg, no diabetic foot exams on patients with bilateral below-the-knee amputations). If needed, clinicians should have access to queries, data extracts, and statistical methods used. Proactive collaboration with stakeholders such as organizational leaders will help ensure that performance measures are valid. To ensure continuous quality improvement, physicians have the responsibility to review the performance feedback they are provided and act on it.
- *Safe EHRs.* EHR software errors and usability issues are increasingly linked to safety hazards that can lead to patient harm (e-iatrogenesis) [35,36]. Clinicians have the right to expect that all EHR-related errors will be reported, investigated, and resolved in a timely manner [37]. EHR vendors and healthcare organizations responsible for maintaining the EHR should make these reports, along with their responses, publically available so that others can learn from them [38]. Clinicians have the responsibility to report, help investigate, and learn from EHR-related safety hazards.
- *Training and assistance.* State-of-the-art EHRs are complex tools designed to facilitate the entry, storage, review, interpretation, and transmission of patient data. Clinicians have the right to receive training—either from their EHR vendor or their healthcare organization—in all EHR features. Ongoing training and support should include access to online instruction and availability of real-time assistance while caring for patients, preferably in-

person [39]. Clinicians have the responsibility to maintain a high level of user proficiency with the same level of diligence as for other clinical skills. To improve efficiency and safety, clinicians must learn to type, complete EHR training, and demonstrate competence in use of all functions required to care for patients (eg, enter orders, add problems, initiate referrals). Finally, clinicians are responsible for asking for help when they reach limits of their EHR proficiency.

- *EHRs that are compatible with real-world clinical workflows.* Clinicians have the right to a safe, effective, and usable EHR that contains evidence-based, problem- and task-specific order sets, documentation templates, and information displays designed to be compatible with their clinical workflows [40]. Clinicians have the responsibility to work with EHR vendors and local information technologists to design, develop, and implement data entry, review, and CDS tools and to modify previous paper-based workflows to overcome limitations of EHRs.
- *EHRs that facilitate communication, coordination, and teamwork.* EHRs fundamentally change the way clinicians coordinate their work activities, communicate, and collaborate to deliver high-quality, safe, and effective healthcare [41]. Most current EHRs are not optimal for team-based care that includes patients and their caregivers [42]. Clinicians have the right to future EHR innovations that facilitate complex communication and coordination tasks across time, space, and people. Clinicians have the responsibility to use EHRs in ways that foster teamwork. They must document their findings, decisions, and actions succinctly, avoid reckless copy-and-paste, and respond to human- and computer-generated requests for information and action in a timely manner.

3.1.4 SETTING THE GROUNDWORK FOR FUTURE DEBATE

Although our essay lays the groundwork for future debate, it has several limitations. First, we do not specifically outline who might enforce these clinician "rights" and responsibilities or what alternatives could be pursued if these conditions are not met. However, we believe it is premature for us to do so at this conceptualization stage without further debate and agreement. Second, we recognize that even with consensus regarding the necessity of these "rights," delivering them in the short-term will be difficult using today's technology and in today's socio-political and economic environments. Our goal, however, is to lay the foundation for a long-term agenda for providing clinicians access to safe, effective and easy- to-use EHRs that support their cognitive and physical work processes. Finally, we recognize that achieving high-quality and affordable healthcare is a complex, socio-technical endeavor. Thus, these clinician rights might not be the perfect solution because there are many competing and often opposing views of the best way to accomplish this endeavor.

A competing view is that other stakeholders in this debate, including payers, administrators, policy makers, and patients, are also entitled to an equally important and valid set of "rights" which may conflict with one or more of the clinicians' "rights". Payers, administrators, and/or policy makers, for example, have the right to mandate use of EHR-related functions that promote patient safety (eg, order entry), prohibit use of EHR-related functions that jeopardize patient safety (eg, use of a non-secure, web-based calendar to facilitate clinician workflow [43] or use of text messaging for order entry [44]), enforce specific rules and regulations (eg, reprimand users for unauthorized access to patient data), create new CDS interventions to encourage efficient, effective, evidence-based care, and evaluate clinicians' performance using EHR data. Likewise, patients have the right to access their data, have any data entry errors corrected, obtain a list of everyone who has viewed their data, confidentially communicate electronically with their providers, and request that certain data not be used for purposes other than research or public health benefit without their written consent [45]. In the event that one group's rights infringe upon those of another group, we are optimistic that organizations and the constituents they represent will participate in an open, constructive debate on these "rights" and reach consensus [46]. Following ratification, relevant stakeholders (eg, EHR vendors, EHR implementers, professional boards, hospital committees, users, patients, and government agencies) can work together to design and implement EHRs and the corresponding policies, procedures and regulations required to ensure these "rights".

REFERENCES

1. HealthConnect Implementation Strategy v2.1 July 6, 2005. Available at: http://www.health.gov.au/internet/hconnect/publishing.nsf/Content/archive-docs/$File/implementation.pdf
2. France FR. eHealth in Belgium, a new "secure" federal network: role of patients, health professions and social security services. Int J Med Inform. 2011 Feb;80(2):e12-6. Epub 2010 Oct 28.

3. EHRS Blueprint: An interoperable EHR framework. Version 2. March 2006. Available at: https://knowledge.infoway-inforoute.ca/EHRSRA/doc/EHRS-Blueprint.pdf
4. Protti D, Johansen I. Widespread adoption of information technology in primary care physician offices in Denmark: a case study. Issue Brief (Commonw Fund). 2010 Mar;80:1-14.
5. House of Commons Public Accounts Committee. The National Programme for IT in the NHS: Progress since 2006. Second Report of Session 2008-09. Available at: http://www.publications.parliament.uk/pa/cm200809/cmselect/cmpubacc/153/153.pdf
6. Blumenthal D. Wiring the health system - origins and provisions of a new federal program. N Engl J Med. 2011 Dec 15;365(24):2323-9.
7. Powsner SM, Wyatt JC, Wright P. Opportunities for and challenges of computerisation. Lancet. 1998 Nov 14;352(9140):1617-22.
8. Bates DW, Leape LL, Cullen DJ, Laird N, Petersen LA, Teich JM, Burdick E, Hickey M, Kleefield S, Shea B, Vander Vliet M, Seger DL. Effect of computerized physician order entry and a team intervention on prevention of serious medication errors. JAMA. 1998 Oct 21;280(15):1311-6.
9. Singh H, Arora HS, Vij MS, Rao R, Khan MM, Petersen LA. Communication outcomes of critical imaging results in a computerized notification system. J Am Med Inform Assoc. 2007 Jul-Aug;14(4):459-66.
10. Singh H, Naik AD, Rao R, Petersen LA. Reducing diagnostic errors through effective communication: harnessing the power of information technology. J Gen Intern Med. 2008 Apr;23(4):489-94.
11. Protti D. Comparison of information technology in general practice in 10 countries. Healthc Q. 2007;10(2):107-16.
12. Westbrook JI, Braithwaite J. Will information and communication technology disrupt the health system and deliver on its promise? Med J Aust. 2010 Oct 4;193(7):399-400.
13. Poissant L, Pereira J, Tamblyn R, Kawasumi Y. The impact of electronic health records on time efficiency of physicians and nurses: a systematic review. J Am Med Inform Assoc. 2005 Sep-Oct;12(5):505-16.
14. Magrabi F, Ong MS, Runciman W, Coiera E. An analysis of computer-related patient safety incidents to inform the development of a classification. J Am Med Inform Assoc. 2010 Nov 1;17(6):663-70.
15. Committee on Patient Safety and Health information Technology Board on Healthcare Services. Health IT and Patient Safety: Building Safer Systems for Better Care. The National Academies Press, Washington, DC, 2011.
16. Singh H, Davis Giardina T, Petersen LA, Smith MW, Paul LW, Dismukes K, Bhagwath G, Thomas EJ. Exploring situational awareness in diagnostic errors in primary care. BMJ Qual Saf. 2011 Sep 2.
17. Sprivulis P, Walker J, Johnston D, Pan E, Adler-Milstein J, Middleton B, Bates DW. The economic benefits of health information exchange interoperability for Australia. Aust Health Rev. 2007 Nov;31(4):531-9.
18. Sittig DF, Singh H. A new sociotechnical model for studying health information technology in complex adaptive healthcare systems. Qual Saf Health Care. 2010 Oct;19 Suppl 3:i68-74.
19. Good medical practice: the duties of a doctor registered with the General Medical Council. Med Educ. 2001 Dec;35 Suppl 1:70-8.

20. World Health Organization. A Human Rights-based Approach to Health. Available at: http://www.who.int/hhr/news/hrba_info_sheet.pdf (Accessed 27 November 2011).
21. The Hippocratic Oath. Available at: http://www.nlm.nih.gov/hmd/greek/greek_oath.html (Accessed 4/7/2011).
22. Stead WW, Searle JR, Fessler HE, Smith JW, Shortliffe EH. Biomedical Informatics: Changing What Physicians Need to Know and How They Learn. Acad Med. 2011 Apr;86(4):429-434.
23. Patient safety and the electronic health record. Committee Opinion No. 472. American College of Obstetricians and Gynecologists. Obstet Gynecol 2010;116:1245–7.
24. Popovits RM. Confidentiality law: Time for change? Behavioral Healthcare 2010 April;30(4):11-13.
25. Watson N. Patients should have to opt out of national electronic care records: FOR. BMJ. 2006 Jul 1;333(7557):39-40.
26. Halamka JD. Patients should have to opt out of national electronic care records: AGAINST. BMJ. 2006 Jul 1;333(7557):41-2.
27. Verghese A. Culture Shock — Patient as Icon, Icon as Patient. N Engl J Med 2008; 359:2748-2751.
28. Gandhi TK, Zuccotti G, Lee TH. Incomplete care--on the trail of flaws in the system. N Engl J Med. 2011 Aug 11;365(6):486-8.
29. Isaac T, Weissman JS, Davis RB, Massagli M, Cyrulik A, Sands DZ, Weingart SN. Overrides of medication alerts in ambulatory care. Arch Intern Med. 2009 Feb 9;169(3):305-11.
30. Strom BL, Schinnar R, Aberra F, Bilker W, Hennessy S, Leonard CE, Pifer E. Unintended effects of a computerized physician order entry nearly hard-stop alert to prevent a drug interaction: a randomized controlled trial. Arch Intern Med. 2010 Sep 27;170(17):1578-83.
31. Wright A, Sittig DF, Ash JS, Bates DW, Feblowitz J, Fraser G, Maviglia SM, McMullen C, Nichol WP, Pang JE, Starmer J, Middleton B. Governance for clinical decision support: case studies and recommended practices from leading institutions. J Am Med Inform Assoc. 2011 Mar 1;18(2):187-94.
32. McCoy AB, Waitman LR, Lewis JB, Wright JA, Choma DP, Miller RA, Peterson JF. A Framework for Evaluating the Clinical Impact of Computerized Medication Safety Alerts . J Am Med Inform Assoc. 2011. doi:10.1136/amiajnl-2011-000185.
33. Ofri D. Quality Measures and the Individual Physician. N Engl J Med 2010; 363:606-607.
34. Department of Health and Human Services Centers for Medicare & Medicaid Services. 42 CFR Part 401, CMS-5059-F, RIN 0938-AQ17. Availability of Medicare Data for Performance Measurement. Available at: http://www.ofr.gov/OFRUpload/OFRData/2011-31232_PI.pdf (Accessed 14 December 2011).
35. Myers RB, Jones SL, Sittig DF. Review of reported clinical information system adverse events in US Food and Drug Administration databases. Appl Clin Inf 2011; 2: 63–74. doi: 10.4338/ACI-2010-11-RA-0064.
36. Institute of Medicine. Health IT and Patient Safety: Building Safer Systems For Better Care. Washington, DC: The National Academies Press, 2012. Available at: http://iom.edu/Reports/2011/Health-IT-and-Patient-Safety-Building-Safer-Systems-for-Better-Care.aspx (Accessed December 16, 2011).
37. Singh H, Classen DC, Sittig DF. Creating an Oversight Infrastructure for Electronic Health Record-Related Patient Safety Hazards. J Patient Saf. 2011 Dec;7(4):169-174.

38. Walker JM, Carayon P, Leveson N, Paulus RA, Tooker J, Chin H, Bothe A Jr, Stewart WF. EHR safety: the way forward to safe and effective systems. J Am Med Inform Assoc. 2008 May-Jun;15(3):272-7.
39. Ash JS, Stavri PZ, Dykstra R, Fournier L. Implementing computerized physician order entry: the importance of special people. Int J Med Inform. 2003 Mar;69(2-3):235-50.
40. Karsh B-T. Clinical practice improvement and redesign: how change in workflow can be supported by clinical decision support. AHRQ Publication No. 09-0054-EF. Rockville, Maryland: Agency for Healthcare Research and Quality. June 2009.
41. Campbell EM, Guappone KP, Sittig DF, Dykstra RH, Ash JS. Computerized provider order entry adoption: implications for clinical workflow. J Gen Intern Med. 2009 Jan;24(1):21-6. Epub 2008 Nov 20.
42. Thomas EJ. Improving teamwork in healthcare: current approaches and the path forward. BMJ Qual Saf. 2011 Jun 28.
43. Department of Veterans Affairs Monthly Report to Congress on Data Incidents. Nov 1-28, 2010. Available at: http://www.va.gov/ABOUT_VA/docs/monthly_rfc_nov2010.pdf (Accessed: 27 November 2011)
44. The Joint Commission - Texting Orders. Released November 10, 2011. Available at: http://www.jointcommission.org/standards_information/jcfaqdetails.aspx?StandardsFaqId=401&ProgramId=1 (Accessed 27 November 2011). Read more: http://www.ihealthbeat.org/articles/2011/11/21/joint-commission-text-messages-should-not-be-used-in-patient-orders.aspx#ixzz1ey8oU1hr
45. Smith M. Patient's Bill of Rights - A Comparative Overview (PRB 01-31E). Government of Canada: Depository Services Program, 2002. Available at: http://dsp-psd.pwgsc.gc.ca/Collection-R/LoPBdP/BP/prb0131-e.htm (Accessed 26 Nov 2011).
46. Beard L, Schein R, Morra D, Wilson K, Keelan J. The challenges in making electronic health records accessible to patients J Am Med Inform Assoc 2011; doi:10.1136/amiajnl-2011-000261

RIGHTS AND RESPONSIBILITIES OF EHR USERS CARING FOR CHILDREN

Dean F. Sittig, Hardeep Singh, and Christopher A. Longhurst

Establishing a safe and effective electronic health record-enabled (EHR) healthcare delivery system is complex and challenging. In addition to support from executive leadership, a robust EHR from a reputable vendor,

---

Reproduced with permission from Archivos Argentinos de Pediatría, *Sittig DF, Singh H, and Longhurst CA. Rights and Responsibilities of EHR Users Caring for Children* **111**,6 *(2013).*

and access to knowledgeable and committed information technology professionals, clinician support is instrumental in overcoming the challenges. While there is an increasing breadth of knowledge about good clinical practices needed to address EHR implementation and use in the general population, clinicians responsible for the care of neonates, children, and adolescents face a unique set of additional challenges. For example, children have unique EHR requirements related to dosing of medications as well as specific needs related to their growth and development that the EHR needs to facilitate [1].

In order to encourage a dialogue between clinicians and other stakeholders to help address and overcome these challenges, we previously proposed that front-line practicing clinicians be given certain "professional rights" for "must have" EHR features, functions, and user privileges that are critical to provide high quality and safe care. We also proposed that each "right" be accompanied by a corresponding user responsibility. Because of the unique circumstances involving the safe and effective care of children and that fact that most children are not cared for in facilities where the EHR has been designed exclusively for children, in this paper we propose "pediatric amendments" to our previously proposed "Rights and Responsibilities of Users of EHRs" [2]. All previously identified rights and responsibilities still apply along with these new pediatric-specific items discussed below.

3.2.1 SUPPORT FOR MEDICATION PRESCRIBING IN CHILDREN

The epidemiology of harm associated with medication prescribing for neonates and children is very different than adult patients. Both hospitalized and ambulatory patients are at higher risk of harm from drug dosing errors than from drug-drug interactions. [3,4] Clinicians seeing pediatric patients have the right to both inpatient and ambulatory electronic prescribing systems that are safer and more effective for children and include weight-based dosing recommendations, age appropriate dosing calculators, dose-range checking, and pediatric-specific drug-drug interaction alerts. [5,6]

Clinicians seeing pediatric patients have the responsibility to consistently and reliably document patient weights, and should maintain familiarity with medication dosing guidelines to mitigate the effect of automation bias [7].

3.2.2 ELECTRONIC DISPLAY OF GROWTH CHARTS

Visual display of patient information is an important decision support tool. Clinicians should have the right to view their young patients' anthropometric data using growth charts [8] that display age-based percentiles for weight, height, head circumference, and body mass index (BMI) within their EHR [9].

All of these age-appropriate displays require up-to-date, accurate data capture; therefore, clinicians have the responsibility to record or facilitate the recording of patient's height, weight, and head circumference. Additionally, they should use this information to apply the appropriate age-specific clinical guidelines and provide copies of these charts to parents.

3.2.3 CHILD-FRIENDLY, EHR-EQUIPPED EXAM ROOM

While not a specific feature or function of the EHR, clinicians caring for children have the right to an EHR-equipped exam room that is designed using appropriate human factors principles [10]. For example, rooms should have a layout that provides adequate room for the patient, a parent and the clinician to move around [11]. In addition, keyboards and touchscreens should be cleaned and disinfected on a regular basis [12]. Finally, the computer, if wall-mounted, should be sturdy enough to withstand a child swinging from the support arm.

Clinicians have the responsibility for positioning the monitor so that he/she, as well as the parent and the patient can see the screen simultaneously. This is particularly important in pediatrics, as children cannot rationalize the use of a computer in the exam room and may unintentionally misinterpret the intention [13].

3.2.4 USER INTERFACE THAT SUPPORTS CORRECT IDENTIFICATION OF PATIENTS

Several studies have suggested that pediatric patients in general and neonates in particular are at higher risk for misidentification because of naming issues during the newborn period and siblings being treated simultaneously at pediatric visits [14]. Clinicians who see these patients have a right to an EHR user interface which minimizes wrong-patient errors. Such functionality may include limiting users to one open chart at a time, availability of patient pictures within the EHR, and including additional patient verification processes with computerized order entry systems. [15,16]

Electronic systems themselves may actually carry the unintended consequence of increasing the risk for wrong-patient errors [17]. Users of these systems have a responsibility to ensure that processes are setup to capture patient photographs in the EHR, and that misidentification errors are appropriately reported and fixed.

3.2.5 AN EHR THAT SUPPORTS ADOLESCENT CONFIDENTIALITY

Although exact legal requirements vary, most countries acknowledge that adolescents have the right to keep mental, behavioral, and sexual healthcare confidential from their parents or guardians. Unfortunately, many commercial EMR's do not yet provide the functionality needed to respect these legal and ethical positions [18]. Pediatric users have the right to EHR software which includes default settings for adolescent privacy, customizable point-of-care privacy controls for clinicians, clear on-screen labeling of confidential data elements, patient-adjustable proxy access capabilities for patient portals, and suppression capabilities for specific items on post-visit summaries, bills, and post-visit surveys. In addition, adolescent privacy standards must be built into health information exchange data sharing agreements.

Clinicians seeing adolescent patients have the responsibility to understand local adolescent confidentiality regulatory requirements. They should also review the entire patient experience from registration to post-clinic surveys to ensure that the adolescent's confidentiality is maintained in light of these requirements.

3.2.6 EHR CONTENT THAT SUPPORTS PEDIATRIC PRACTICE

To deliver appropriate preventative well-child care, pediatricians have the right to an EHR with content that supports the care of children. This includes appropriate decision support rules for timely preventive care such as administration of immunizations and linkages to immunization registries as well as content for pediatric normative values (e.g. laboratory test values) that frequently change with age [19]. Furthermore, EHRs must be optimized to support recording of quality measures for pediatrics.

Pediatricians have the responsibility to review decision support rules (e.g. do they match local vaccination schedules) and record key data that would lead to the generation of appropriate decision support.

3.2.7 SUMMARY

The care of children and neonates presents complex challenges for the design and operation of healthcare facilities and EHRs worldwide. For clinicians to provide the highest quality, safe and effective care to children, EHRs providing care to children must be properly designed and configured and clinicians must use them correctly. Organizations that provide their clinicians with state-of-the-art EHRs and grant them the "professional rights" we previously identified along with these "pediatric amendments" could see dramatic improvements in clinician usage of their EHRs. This will lead us closer to the ultimate goal of improving the quality, safety, and effectiveness of care delivered to children.

REFERENCES

1. Spooner SA; Council on Clinical Information Technology, American Academy of Pediatrics. Special requirements of electronic health record systems in pediatrics. Pediatrics. 2007 Mar;119(3):631-7.
2. Sittig DF, Singh H. Rights and responsibilities of users of electronic health records. CMAJ. 2012 Sep 18;184(13):1479-83. doi: 10.1503/cmaj.111599.
3. Kaushal R, Bates DW, Landrigan C, McKenna KJ, Clapp MD, Federico F, Goldmann DA. Medication errors and adverse drug events in pediatric inpatients. JAMA. 2001 Apr 25;285(16):2114-20.
4. Kaushal R, Goldmann DA, Keohane CA, Christino M, Honour M, Hale AS, Zigmont K, Lehmann LS, Perrin J, Bates DW. Adverse drug events in pediatric outpatients. Ambul Pediatr. 2007 Sep-Oct;7(5):383-9.
5. Harper MB, Longhurst CA, McGuire T, Tarrago R, Patterson A, CHA CDS Working Group. Core drug-drug interaction alerts for inclusion in pediatric electronic health

records with computerized prescriber order entry. Journal of Patient Safety. (2013 in press).
6. Stevens LA, Palma JP, Pander KK, Longhurst CA. Immunization registries in the EMR Era. Online J Public Health Inform. 2013;5(2); 1-11. Available at: http://ojphi.org/ojs/index.php/ojphi/article/view/4696/3717
7. Goddard K, Roudsari A, Wyatt JC. Automation bias: a systematic review of frequency, effect mediators, and mitigators. J Am Med Inform Assoc. 2012 Jan-Feb;19(1):121-7. doi: 10.1136/amiajnl-2011-000089. Epub 2011 Jun 16.
8. Rosenbloom ST, Qi X, Riddle WR, Russell WE, DonLevy SC, Giuse D, Sedman AB, Spooner SA. Implementing pediatric growth charts into an electronic health record system. J Am Med Inform Assoc. 2006 May-Jun;13(3):302-8.
9. Lowry S, Quinn M, Ramaiah M, Brick D, Patterson E, Zhang J, Abbott P, Gibbons M. A Human Factors Guide to Enhance EHR Usability of Critical User Interactions when Supporting Pediatric Patient Care. National Institutes of Standards and Technology: US Department of Commerce. 06/28/2012. NISTIR 7865. Available at: http://www.nist.gov/healthcare/usability/upload/NIST-IR-7865.pdf
10. Freihoefer K, Nyberg G, Vickery C. Clinic exam room design: present and future. HERD. 2013 Spring;6(3):138-56.
11. Henriksen K, Dayton E, Keyes MA, et al. Understanding Adverse Events: A Human Factors Framework. In: Hughes RG, editor. Patient Safety and Quality: An Evidence-Based Handbook for Nurses. Rockville (MD): Agency for Healthcare Research and Quality (US); 2008 Apr. Chapter 5. Available from: http://www.ncbi.nlm.nih.gov/books/NBK2666/
12. Neely AN, Sittig DF. Basic microbiologic and infection control information to reduce the potential transmission of pathogens to patients via computer hardware. J Am Med Inform Assoc. 2002 Sep-Oct;9(5):500-8.
13. Toll E. A piece of my mind. The cost of technology. JAMA. 2012 Jun 20;307(23):2497-8. doi: 10.1001/jama.2012.4946.
14. Gray JE, Suresh G, Ursprung R, Edwards WH, Nickerson J, Shiono PH, Plsek P, Goldmann DA, Horbar J. Patient misidentification in the neonatal intensive care unit: quantification of risk. Pediatrics. 2006 Jan;117(1):e43-7.
15. McCoy AB, Wright A, Kahn MG, Shapiro JS, Bernstam EV, Sittig DF. Matching identifiers in electronic health records: implications for duplicate records and patient safety. BMJ Qual Saf. 2013 Mar;22(3):219-24. doi: 10.1136/bmjqs-2012-001419.
16. Hyman D, Laire M, Redmond D, Kaplan DW. The use of patient pictures and verification screens to reduce computerized provider order entry errors. Pediatrics. 2012 Jul;130(1):e211-9. doi: 10.1542/peds.2011-2984. Epub 2012 Jun 4.
17. Levin HI, Levin JE, Docimo SG. "I meant that med for Baylee not Bailey!": a mixed method study to identify incidence and risk factors for CPOE patient misidentification. AMIA Annu Symp Proc. 2012;2012:1294-301. Epub 2012 Nov 3.
18. Anoshiravani A, Gaskin GL, Groshek MR, Kuelbs C, Longhurst CA. Special requirements for electronic medical records in adolescent medicine. J Adolesc Health. 2012 Nov;51(5):409-14. doi: 10.1016/j.jadohealth.2012.08.003.
19. Spooner SA, Classen DC. Data standards and improvement of quality and safety in child health care. Pediatrics. 2009 Jan;123 Suppl 2:S74-9. doi: 10.1542/peds.2008-1755E.

CHAPTER 4

CONCEPTUAL FOUNDATION OF SAFER GUIDES

A NEW SOCIO-TECHNICAL MODEL FOR STUDYING HEALTH INFORMATION TECHNOLOGY IN COMPLEX ADAPTIVE HEALTHCARE SYSTEMS

Dean F. Sittig and Hardeep Singh

4.1.1 INTRODUCTION

An ongoing challenge to the design, development, implementation, and evaluation of health information technology (HIT) interventions is to operationalize their use within the complex adaptive health care system that consists of high-pressured, fast-paced, and distributed settings of care delivery. Many conceptual models of user interaction, acceptance, and evaluation exist [1,2], but most are relatively limited in scope. Given the dearth of models that are specifically designed to address safe and effective HIT

A New Sociotechnical Model for Studying Health Information Technology in Complex Adaptive Healthcare Systems. Reproduced from Quality and Safety in Health Care, *Sittig DF and Singh H,* **19**, *pp. i68-i74, copyright 2010 with permission from BMJ Publishing Group Ltd.*

development and use, we have developed a comprehensive, socio-technical model that provides a multi-dimensional framework within which any HIT innovation, intervention, application, or device implemented within a complex adaptive healthcare system can be studied. This model builds upon and bridges previous frameworks and is further informed by our own work to study the safe and effective implementation and use of HIT interventions. In this paper we describe the conceptual foundations of our model and provide several examples of its utility for studying HIT interventions within real-world clinical contexts.

4.1.2 BACKGROUND

Previous analyses of HIT interventions have been limited by a lack of conceptual models that have been specifically developed for this purpose. Examples of models previously applied by HIT investigators include Rogers' diffusion of innovations theory [3,4,5], Venkatesh's unified theory of acceptance and use of technology [6,7,8,9], Hutchins' theory of distributed cognition [10–14], Reason's Swiss Cheese Model [15–17], and Norman's 7-step human-computer interaction model [18–20]. Although all of these models account for one or more important facets of technology implementation, we believe that the scope of each model limits its utility to address the full range of factors that should be considered in the design, development, implementation, use, and evaluation of HIT interventions. For example, these models were not specifically designed to address the complex relationships between the HIT hardware, software, information content, and the human-computer interface. Furthermore, while most of these models provide general guidance to study the high-level aspects of HIT implementation within a given clinical environment, none of them includes a measurement and monitoring infrastructure (e.g., methods to routinely collect data, create or review reports or conduct surveillance of outcomes). Based on these limitations, our aim was to develop a more comprehensive model to integrate specific technological and measurement dimensions of HIT with other socio-technical dimensions (e.g., people, workflow, communication, organizational policies, external rules and regulations).

Conceptual Foundation of SAFER Guides

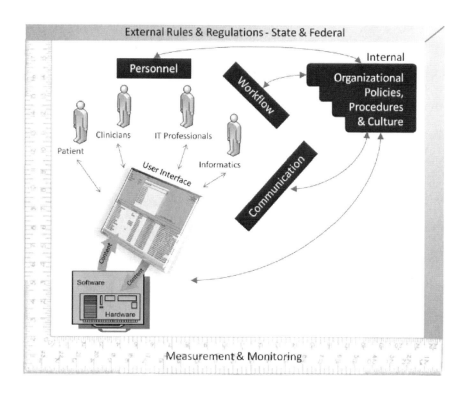

FIGURE 4.1.1 Illustration of the complex inter-relationships between the 8 dimensions of the new Socio-technical model.

4.1.3 PREVIOUS HEALTH INFORMATION TECHNOLOGY-FOCUSED SOCIO-TECHNICAL SYSTEMS MODELS.

Four related socio-technical models have been particularly influential in providing the foundation of our proposed model. First, Henriksen's model addresses (1) individual provider characteristics; (2) the nature or complexity of the work or task performed; (3) the physical environment where care takes place; (4) the human-system interfaces involved; and (5) various characteristics of the organization (social, environment, and management) [21]. Second, Vincent's framework for analyzing risk and safety proposes a hierarchy of factors that can potentially influence clinical practice [22]. Third, Carayon's Systems Engineering Initiative for Patient Safety (SEIPS) model [23] identifies three domains: (1) characteristics of providers, their tools and resources, and the physical/organizational setting; (2) interpersonal and technical aspects of health care activities; and (3) change in the patient's health status or behavior. Finally, Harrison et al.'s Interactive Socio-technical Analysis (ISTA) framework provides an excellent broad overview of the complex, emergent, inter-relationships between the HIT, clinicians, and workflows within any healthcare system [24].

While these socio-technical models include a "technology" component, none break down the "technology" into its individual components to enable researchers to dissect out the causes of particular HIT implementation or use problems, or to help identify specific solutions. We have found that many HIT problems we are studying revolve around the interplay of hardware, software, content (e.g., clinical data and computer-generated decision support), and user interfaces. Failing to acknowledge these specific technology-specific elements or attempting to treat them separately can hinder overall understanding of HIT-related challenges. For example, the "content" dimension of our model accounts for much of what informaticians do, that is, studying the intricacies of controlled clinical vocabularies that provide the cognitive interface between the inexact, subjective, highly variable world of biomedicine and the highly structured, tightly controlled, digital world of computers [25]. A well-constructed, robust user interface vocabulary can make all the difference in the world to a busy clinician struggling to quickly and accurately enter a complex clinical order for a critically ill patient [26], and it is important to distinguish this aspect of technology from others that may contribute to additional challenges (e.g., a user interface that is difficult to navigate, an order entry application that is slow to respond, or computers that are only available at the main nursing station). Failure to do so, for example, leads to general statements such as "clinicians struggled with the new technology" or "it takes clinicians longer to complete their tasks using the new technology" without providing any insight into specific causes of the problems or their solutions. In this example, without a multidimensional understanding of the technological dimensions of the failed IT application, the researcher may incorrectly conclude that the hardware, application software, or user was responsible, when in fact a poorly designed or implemented clinical vocabulary might have been the root of the problem.

Finally the preceding models do not account for the special monitoring processes and governance structures that must be put in place while designing and developing, implementing, or using HIT. For example, identifying who will make the decision on what, when, and how clinical decision support (CDS) interventions will be added [27]; developing a process for monitoring the effect of new CDS on the systems' response time [28]; building tools to track the CDS that is in place [29]; developing an approach for testing CDS; defining approaches for identifying rules that interact; developing robust processes for collecting feedback from users and communicating new system fixes, features, and functions; and building tools for monitoring the CDS system itself [30].

4.1.4 MOVING TOWARDS A NEW SOCIO-TECHNICAL MODEL FOR HIT

To overcome the limitations of previous models, we propose a new sociotechnical model to study the design, development, use, implementation, and evaluation of HIT (Figure 4.1.1). Our comprehensive 8-dimensional model accounts for key factors that influence the success of HIT interventions. A major assumption of our model is that the 8 dimensions cannot be viewed as a series of independent, sequential steps. As with other components of complex adaptive systems, these 8 interacting dimensions must be studied in relationship to each other. Clearly, several of our model's components are more tightly coupled than others, for example, the hardware, software, content, and user interface are all completely dependent on one another. However, all the other social components also exert strong influences on these technical components.

In our model, one cannot expect to gain an in-depth understanding of the intricacies of complex HIT interventions simply by integrating the results of studies performed within any single dimension of the model [31]. Rather, HIT interventions must be understood in the context of their simultaneous effects across multiple dimensions of the model. For instance, a recent evaluation of a national program to develop and implement centrally stored electronic summaries of patients' medical records in the UK revealed their benefits to be lower than anticipated and cautioned that complex interdependencies between many socio-technical factors at the clinical encounter-, organizational- and the national-level are to be expected in such evaluations [32]. These study findings are illustrative of how and why our proposed model could be useful.

The 8 dimensions include:

1. *Hardware and Software Computing Infrastructure.* This dimension of the model focuses solely on the hardware and software required to run the applications. The most visible part of this dimension is the computer, including the monitor, printer, and other data display devices along with the keyboard, mouse, and other data entry devices used to access clinical applications and medical or imaging devices. This dimension also includes the centralized (network-attached) data storage devices and all of the networking equipment required to allow applications or devices to retrieve and store patient data. Also included in this dimension is software at both the operating system and application levels. Finally, this dimension of the model subsumes all the machines, devices, and software required to keep the computing infrastructure functioning such as the high-capacity air conditioning system, the batteries that form the uninterruptable power supply (UPS) that provides short-term electrical power in the event of an electrical failure, and the diesel-powered backup generators that supply power during longer outages. In short, this dimension is purely technical; it is only composed of the physical devices and the software required keeping these devices running. One of the key aspects of this dimension is that, for the most part, the user is not aware that most of this infrastructure exists until it fails [33]. For example, in 2002 the Beth Israel Deaconess Medical Center in Boston experienced a four-day computer outage due to old, out-of-date computer equipment coupled with an outdated software program designed to direct traffic on a much less complex network. Furthermore, their network diagnostic tools were ineffective because they could only be used when the network was functioning [34].

2. *Clinical Content.* This dimension includes everything on the data-information-knowledge continuum that is stored in the system (i.e., structured and unstructured textual or numeric data and images that are either captured directly from imaging devices or scanned from paper-based sources) [35]. Clinical content elements can be used to configure certain software requirements. Examples include controlled vocabulary items that are selected from a list while ordering a medication or a diagnostic test, and the logic required to generate an alert for certain types of medication interactions). These elements may also describe certain clinical aspects of the patients' condition (e.g., laboratory test results, discharge summaries, or radiographic images). Other clinical content, such as demographic data and patient location, can be used to manage administrative aspects of a patient's care. These data can be entered (or created), read, modified, or deleted by authorized users and stored either on the local computer or on a network. Certain elements of the clinical content, such as that which informs clinical decision support (CDS) interventions, must be managed on a regular basis [36].
3. *Human Computer Interface.* An interface enables unrelated entities to interact with the system and includes aspects of the system that users can see, touch, or hear. The hardware and software "operationalize" the user interface; provided these are functioning as designed, any problems with using the system are likely due to human-computer interaction (HCI) issues. The HCI is guided by a user interaction model created by the software designer and developer [37]. During early pilot testing of the application in the target clinical environment, both the user's workflow and the interface are likely to need revisions. This process of iterative refinement, wherein both the user and user interface may need to change, must culminate in a human computer interaction model that matches the user's modified clinical workflow. For example, if a clinician wants to change the dose of a medication, the software requires the clinician to discontinue the old order and enter a new one, but the user interface should hide this complexity. This dimension also includes the ergonomic aspects of the interface [38]. If users are forced to use a computer mouse while standing, they may have

difficulty controlling the pointer on the screen because they are moving the mouse using the large muscles of their shoulder rather than the smaller muscles in the forearm. Finally, the lack of a feature or function within the interface represents a problem with both the interface and with the software or hardware that implements the interface.

4. *People.* This dimension represents the humans (e.g., software developers, system configuration and training personnel, clinicians, and patients) involved in all aspects of the design, development, implementation, and use of HIT. It also includes the ways that systems help users think and make them feel [39]. Although user training is clearly an important component of the user portion of the model, it may not by itself overcome all user-related problems. Many "user" problems actually result from poor system design or errors in system development or configuration. In addition to the users of these systems, this dimension includes the people who design, develop, implement, and evaluate these systems. For instance, these people must have the proper knowledge, skills, and training required to develop applications that are safe, effective, and easy to use. This is the first aspect of the model that is purely on the social end of the socio-technical spectrum. In most cases, users will be clinicians or employees of the health system. However, with recent advances in patient-centered care and development of personal health record systems and "home monitoring" devices, patients are increasingly becoming important users of HIT. Patients and/or their caregivers may not possess the knowledge or skills to manage new health information technologies, and this is of specific concern as more care shifts to the patient's home [40].

5. *Workflow and Communication.* This is the first portion of the model that acknowledges that people often need to work cohesively with others in the health care system to accomplish patient care. This collaboration requires significant two-way communication. The workflow dimension accounts for the steps needed to ensure that each patient receives the care they need at the time they need it. Often, the clinical information system does not initially match the actual "clinical" workflow. In this case, either the workflow must

be modified to adapt to the HIT, or the HIT system must change to match the various workflows identified.

6. *Internal Organizational Policies, Procedures, and Culture.* The organization's internal structures, policies, and procedures affect every other dimension in our model. For example, the organization's leadership allocates the capital budgets that enable the purchase of hardware and software, and internal policies influence whether and how offsite data backups are accomplished. The organizational leaders and committees who write and implement IT policies and procedures are responsible for overseeing all aspects of HIT system procurement, implementation, use, monitoring, and evaluation. A key aspect of any HIT project is to ensure that the software accurately represents and enforces, if applicable, organizational policies and procedures. Likewise, it is also necessary to ensure that the actual clinical workflow involved with operating these systems is consistent with policies and procedures. Finally, internal rules and regulations are often created in response to the external rules and regulations that form the basis of the next dimension of the model.

7. *External Rules, Regulations, and Pressures.* This dimension accounts for the external forces that facilitate or place constraints on the design, development, implementation, use, and evaluation of HIT in the clinical setting. For example, the recent passage of the American Recovery and Reinvestment Act (ARRA) of 2009, which includes the Health Information Technology for Economic and Clinical Health (HITECH) Act, makes available over $20 billion dollars for health care practitioners who become "meaningful users" of health IT. Thus, ARRA introduces the single largest financial incentive ever to facilitate electronic health record (EHR) implementation. Meanwhile, a host of federal, state, and local regulations regulate the use of HIT. Examples include the 1996 Health Insurance Portability and Accountability Act (HIPAA), recent changes to the Stark Laws, and restrictions on secondary use of clinical data. Finally, there are three recent national developments that have the potential to affect the entire health care delivery system in the context of HIT. These include: 1) the initiative

to develop the data and information exchange capacity to create a national health information network [41]; 2) the initiative to enable patients to access copies of the clinical data via personal health records [42]; and 3) clinical and IT workforce shortages [43].

8. *System Measurement and Monitoring.* This dimension has largely been unaccounted for in previous models. We posit that the effects of HIT must be measured and monitored on a regular basis. An effective system measurement and monitoring program must address four key issues related to HIT features and functions [44]. First is the issue of availability—the extent to which features and functions are available and ready for use. Measures of system availability include response times and percent uptime of the system. A second measurement objective is to determine how the various features and functions are being used by clinicians. For instance, one such measure is the rate at which clinicians override CDS warnings and alerts. Third, the effectiveness of the system on health care delivery and patient health should be monitored to ensure that anticipated outcomes are achieved. For example, the mean HbA1c value for all diabetic patients in a practice may be measured before and after implementation of a system with advanced CDS features. Finally, in addition to measuring the expected outcomes of HIT implementation, it is also vital to identify and document unintended consequences that manifest themselves following use of these systems [45]. For instance, it may be worthwhile to track practitioner efficiency before and after implementation of a new clinical charting application [46]. In addition to measuring the use and effectiveness of HIT at the local level, we must develop the methods to measure and monitor these systems and assess the quality of care resulting from their use on a state, regional, or even national level [47,48].

4.1.5 RELATIONSHIPS AND INTERACTIONS BETWEEN OUR MODEL'S COMPONENTS

Our research and experience has led us, and others, to conclude that HIT-enabled healthcare systems are best treated as complex adaptive systems

[49]. The most important result of this conclusion is that hierarchical decomposition (i.e., breaking a complex system, process, or device down into its components, studying them, and then integrating the results in an attempt to understand how the complete system functions) cannot be used to study HIT [50]. As illustrated by the evaluation of centrally stored electronic summaries in the UK, complex interdependencies between various socio-technical dimensions are to be expected and our HIT model (had it existed at the time) might have potentially predicted some of them and allowed them to address them prior to go-live rather than in the evaluation stages of the project. Therefore, one should not view or use our model as a set of independent components which can be studied in isolation and then synthesized to develop a realistic picture of how HIT is used within the complex adaptive healthcare system. Rather, the key to our model is how the eight dimensions interact and depend on one another. They must be studied as multiple, interacting components with non-linear, emergent, dynamic behavior (i.e., small changes in one aspect of the system lead to small changes in other parts of the system under some conditions, but large changes at other times) that often appear random or chaotic. This is typical of complex adaptive systems, and our model reflects these interactions.

For example, a computer-based provider order entry (CPOE) system that works successfully on an adult, surgical nursing unit within a hospital may not work at all in the nearby pediatric unit for any number of potential reasons, including: 1) hardware/software (e.g., fewer computers, older computers, poor wireless reception, poor placement); 2) content (e.g., no weight- or age-based dosing, no customized order sets or documentation templates); 3) user-interface (e.g., older workforce that has trouble seeing the small font on the screen); or 4) personnel (e.g., no clinical champion within the medical staff). However, each of these dimensions has a potential relationship with one or more of the other dimensions. For instance, computers may have been few or old because of some organizational limitations, there may be no customized order sets because clinician-users did not agree on how best to do it, and there was no clinical champion because the organization did not provide any incentive for the additional time this role would entail. Other reasons could include problems with the user interface and the communication and workflow related to how nurses process new medication orders using the EHR and record administration of medications. These issues, in turn, may have been due to organizational policies and procedures. For example, the unit governance committee may have decided not to approve a request for mobile computers, with the result that nurses spent more time away from patients and therefore had a slower workflow related to processing new orders. The preceding example illustrates the interaction of six dimensions of our model: hardware/software, clinical content, user interface, people, workflow, and organizational policies. Additionally, some form of monitoring could have detected these issues. In summary, our model provides HIT researchers with several new avenues of thinking about key technology components and how these dimensions can be accounted for in future research.

4.1.6 THE NEW HIT MODEL IN ACTION IN REAL-WORLD SETTINGS

The following sections illustrate how we have used the socio-technical model of safe and effective HIT use within our research. In an attempt to describe how the model can be applied across the breadth of HIT research and development, and to provide examples of different systems and interventions that can be analyzed within this new paradigm, we highlight key elements of our model in the context of several recent projects.

4.1.6.1 HIT DESIGN AND DEVELOPMENT

The design and development of CDS interventions within clinicians' workflow presents several challenges. We conducted several qualitative studies to gain insight into the 8 dimensions of our model during the development of a CDS tool within a CPOE application. This CDS intervention was designed to alert clinicians whenever they attempted to order a medication that was contraindicated in elderly patients or one that had known serious interactions with Warfarin. We used several methods, including focus groups, usability testing, and educational sessions with clinician users [51], to identify issues related to hardware/software, content, interface, people, measurement, workflow/communication, and internal policies and

Conceptual Foundation of SAFER Guides 117

procedures. These efforts helped us, for example, to understand the need to meet with the organization's Pharmacy and Therapeutics (P & T) committee (i.e., internal policy) to convince them to modify the medication formulary as well as the information technology professional (i.e., people) who was responsible for maintaining the textual content of the alerts (i.e., font size, contents and order of the messages) to fit within the constraints of the alert notification window (i.e., user interface) which eliminated the need to train clinicians to use the horizontal scrolling capability. This is just one simple example of how use of the 8 dimensional model paid huge dividends during the development and implementation stages of this highly successful project [52,53].

4.1.6.2 HIT IMPLEMENTATION

In a recent article we described lessons that could be learned from CPOE implementation at another site [54]. One of the most important conclusions from this implementation was that problems could, and often do, occur in all 8 dimensions of the model (see Table 4.1.1) [55].

4.1.6.3 HIT USE

Safe and effective use of an EHR-based notification system involves many factors that are addressed by almost all dimensions of our model [56,57]. This CDS system generates automated asynchronous "alerts" to notify clinicians of important clinical findings. We examined communication outcomes of over 2500 such alerts that were specifically related to abnormal test results. We found that 18.1% of abnormal lab alerts and 10.2% of abnormal imaging alerts were never acknowledged (i.e., were unread by the receiving provider). Additionally, 7-8% of these alerts lacked timely follow-up, which was unrelated to acknowledgment of the alert.

Despite a notification system that ensured transmission of results, it was concerning that abnormal test results did not always receive timely follow-up, even when acknowledged. This study revealed complex interactions between users, the user interface, software, content, workflow/

communication, and organizational policies related to who was responsible for abnormal test follow-up. Our findings thus highlighted the multiple dimensions of our model that need to be addressed to improve the safety of EHR-based notification systems and perhaps other forms of CDS (see Table 4.1.1) [59–62]. We are now applying the socio-technical model to study barriers, facilitators, and interventions for safe and effective test result notification through EHRs.

4.1.6.4 HIT EVALUATION

Our model recently provided us guidance in HIT evaluation, reminding us that however technologically savvy we make our patient care processes, we must also carefully monitor their impact, effectiveness, and unintended consequences. We recently evaluated why, despite implementation of an automated notification system to enhance communication of fecal occult blood test (FOBT) results, providers did not take follow-up actions in almost 40% of cases [63]. Again, our findings highlighted multiple dimensions corresponding to our socio-technical model. For instance, we found that clinician non-response to automated notifications was related to a software configuration error that prevented transmission of a subset of test results but we also found that if the institution was using certain types of workflows related to test performance and that if organizational procedures for computerized order-entry of FOBTs were different, the problem may not have occurred. Thus, we found our multi-dimensional approach, which accounted for interactions, to be useful for comprehensive evaluation of HIT after implementation.

4.1.7 CONCLUSIONS

The 8 dimensions of the safe and effective HIT use model introduced in this manuscript establish a new paradigm for the study of HIT. We have successfully applied this model to study several HIT interventions at different levels of design, development, implementation, use and evaluation. We anticipate that additional study of the 8 dimensions and their complex

Conceptual Foundation of SAFER Guides

TABLE 4.1.1: Illustration of how the 8-dimensions of our socio-technical model have been used to analyze different HIT-related interventions and how other dimensions might need to be addressed for every dimension

Socio-technical model dimension	Lessons Learned from Implementation of Computer-based Provider Order Entry	Follow-up of Alerts related to Abnormal Diagnostic Imaging Results
Hardware and Software	The majority of computer terminals were linked to the hospital computer system via wireless signal, communication bandwidth was often exceeded during peak operational periods, which created additional delays between each click on the computer mouse.	Alerts should be retracted when the patient dies or if the radiologist calls, or the patient is admitted before the alert is acknowledged. However, this can be done only through a centralized organizational policy.
Clinical Content	No ICU-specific order sets were available at the time of CPOE implementation. The hurried implementation timeline established by the leaders in the organization prohibited their development.	Interventions to reduce alert overload and improve the signal to noise ratio should be explored. Unnecessary alerts should be minimized. However, people (physicians) may not agree which alerts are essential and which ones are not [ ].
Human Computer Interface	The process of entering orders often required an average of 10 clicks on the computer mouse per order, which translated to 1 to 2 minutes to enter a single order. Organizational leaders eventually hired additional clinicians to "work the CPOE system" while others cared for the patients.	Unacknowledged alerts must stay active on the EMR screen for longer periods, perhaps even indefinitely, and should require the provider's signature and statement of action before they are allowed to drop off the screen. However, providers might not want to spend additional time stating their actions; who will make this decision?
People	Leaders at all levels of the institution made implementation decisions (re: hardware placement, software configuration, content development, user interface design, etc.) that placed patient care in jeopardy.	Many clinicians did not know how to use many of the EMR's advanced features that greatly facilitated the processing of alerts so training should be revamped. However, providers are only given 4 hours of training time by the institution
Workflow and Communication	Rapid implementation timeline did not allow time for clinicians to adapt to their new routines and responsibilities. In addition, poor hardware and software design and configuration decisions complicated the workflow issues.	Communicating alerts to 2 recipients, which occurred when tests were ordered by a healthcare practitioner other than the patient's regular PCP, significantly increased the odds that the alert would not be read and would not receive timely follow-up action. No policy was available that states who is responsible for follow-up. Additionally, back-up notification required by the institution to improve critical test result follow-up, a Joint Commission goal.

TABLE 4.1.1: *Cont.*

Socio-technical model dimension	Lessons Learned from Implementation of Computer-based Provider Order Entry	Follow-up of Alerts related to Abnormal Diagnostic Imaging Results
Organizational Policies and Procedures	Order entry was not allowed until after the patient had physically arrived at the hospital and been fully registered into the clinical information system.	Every institution must develop and publicize a policy regarding who is responsible (PCP vs the ordering provider, who may be a consultant) for taking action on abnormal results. Also meets External Joint commission requirements.
External Rules, Regulations, and Pressures	Following the IOM's report "To Err is Human: Building a Safer Health System" and subsequent congressional hearings the issue of patient safety has risen to a position of highest priority among health care organizations.	Poor reimbursement and heavy workload of patients puts productivity pressure on providers The nature of high-risk transitions between health care practitioners, settings, and systems of care makes timely and effective electronic communication particularly challenging.
System Measurement and Monitoring	Monitoring identified a significant increase in patient mortality following CPOE implementation.	An audit and performance feedback system should be established to give providers information on timely follow-up of patients' test results on a regular basis. However, providers may not want feedback or the institution does not have the persons required to do so.

interactions will yield further refinements to this model and, ultimately, improvements in the quality and safety of the HIT applications that translate to better health and welfare for our patients.

REFERENCES

1. Beuscart-Zéphir MC, Aarts J, Elkin P. Human factors engineering for healthcare IT clinical applications. Int J Med Inform. 2010 Feb 16.
2. Holden RJ, Karsh B. A theoretical model of health information technology usage behaviour with implications for patient safety. Behaviour & Information Technology, (2009; 28: 21-38.
3. Rogers EM. Diffusion of Innovations, 5th Edition. Free Press, 2003 512pgs.
4. Ash J. Organizational factors that influence information technology diffusion in academic health sciences centers. J Am Med Inform Assoc. 1997 Mar-Apr;4(2):102-11.

5. Gosling AS, Westbrook JI, Braithwaite J. Clinical team functioning and IT innovation: a study of the diffusion of a point-of-care online evidence system. J Am Med Inform Assoc. 2003 May-Jun;10(3):244-51.
6. Venkatesh, V., Morris, M.G., Davis, F.D., and Davis, G.B. "User Acceptance of Information Technology: Toward a Unified View," MIS Quarterly, 27, 2003, 425-478.
7. Holden RJ, Karsh BT. The technology acceptance model: its past and its future in health care. J Biomed Inform. 2010 Feb;43(1):159-72.
8. Duyck P, Pynoo B, Devolder P, Voet T, Adang L, Vercruysse J. User acceptance of a picture archiving and communication system. Applying the unified theory of acceptance and use of technology in a radiological setting. Methods Inf Med. 2008;47(2):149-56.
9. Kijsanayotin B, Pannarunothai S, Speedie SM. Factors influencing health information technology adoption in Thailand's community health centers: applying the UTAUT model. Int J Med Inform. 2009 Jun;78(6):404-16.
10. Hutchins E. Cognition in the Wild. MIT Press, Cambridge, MA 1996; 401pp.Hazlehurst B, McMullen C, Gorman P, Sittig D. How the ICU follows orders: care delivery as a complex activity system. AMIA Annu Symp Proc. 2003:284-8.
11. Cohen T, Blatter B, Almeida C, Shortliffe E, Patel V. A cognitive blueprint of collaboration in context: distributed cognition in the psychiatric emergency department. Artif Intell Med. 2006 Jun;37(2):73-83.
12. Hazlehurst B, McMullen CK, Gorman PN. Distributed cognition in the heart room: how situation awareness arises from coordinated communications during cardiac surgery. J Biomed Inform. 2007 Oct;40(5):539-51.
13. Patel VL, Zhang J, Yoskowitz NA, Green R, Sayan OR. Translational cognition for decision support in critical care environments: a review. J Biomed Inform. 2008 Jun;41(3):413-31.
14. Reason J. Human error: models and management. BMJ. 2000 Mar 18;320(7237):768-70.
15. van der Sijs H, Aarts J, Vulto A, Berg M. Overriding of drug safety alerts in computerized physician order entry. J Am Med Inform Assoc. 2006 Mar-Apr;13(2):138-47.
16. Lederman RM, Parkes C. Systems failure in hospitals--using Reason's model to predict problems in a prescribing information system. J Med Syst. 2005 Feb;29(1):33-43.
17. Norman, D. (1988). The Psychology of Everyday Things. New York: Basic Books.
18. Malhotra S, Jordan D, Shortliffe E, Patel VL. Workflow modeling in critical care: piecing together your own puzzle. J Biomed Inform. 2007 Apr;40(2):81-92.
19. Sheehan B, Kaufman D, Stetson P, Currie LM. Cognitive analysis of decision support for antibiotic prescribing at the point of ordering in a neonatal intensive care unit. AMIA Annu Symp Proc. 2009 Nov 14;2009:584-8.
20. Henriksen K, Kaye R, Morisseau D. Industrial ergonomic factors in the radiation oncology therapy environment. In: Nielsen R, Jorgensen K, eds. Advances in industrial ergonomics and safety V. Washington, DC: Taylor and Francis; 1993. p. 325-335
21. Vincent C, Taylor-Adams S, Stanhope N. Framework for analysing risk and safety in clinical medicine. BMJ. 1998 Apr 11;316(7138):1154-7.
22. Carayon P, Schoofs Hundt A, Karsh BT, Gurses AP, Alvarado CJ, Smith M, Flatley Brennan P. Work system design for patient safety: the SEIPS model. Qual Saf Health Care. 2006 Dec;15 Suppl 1:i50-8.

23. Harrison MI, Koppel R, Bar-Lev S. Unintended consequences of information technologies in health care--an interactive sociotechnical analysis. J Am Med Inform Assoc. 2007 Sep-Oct;14(5):542-9.
24. Rector AL. Clinical terminology: why is it so hard? Methods Inf Med. 1999 Dec;38(4-5):239-52.
25. Rosenbloom ST, Miller RA, Johnson KB, Elkin PL, Brown SH. Interface terminologies: facilitating direct entry of clinical data into electronic health record systems. J Am Med Inform Assoc. 2006 May-Jun;13(3):277-88.
26. Wright A, Sittig DF, Ash JS, Bates DW, Fraser G, Maviglia SM McMullen C, Nicol WP. Pang JE, Starmer J, Middleton B. Governance for Clinical Decision Support: Case Studies and Best Practices of Exemplary Institutions. J Amer Med Inform Assoc. 2010 (under review)
27. Sittig DF, Campbell EM, Guappone KP, Dykstra RH, Ash JS. Recommendations for Monitoring and Evaluation of In-Patient Computer-based Provider Order Entry Systems: Results of a Delphi Survey. Proc. Amer Med Informatics Assoc Fall Symposium (2007) p 671-675.
28. Sittig DF, Simonaitis, L, Carpenter, JD, Allen, GO, Doebbeling, BN, Sirajuddin, AM, Ash, SJ, Middleton, B. The state of the art in clinical knowledge management: An inventory of tools and techniques. Int J Med Inform. 2010 Jan;79(1):44-57.
29. Hripcsak G. Monitoring the monitor: automated statistical tracking of a clinical event monitor. Comput Biomed Res. 1993 Oct;26(5):449-66.
30. Rasmussen J. Risk management in a dynamic society: a modelling problem. Safety Science, 27(2): 183-213; 1997.
31. Greenhalgh T, Stramer K, Bratan T, Byrne E, Russell J, Potts HW. Adoption and non-adoption of a shared electronic summary record in England: a mixed-method case study. BMJ. 2010 Jun 16;340:c3111. doi: 10.1136/bmj.c3111.
32. Leveson NG, Turner CS. An Investigation of the Therac-25 Accidents. IEEE Computer, 1993; 26 (7): 18-41. Updated version available at: http://sunnyday.mit.edu/papers/therac.pdf
33. Kilbridge P. Computer crash--lessons from a system failure. N Engl J Med. 2003 Mar 6;348(10):881-2.
34. Bernstam EV, Smith JW, Johnson TR. What is biomedical informatics? J Biomed Inform. 2010 Feb;43(1):104-10.
35. Sittig DF, Wright A, Simonaitis L, Carpenter JD, Allen GO, Doebbeling BN, Sirajuddin AM, Ash JS, Middleton B. The state of the art in clinical knowledge management: an inventory of tools and techniques. Int J Med Inform. 2010 Jan;79(1):44-57.
36. Shneiderman B, Plaisant C, Cohen M, Jacobs S. Designing the User Interface: Strategies for Effective Human-Computer Interaction, 5th ed. Pearson Educaiton, 2009. 672 Pgs.
37. Svanæs D, Alsos OA, Dahl Y. Usability testing of mobile ICT for clinical settings: Methodological and practical challenges. Int J Med Inform. 2008 Sep 10.
38. Sittig DF, Krall M, Kaalaas-Sittig J, Ash JS. Emotional aspects of computer-based provider order entry: a qualitative study. J Am Med Inform Assoc. 2005 Sep-Oct;12(5):561-7.
39. Henriksen K, Joseph A, Zayas-Caban T. The Human Factors of Home Health Care: A Conceptual

40. Model for Examining Safety and Quality Concerns. J Patient Safety; 5(4), December 2009.
41. American Recovery and Reinvestment Act of 2009, State Grants to Promote Health Information Technology Planning and Implementation Projects. Available at: https://www.grantsolutions.gov/gs/preaward/previewPublicAnnouncement.do?id=10534
42. Sittig DF. Personal health records on the internet: a snapshot of the pioneers at the end of the 20th Century. Int J Med Inform. 2002 Apr;65(1):1-6.
43. Detmer DE, Munger BS, Lehmann CU. Medical Informatics Board Certification: History, Current Status, and Predicted Impact on the Medical Informatics Workforce. Applied Clinical Informatics 1(1):11-18; 2010. Available: http://www.schattauer.de/nc/en/magazine/subject-areas/journals-a-z/applied-clinical-informatics/issue/special/manuscript/12624/download.htmlLeonard KJ, Sittig DF.Improving information technology adoption and implementation through the identification of appropriate benefits: creating IMPROVE-IT. J Med Internet Res. 2007 May 4;9(2):e9.
44. Ash JS, Berg M, Coiera E. Some unintended consequences of information technology in health care: the nature of patient care information system-related errors. J Am Med Inform Assoc. 2004 Mar-Apr;11(2):104-12.
45. Bradshaw KE, Sittig DF, Gardner RM, Pryor TA, Budd M. Computer-based data entry for nurses in the ICU. MD Comput. 1989 Sep-Oct;6(5):274-80.
46. Sittig DF, Shiffman RN, Leonard K, Friedman C, Rudolph B, Hripcsak G, Adams LL, Kleinman LC, Kaushal R. A draft framework for measuring progress towards the development of a National Health Information Infrastructure. BMC Med Inform Decis Mak. 2005 Jun 13;5:14.
47. Sittig DF, Classen DC. Safe electronic health record use requires a comprehensive monitoring and evaluation framework. JAMA. 2010 Feb 3;303(5):450-1.
48. Begun JW, Zimmerman B, Dooley K. Health Care Organizations as Complex Adaptive Systems. In: Mick SM, Wyttenbach M (eds.), Advances in Health Care Organization Theory San Francisco: Jossey-Bass, 2003; pp 253-288.
49. Rouse WB. Health Care as a Complex Adaptive System: Implications for Design and Management. The Bridge, Spring 2008; pgs. 17-25.
50. Feldstein A, Simon SR, Schneider J, Krall M, Laferriere D, Smith DH, Sittig DF, Soumerai SB. How to design computerized alerts to safe prescribing practices. Jt Comm J Qual Saf. 2004 Nov;30(11):602-13
51. Feldstein AC, Smith DH, Perrin N, Yang X, Simon SR, Krall M, Sittig DF, Ditmer D, Platt R, Soumerai SB. Reducing Warfarin medication interactions: an interrupted time series evaluation. Arch Intern Med. 2006 May 8;166(9):1009-15.
52. Smith DH, Perrin N, Feldstein A, Yang X, Kuang D, Simon SR, Sittig DF, Platt R, Soumerai SB. The impact of prescribing safety alerts for elderly persons in an electronic medical record: an interrupted time series evaluation. Arch Intern Med. 2006 May 22;166(10):1098-104.
53. Sittig DF, Ash JS, Zhang J, Osheroff JA, Shabot MM. Lessons from "Unexpected increased mortality after implementation of a commercially sold computerized physician order entry system". Pediatrics. 2006 Aug;118(2):797-801.
54. Sittig DF, Ash JS. Clinical information systems: Overcoming adverse consequences. Sudbury, MA: Jones and Bartlett. 2010.

55. Singh H, Thomas EJ, Mani S, Sittig D, Arora H, Espadas D, Khan MM, Petersen LA. Timely follow-up of abnormal diagnostic imaging test results in an outpatient setting: are electronic medical records achieving their potential? Arch Intern Med. 2009 Sep 28;169(17):1578-86.
56. Singh H, Thomas EJ, Sittig DF, Wilson L, Espadas D, Khan MM, Petersen LA. Notification of Abnormal Laboratory Test Results in an Electronic Medical Record: Do Any Safety Concerns Remain? Am J Med. 2010 Mar;123(3):238-44..
57. van der Sijs H, Aarts J, van Gelder T, Berg M, Vulto A. Turning off frequently overridden drug alerts: limited opportunities for doing it safely. J Am Med Inform Assoc. 2008 Jul-Aug;15(4):439-48
58. Hysong SJ, Sawhney MK, Wilson L, Sittig DF, Esquivel A, Watford M, Davis T, Espadas D, Singh H. Improving outpatient safety through effective electronic communication: A study protocol. Implement Sci. 2009 Sep 25;4(1):62. PMID: 19781075
59. Hysong SJ, Sawhney MK, Wilson L, Sittig DF, Espadas D, Davis TL, Singh H. Provider Management Strategies of Abnormal Test Result Alerts: A Cognitive Task Analysis. J Am Med Inform Assoc 17:71-77, 2010. PMID: 20064805
60. Singh H, Wilson L, Reis B, Sawhney MK, Espadas D, Sittig DF. Ten Strategies to Improve Management of Abnormal Test Result Alerts in the Electronic Health Record. In press. Journal of Patient Safety. 2010 Jun;6(2):121-123.
61. Singh H, Vij M. Eight Recommendations for Policies for Communication of Abnormal Test Results. In press. Joint Commission Journal on Quality and Patient Safety. 2010.
62. Singh H, Wilson L, Petersen LA, Sawhney MK, Reis B, Espadas D, Sittig DF. Improving Follow-up of Abnormal Cancer Screens using Electronic Health Records: Trust but Verify Test Result Communication. BMC Med Inform Decis Mak. 2009 Dec 9;9:49.

ELECTRONIC HEALTH RECORDS AND NATIONAL PATIENT SAFETY GOALS

Dean F. Sittig and Hardeep Singh

Electronic health records (EHRs) are essential to improving patient safety [1]. Hospitals and healthcare providers are implementing electronic health records (EHRs) at an unprecedented pace in response to the American Recovery and Reinvestment Act of 2009 (ARRA) [2–4]. Meanwhile, the number of certified EHR vendors in the US has increased from 60 [5,6],

---

From The New England Journal of Medicine, *Sittig DF and Singh H, Electronic Health Records and National Patient Safety Goals. Copyright © 2014 Massachusetts Medical Society. Reprinted with permission from Massachusetts Medical Society.*

to over 1000 [7] since mid-2008. Recent evidence has highlighted significant and often unexpected risks resulting from the use of EHRs and other health information technology [8–12]. These concerns are compounded by the extraordinary pace of EHR development and implementation. Thus, the unique safety risks posed by the use of EHRs should be considered alongside the potential benefits of these systems.

At a time when institutions are focused heavily on achieving meaningful use requirements, we propose that clearer guidance be provided for them to align their patient safety activities with those required for an EHR-enabled healthcare system [13]. A set of EHR-specific safety goals, modeled after the Joint Commission's National Patient Safety Goals (NPSGs), may provide organizations with unique focus areas for sustained improvements in organizational infrastructure, processes, and culture as they adapt to new technology.

EHR implementation is still highly variable across healthcare systems and providers, with equally variable implications for patient safety. For instance, the key patient safety priorities for an organization in the midst of an EHR rollout are somewhat distinct from those of an organization that has used a fully integrated EHR for five or more years. To account for variation in stages of implementation and levels of complexity across clinical practice settings, we propose a three-phase framework for development of EHR-specific patient safety goals (e-PSGs). The first phase of the framework, aimed at all EHR users but especially at recent and future adopters, includes goals to mitigate risks that are unique and specific to the use of technology [14] (e.g., unavailable or malfunctioning hardware or software). The second phase addresses issues created by failure to use or misuse of appropriate technology [15]. The final phase focuses on uses of technology to monitor safety events and identify potential safety issues before they can harm patients [16]. In the following sections, we illustrate how this framework can lay the foundation for development of e-PSGs within the context of EHR-enabled healthcare.

4.2.1 PHASE 1: DEVELOP GOALS TO ADDRESS SAFETY CONCERNS UNIQUE TO TECHNOLOGY

Device failures and natural or man-made disasters are inevitable. The potential consequences of an EHR failure become increasingly significant as large-scale systems are deployed across multiple facilities within a health care system, often across a wide geographical area. These broadly distributed systems may be tightly coupled and lightning fast; hence, a malfunction can rapidly affect not only a single department or institution but possibly an entire community [17]. Furthermore, the operations of such systems are often decentralized and relatively opaque to end users [18] such that problems evade easy detection and solution. As a recent example, on April 21, 2010, one-third of hospitals in Rhode Island were forced to postpone elective surgeries and divert non-life-threatening emergencies [19] when an erroneous automatic anti-virus software update set off a chain of events that caused "uncontrolled [computer] restarts and loss of networking functionality." [20] A potential goal, therefore, should be to reduce the impact of EHR downtime on clinical operations and patient safety. Table 4.2.1 lists some of the activities that organizations could undertake to achieve this goal.

Safety can also be compromised as the result of miscommunication between components of an EHR system. For example, it is not uncommon for data translation tables, used to encode and decode orders between disparate systems, to have mismatched data fields [34]. These errors may result in inadvertent changes to orders that are virtually undetectable by the computer or by humans not privy to the original sender's intentions. An example of such an error is an order for 30 mg oxycodone sustained release that is correctly entered in the computer-based provider order entry (CPOE) system but erroneously mapped to 30 mg oxycodone immediate release in the pharmacy management system and incorrectly dispensed. Errors related to system-to-system information transfer may be detected by testing interacting components within the "live EHR" environment. However, this process is resource intensive and therefore may not receive adequate effort and attention. Therefore, an e-PSG could focus on reducing miscommunication of data transmitted between different safety-critical components of the EHR.

Recent evidence has identified both problem areas above (EHR accessibility and information transfer) as the most common issues in reported EHR-related safety events [9,11,12].

Conceptual Foundation of SAFER Guides

TABLE 4.2.1: Framework for Potential EHR-Related National Patient Safety Goals (e-PSGs).

Potential Goal	Rationale	Suggestions to Achieve the Goal
Phase 1: Safety Concerns Unique to Technology		
Reduce the impact of EHR downtime on patient safety	A robust computing infrastructure should include a plan for when the computer is unexpectedly unavailable.	• Maintain backup paper forms for ordering and clinical documentation in clinical areas
		• Employ clearly marked, easily activated, password protected, read-only backup systems that contain the most recent clinical results and orders
		• Ensure complete, encrypted, daily, off-site storage of all patient data
		• Use redundant hardware (e.g., database servers) for mission-critical applications
		• Maintain uninterrupted power supplies capable of maintaining computer operations until generators come on-line
		• Develop downtime (and re-activation) policies and procedures to operationalize plans and train personnel on these plans
		• Report EHR uptime rates to organization's board of directors on regular basis
Reduce miscommunication of data transmitted between different components of EHRs	Miscommunication can be problematic when sending remotely generated, "asynchronous" orders through multiple components of an EHR system.	• Mandate regression testing (i.e., testing to ensure that intended changes are correct and did not corrupt any other parts of the system) of all mission-critical applications after every modification
		• Reduce the number of interfaces between mission-critical systems (e.g., between CPOE and pharmacy management systems) developed by different software vendors
Phase 2: Address Failure to Use EHRs Appropriately		
Mandate computer-based provider order entry for all medications, laboratory, and radiology test orders	CPOE with advanced clinical decision support has been shown to reduce errors of omission and commission.	• Create order sets for the most common condition-, task-, and service-specific clinical scenarios
		• Make clinician login privileges conditional on training and testing in order entry
		• Report CPOE rates to organization's board of directors on regular basis

TABLE 4.2.1: *Cont.*

Potential Goal	Rationale	Suggestions to Achieve the Goal
Reduce alert fatigue	Alerts with low specificity result in a high rate of clinician overrides and lead to "alert fatigue." Clinicians thus may inadvertently ignore important information.	• Implement drug-drug interaction checking only for life-threatening combinations • Focus CDS interventions on key organizational safety goals • Ensure that timing, content, and delivery of CDS interventions are appropriate to recipients and workflows • Monitor the number and override rate of all alerts • Report Alert override rates to organization's board of directors on regular basis
Enter all medications, allergies, diagnostic test results, and clinical problems as structured or coded data	Structured data is needed to realize the full potential of computer-generated decision support (e.g., drug-allergy checking, automated abnormal test result notification, or drug-condition reminders)	• Use standard clinical vocabularies • Implement two-way, system-system interfaces with all ancillary information systems both within and outside the organization to facilitate the capture and use of coded data • Develop order entry templates
Phase 3: Use EHRs to Monitor and Improve Patient Safety		
Use EHR-based "triggers" to monitor and improve patient safety	Current incident reporting systems capture a small proportion of events or only specific types of events. Safety trends cannot be measured reliably at present.	• Identify high-risk target conditions relevant to their clinical contexts • Develop search criteria to identify them (e.g., patients in need of particular tests, follow-up actions, or those experiencing specific safety events) • Query the EHR regularly to detect events based on search criteria • Assign staff to take action on identified events

4.2.2 PHASE 2: DEVELOP GOALS TO MITIGATE SAFETY CONCERNS FROM FAILURE TO USE EHRS APPROPRIATELY

One rationale for widespread use of EHRs is that certain types of patient harm can be prevented when EHRs are used appropriately. For instance, EHRs facilitate and/or standardize the transfer of information between providers and help close the communication loop, such as when providers who order tests are notified promptly of abnormalities. However, these benefits are predicated on the assumption that EHRs are used correctly and as intended in routine practice [35]. For example, if CPOE were used on some nursing units but not others, clinicians would need to check for orders and test results in multiple locations, increasing the opportunity to miss information. Other partial uses of CPOE (e.g., used for ordering medications but not laboratory tests) could leave the non-computerized processes more vulnerable to error, with no way of ensuring closed-loop electronic communication of test results to the ordering providers and potentially leading to more missed results [36]. Another hazard can arise if providers bypass structured data fields in CPOE and instead use EHR-based free-text communication to prescribe or discontinue medications, since free-text orders are not standardized and vulnerable to miscommunication [37]. To reduce these safety concerns, another e-PSG could mandate use of CPOE for all medication, laboratory, and radiology test orders. Table 4.2.1 lists several potential strategies to help achieve this goal.

Second, implementation and use of complex clinical decision support (CDS) embedded within EHR systems are prone to human error and cognitive constraints [38,39]. Thus, decisions related to various aspects of CDS interventions must be periodically evaluated [40]. For example, although point-of-care, CDS interventions are necessary to achieve the full benefits of EHRs and "meaningful use" payments [41], interruptive alerts must be used judiciously. Many organizations turn on alerts with low specificity, resulting in high rates of clinician overrides [24]. Frequent overrides are associated with "alert fatigue," which may cause clinicians to inadvertently ignore important information. Thus, another potential e-PSG could be to reduce alert fatigue. Alerts with override rates above a certain threshold should be discontinued or modified to increase their specificity [42]. Similarly, hard stops (i.e., when users cannot proceed with the desired action) must be used only for the most egregious errors [43]. Having such a goal will stimulate a multidisciplinary approach to reducing alerts that involves bringing cognitive scientists, human factors engineers and informaticians [44,45], to work on these complex issues with the clinicians. Some additional suggestions to achieve this goal are listed in Table 4.2.1.

Third, although there are safety benefits of managing free text, dictated reports and scanning images of test results into EHRs including improved legibility and rapid access [46], many institutions are not currently coding certain critical data. A lack of structured or coded data prevents the system from being able to provide meaningful feedback or interpretation of results to the user (i.e. no alert for lisinopril will be generated if captopril angioedema was not previously entered as coded allergy data). Therefore, to realize the full safety benefits of complex CDS tools [47] (e.g., drug-allergy checking [48], automated abnormal test result notification [28], or drug-condition reminders [29]) another e-PSG could focus on ensuring that critical data such as medications, allergies, diagnostic test results, and clinical problems are entered as structured or coded data in the EHR [49]. Strategies to help achieve this goal are summarized in Table 4.2.1.

4.2.3 PHASE 3. DEVELOP GOALS RELATED TO USE OF EHRS TO MONITOR AND IMPROVE PATIENT SAFETY

To achieve the goals of many national stakeholders and initiatives to improve patient safety, including Agency for Healthcare Research and Quality (AHRQ), The Joint Commission and the recent "Partnership for Patients" [50]. Current methods to measure safety events over rely on incident reports, which have several limitations including detection of only a small proportion of events [32]. In contrast, systems can be programmed to automatically detect easily overlooked and underreported errors of omission, such as patients who are overdue for medication monitoring, patients who lack appropriate surveillance after treatment, and patients who do not receive follow-up for abnormal laboratory or radiology tests [51]. EHR-based trigger approaches [33] can also be used to detect errors of commission such as adverse drug events [52], postoperative complications [53], and errors related to misidentification of patients [54]. Organizations must leverage EHRs for purposes of improving rapid detection of common errors (including EHR-related errors), to monitor for high-priority safety events and to more reliably track trends over time. EHRs could also play a role in improving the existing infrastructure of reporting and analysis by

Patient Safety Organizations by facilitating the generation of data files describing particular safety events using the AHRQ common format v1.2 [55]. Thus, an e-PSG could relate to the use of the EHR to monitor, report, and identify potential safety issues and events. This would make detection and reporting more efficient and help shift resources towards investigation and action. Strategies to help achieve this goal are summarized in Table 4.2.1.

4.2.4 APPLICATION OF THE THREE-PHASE E-PSG FRAMEWORK

Given that only 48% of all eligible hospitals and only 20% of eligible physicians have currently received Stage 1 meaningful use payments [56], the development and application of e-PSGs could partially address the Institute of Medicine's recent recommendation to the ONC to create an EHR safety action and surveillance plan [8]. Such a plan should be tailored to the appropriate stage of EHR implementation. Recent adopters of EHRs could focus on Phase 1 goals in our safety framework, making sure that the technology is safe to use, whereas organizations that have already achieved stage 1 meaningful use criteria and have been using EHRs for several years would aim for goals from all phases. Measurements related to e-PSGs would allow tracking and benchmarking of EHR-related safety performance nationally [57]. Policymakers and EHR vendors could collaborate on development and certification of automated methods to measure and report new indicators from "meaningful use" certified EHRs in eligible hospitals annually. Examples of potential measures for e-PSGs might include EHR uptime rate (e.g., minutes the EHR was available to clinicians divided by number of minutes in a year [23]), CPOE rate (e.g., number of orders electronically entered divided by the total number of orders during the year [23], and alert override rate (e.g., number of point-of-care alerts ignored divided by the total number of point-of-care alerts generated [23]).

These goals will also need to be reviewed regularly and updated as needed based on national priorities and research on EHR-related patient safety. In addition, many strategies not addressed in this paper could be considered as recommendations or good clinical practices and progress in a step-wise fashion to future e-PSGs.

REFERENCES

1. Blumenthal D, Glaser JP. Information technology comes to medicine. N Engl J Med. 2007 Jun 14;356(24):2527-34.
2. Centers for Medicare and Medicaid Services, Medicare and Medicaid EHR Incentive Programs and the Office of the National Coordinator for Health IT, Certified Health IT Products List. 2012 [cited 2012 April 28]; Available from: https://explore.data.gov/d/eybk-7w2b
3. Tagalicod R AR, Kahn J. Medicare & Medicaid EHR Incentive Programs. 2012 [cited 2012 January 16]; Available from: http://healthit.hhs.gov/portal/server.pt/document/956320/ehr_incentiveprogramanalysis_1_10_12_pdf
4. Blumenthal D. Stimulating the adoption of health information technology. N Engl J Med. 2009 Apr 9;360(15):1477-9. Epub 2009 Mar 25.
5. The list of CCHIT Certified Ambulatory EHRs certified under the 2007 criteria http://web.archive.org/web/20090123165327/http://cchit.org/choose/ambulatory/2007/index.asp
6. The list of CCHIT Certified Inpatient EHRs certified under the 2007 criteria http://web.archive.org/web/20090221100419/http://cchit.org/choose/inpatient/2007/index.asp
7. EHR products classified as Complete Ambulatory and Inpatient EHRs. Available at: oncchpl.force.com/ehrcert   Accessed 4/28/2012.
8. Committee on Patient Safety and Health Information Technology; Institute of Medicine. "Health IT and patient safety: building safer systems for better care". Washington, DC: The National Academies Press, 2012.
9. Myers RB, Jones SL, Sittig DF. Review of reported clinical information system adverse events in US Food and Drug Administration databases. Appl Clin Inform 2011; 2: 63–74. doi: 10.4338/ACI-2010-11-RA-0064.
10. Harrington, L., D. Kennerly, and C. Johnson. 2011. Safety issues related to the electronic medical record (emr): Synthesis of the literature from the last decade, 2000–2009. Journal of Healthcare Management 56(1):31-44.
11. Magrabi F, Ong MS, Runciman W, Coiera E. Using FDA reports to inform a classification for health information technology safety problems. J Am Med Inform Assoc 2012; 19: 45-53.
12. Warm D; Edwards P: Classifying health information technology patient safety related incidents – an approach used in Wales. Appl Clin Inf 2012; 3:248–257. Available at: http://dx.doi.org/10.4338/ACI-2012-03-RA-0010.
13. Radecki RP, Sittig DF. Application of electronic health records to the Joint Commission's 2011 National Patient Safety Goals. JAMA 2011 Jul 6;306(1):92-3.
14. Kilbridge P. Computer crash--lessons from a system failure. N Engl J Med. 2003 Mar 6;348(10):881-2.
15. Sittig DF, Singh H. Defining health information technology-related errors: new developments since to err is human. Arch Intern Med. 2011 Jul 25;171(14):1281-4.
16. Jha AK, Classen DC. Getting moving on patient safety--harnessing electronic data for safer care. N Engl J Med. 2011 Nov 10;365(19):1756-8.

17. Perrow, Charles. Normal Accidents: Living with High-Risk Technologies. Princeton University Press, 1999.
18. The Menlo Report: Ethical Principles Guiding Information and Communication Technology Research. Available at: http://www.cyber.st.dhs.gov/wp-content/uploads/2011/12/MenloPrinciplesCORE-20110915-r560.pdf
19. NPR Staff. Anti-Virus Program Update Wreaks Havoc With PCs. April 21, 2010 Available at: http://www.npr.org/templates/story/story.php?storyId=126168997&sc=17&f=1001
20. Patel N. Botched McAfee update shutting down corporate XP machines worldwide. From Engadget.com April 21, 2010. Available at: http://www.engadget.com/2010/04/21/mcafee-update--shutting-down-xp-machines/
21. Sittig DF, Ash JS, Jiang Z, Osheroff JA, Shabot MM. Lessons from "unexpected increased mortality after implementation of a commercially sold computerized physician order entry system". Pediatrics. 2006 Aug;118(2):797-801.
22. Wright A, Feblowitz JC, Pang JE, Carpenter JD, Krall MA, Middleton B, Sittig DF. Use of order sets in inpatient computerized provider order entry systems: a comparative analysis of usage patterns at seven sites. Int J Med Inform 2012 Nov;81(11):733-45. doi: 10.1016/j.ijmedinf.2012.04.003.
23. Sittig DF, Campbell E, Guappone K, Dykstra R, Ash JS. Recommendations for monitoring and evaluation of in-patient Computer-based Provider Order Entry systems: results of a Delphi survey. AMIA Annu Symp Proc. 2007 Oct 11:671-5.
24. Lin CP, Payne TH, Nichol WP, Hoey PJ, Anderson CL, Gennari JH. Evaluating clinical decision support systems: monitoring CPOE order check override rates in the Department of Veterans Affairs' Computerized Patient Record System. J Am Med Inform Assoc. 2008 Sep-Oct;15(5):620-6.
25. Phansalkar S, Desai AA, Bell D, Yoshida E, Doole J, Czochanski M, Middleton B, Bates DW. High-priority drug-drug interactions for use in electronic health records. J Am Med Inform Assoc. 2012 Apr 26. [Epub ahead of print] PMID: 22539083
26. Sittig DF, Teich JM, Osheroff JA, Singh H. Improving clinical quality indicators through electronic health records: it takes more than just a reminder. Pediatrics. 2009 Jul;124(1):375-7.
27. Osheroff JA, Teich JM, Levick D, Saldana L, Velasco FT, Sittig DF, Rogers KM, Jenders RA. Improving Outcomes with Clinical Decision Support: An Implementer's Guide, Second Edition Healthcare Information and Management Systems Society, 2012.
28. Kuperman GJ, Teich JM, Tanasijevic MJ, Ma'Luf N, Rittenberg E, Jha A, Fiskio J, Winkelman J, Bates DW. Improving response to critical laboratory results with automation: results of a randomized controlled trial. J Am Med Inform Assoc. 1999 Nov-Dec;6(6):512-22.
29. Gandhi TK, Zuccotti G, Lee TH. Incomplete care--on the trail of flaws in the system. N Engl J Med. 2011 Aug 11;365(6):486-8.
30. American Society of Health-System Pharmacists. ASHP guidelines on pharmacy planning for implementation of computerized provider- order entry systems in hospitals and health systems. Am J Health-Syst Pharm. 2011; 68:e9-31. Available at: http://www.ashp.org/DocLibrary/BestPractices/AutoITGdlCPOE.aspx Accessed: 7/2/12

31. Haynes K, Linkin DR, Fishman NO, Bilker WB, Strom BL, Pifer EA, Hennessy S. Effectiveness of an information technology intervention to improve prophylactic antibacterial use in the postoperative period. J Am Med Inform Assoc. 2011 Mar-Apr;18(2):164-8. Epub 2011 Jan 24.
32. Shojania KG. The elephant of patient safety: what you see depends on how you look. Jt Comm J Qual Patient Saf. 2010 Sep;36(9):399-401.
33. Classen DC, Resar R, Griffin F, Federico F, Frankel T, Kimmel N, Whittington JC, Frankel A, Seger A, James BC.'Global trigger tool' shows that adverse events in hospitals may be ten times greater than previously measured. Health Aff (Millwood). 2011 Apr;30(4):581-9.
34. Hamblin JF, Bwitit PT, Moriarty HT. Pathology Results in the Electronic Health Record. The electronic Journal of Health Informatics. 2010; Vol 5(2): e15. Available at: http://www.ejhi.net/ojs/index.php/ejhi/article/view/131
35. Sittig DF, Singh H. A new sociotechnical model for studying health information technology in complex adaptive healthcare systems. Qual Saf Health Care. 2010 Oct;19 Suppl 3:i68-74.
36. Singh H, Wilson L, Petersen LA, Sawhney MK, Reis B, Espadas D, Sittig DF.Improving follow-up of abnormal cancer screens using electronic health records: trust but verify test result communication. BMC Med Inform Decis Mak. 2009 Dec 9;9:49.
37. Singh H, Mani S, Espadas D, Petersen N, Franklin V, Petersen LA. Prescription errors and outcomes related to inconsistent information transmitted through computerized order entry: a prospective study. Arch Intern Med. 2009 May 25;169(10):982-9.
38. Koppel R, Metlay JP, Cohen A, Abaluck B, Localio AR, Kimmel SE, Strom BL.Role of computerized physician order entry systems in facilitating medication errors. JAMA. 2005 Mar 9;293(10):1197-203.
39. Wetterneck TB, Walker JM, Blosky MA, Cartmill RS, Hoonakker P, Johnson MA, Norfolk E, Carayon P. Factors contributing to an increase in duplicate medication order errors after CPOE implementation. J Am Med Inform Assoc. 2011 Nov-Dec;18(6):774-82.
40. McCoy AB, Waitman LR, Lewis JB, Wright JA, Choma DP, Miller RA, Peterson JF. A framework for evaluating the appropriateness of clinical decision support alerts and responses. J Am Med Inform Assoc. 2012 May-Jun;19(3):346-52.
41. Blumenthal D, Tavenner M. The "meaningful use" regulation for electronic health records. N Engl J Med. 2010 Aug 5;363(6):501-4.
42. Paterno MD, Maviglia SM, Gorman PN, Seger DL, Yoshida E, Seger AC, Bates DW, Gandhi TK. Tiering drug-drug interaction alerts by severity increases compliance rates. J Am Med Inform Assoc. 2009 Jan-Feb;16(1):40-6.
43. Sittig DF, Singh H. Rights and responsibilities of users of electronic health records. CMAJ. 2012 Feb 13. [Epub ahead of print]
44. Gardner RM, Overhage JM, Steen EB, Munger BS, Holmes JH, Williamson JJ, Detmer DE; AMIA Board of Directors. Core content for the subspecialty of clinical informatics. J Am Med Inform Assoc. 2009 Mar-Apr;16(2):153-7.
45. Kulikowski CA, Shortliffe EH, Currie LM, et al. AMIA Board white paper: definition of biomedical informatics and specification of core competencies for graduate education in the discipline. J Am Med Inform Assoc 2012 Jun 21.

46. Powsner SM, Wyatt JC, Wright P. Opportunities for and challenges of computerisation. Lancet. 1998 Nov 14;352(9140):1617-22.
47. Wright A, Sittig DF, Ash JS, Feblowitz J, Meltzer S, McMullen C, Guappone K, Carpenter J, Richardson J, Simonaitis L, Evans RS, Nichol WP, Middleton B. Development and evaluation of a comprehensive clinical decision support taxonomy: comparison of front-end tools in commercial and internally developed electronic health record systems. J Am Med Inform Assoc. 2011 May 1;18(3):232-42.
48. Kuperman GJ, Gandhi TK, Bates DW. Effective drug-allergy checking: methodological and operational issues. J Biomed Inform. 2003 Feb-Apr;36(1-2):70-9.
49. Wright A, Goldberg H, Hongsermeier T, Middleton B. A description and functional taxonomy of rule-based decision support content at a large integrated delivery network. J Am Med Inform Assoc. 2007 Jul-Aug;14(4):489-96.
50. US Department of Health & Human Services, Partnership for Patients: Better care, Lower costs. Available at: http://www.healthcare.gov/compare/partnership-for-patients
51. Singh H, Thomas EJ, Mani S, Sittig D, Arora H, Espadas D, Khan MM, Petersen LA. Timely follow-up of abnormal diagnostic imaging test results in an outpatient setting: are electronic medical records achieving their potential? Arch Intern Med. 2009 Sep 28;169(17):1578-86.
52. Nwulu U, Nirantharakumar K, Odesanya R, McDowell SE, Coleman JJ. Improvement in the detection of adverse drug events by the use of electronic health and prescription records: An evaluation of two trigger tools. Eur J Clin Pharmacol. 2012 Jun 17. [Epub ahead of print].
53. Griffin FA, Classen DC. Detection of adverse events in surgical patients using the Trigger Tool approach. Qual Saf Health Care. 2008 Aug;17(4):253-8.
54. Adelman JS, Kalkut GE, Schechter CB, Weiss JM, Berger MA, Reissman SH, Cohen HW, Lorenzen SJ, Burack DA, Southern WN. Understanding and preventing wrong-patient electronic orders: a randomized controlled trial. J Am Med Inform AssocJune 2012 doi:10.1136/amiajnl-2012-001055.
55. The Patient Safety Organization Privacy Protection Center. Available at: https://www.psoppc.org/web/patientsafety
56. CMS Office of Public Affairs. More than 100,000 health care providers paid for using electronic health records: CMS and ONC surpass 2012 goals for EHR adoption and use. Available at: http://tinyurl.com/CMS-EHR-Users
57. Singh H, Classen DC, Sittig DF. Creating an oversight infrastructure for electronic health record-related patient safety hazards. J Patient Saf. 2011 Dec;7(4):169-74. doi: 10.1097/PTS.0b013e31823d8df0.
58. Department of Health and Human Services, Office of the Secretary. 45 CFR Part 170, RIN 0991–AB82 — Health information technology: standards, implementation specifications, and certification criteria for electronic health record technology, 2014 edition; revisions to the Permanent Certification Program for Health Information Technology. Fed Regist 2012;77(171):54163-260.

CHAPTER 5

SAFER GUIDE DEVELOPMENT METHODS

SAFETY ASSURANCE FACTORS FOR ELECTRONIC HEALTH RECORD RESILIENCE (SAFER): STUDY PROTOCOL

Hardeep Singh, Joan S. Ash, and Dean F. Sittig

5.1 BACKGROUND

Several countries have made recent multi-billion dollar investments in electronic health record (EHR) infrastructure to transform their health care delivery systems. However, implementation of EHR-related initiatives has encountered greater than expected challenges [1-4]. Although successful transformations have occurred in a few pioneering healthcare organizations across the globe, [5,6] the vast majority of organizations are still in the process of implementing their EHRs and modifying their work processes [7,8].

In some instances, reports warn of unintended consequences of health information technology (HIT) adoption, including new safety problems

---

© Singh H, Ash JS, and Sittig DF; licensee BioMed Central Ltd. Safety Assurance Factors for Electronic Health Record Resilience (SAFER): Study Protocol. BMC Medical Informatics and Decision Making 13,46 (2013), doi:10.1186/1472-6947-13-46. Published under the Creative Commons Attribution 2.0 Generic License, http://creativecommons.org/licenses/by/2.0/.

and reduced provider efficiency resulting from the implementation and use of EHRs [9-20]. EHR-related errors occur in a sociotechnical environment, described more fully below, which includes the hardware and software required to implement the heath IT, as well as the social environment in which it is implemented. EHR-related errors should be defined from the sociotechnical viewpoint of end users, rather than from the purely technical viewpoint of manufacturers, developers, vendors, and personnel responsible for implementation. In this context, EHR-related errors could occur anytime the EHR is unavailable for use, malfunctions, is used incorrectly, or when EHR components interact incorrectly, resulting in data being lost or incorrectly entered, displayed, or transmitted [20]. Examples could include technology errors, such as lack of transmission of test results due to software configuration problems, or technology use problems such as juxtaposition errors due to clicking the wrong item in a drop-down menu when ordering a medication. EHR-related errors are complex, and the roots of these errors are often multifaceted. Risks for EHR-related errors and breakdowns may be related to features of the technology itself, user behaviors, and organizational influences on how the EHR is routinely used, maintained, and monitored.

To respond to the safety challenges described herein, the Office of the National Coordinator for Health Information Technology (ONC) sponsored an Institute of Medicine report, *Health IT and Patient Safety: Building Safer Systems for Better Care* [21]. In addition, ONC leveraged another ongoing project on "Anticipating the Unintended Consequences of Health IT" and requested development of health IT patient safety self-assessment guides to address these safety concerns.

To ensure that EHRs fulfill their promise of making healthcare safer and enhancing health care quality, hospitals and other clinical entities require proactive monitoring strategies to detect new, unexpected EHR-related errors. This would enable them to transform into safe and effective "EHR-enabled clinical work systems" by building resilience into their systems and processes. Resilience is the "degree to which a system continuously prevents, detects, mitigates or ameliorates hazards or incidents so that an organization can bounce back to its original ability to provide care" [22]. Through proactive, systematic assessment of risks and vulnerabilities, health care organizations can address potential EHR-related safety hazards before harmful incidents occur. The overall objective of this project is to develop and validate proactive self-assessment tools to ensure that EHR-enabled clinical work systems are safe and effective.

5.2 METHODS/DESIGN

The project will be conducted in two main phases. The first phase develops self-assessment guides that can be used by clinicians and health care organizations to evaluate certain high-risk components of their EHR-enabled clinical work systems. The second phase will examine the usefulness of the self-assessment guides by beta testing the guides at selected sites with EHR-enabled systems and by conducting on-site evaluations.

This work will be guided by an 8-dimension, sociotechnical model of safe and effective health IT use [23] that we developed in response to difficulties clinicians and organizations have encountered with EHR implementation. This model (see Figure 5.1 and Table 5.1) provides a comprehensive framework for studying all aspects of health IT design, development, implementation, use, and evaluation within complex, adaptive health care systems.

TABLE 5.1: Dimensions of the sociotechnical model[23]

Hardware and software	The computing infrastructure used to power, support, and operate clinical applications and devices.
Clinical content	The text, numeric data, and images that constitute the "language" of clinical applications.
Human-computer interface	All aspects of technology that users can see, touch, or hear as they interact with it.
People	Everyone who interacts in some way with technology, including developers, users, IT personnel, and informaticians.
Workflow and communication	Processes to ensure that patient care is carried out effectively.
Internal organizational features	Policies, procedures, work-environment and culture.
External rules and regulations	Federal or state rules that facilitate or constrain preceding dimensions
Measurement and monitoring	Processes to evaluate both intended and unintended consequences of health IT implementation and use.

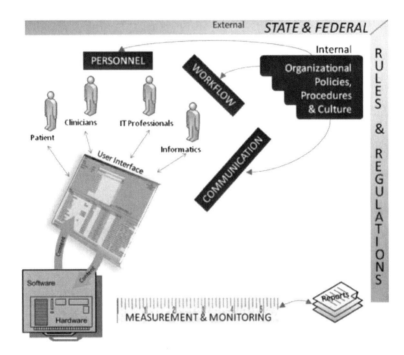

FIGURE 5.1: Sociotechnical model for safe and effective health IT use (reproduced with permission from BMJ Quality and Safety)[23].

We will use the eight dimensions of this sociotechnical model (Figure 5.1) to develop self-assessment guides to address certain high-risk aspects of the EHR-enabled clinical work system. These dimensions include the hardware and software that "run" the EHR; the clinical content that is used to configure the various EHR modules; the user interface that allows clinicians to interact with the EHR application; the people who are required to configure the system, train the users, and use the system; the clinical workflow and communication processes that enable clinicians to provide patient care; the internal organizational policies, procedures, and culture that "govern" all the activities associated with using the EHR; the external rules and regulations that affect the healthcare delivery system; and, finally, the measurement and monitoring that is required to determine what is working and what is not. Within each of these 8 dimensions we will identify multiple themes that affect safety and effectiveness of EHR-enabled work systems in order to provide guidance about what an organization could do to address potential problem areas.

Based upon previous work, current literature, and expert opinion, [1,3,14,15,24-29] we have identified 9 high-risk areas around which the content of the self-assessment guides will be organized. These areas include features, functions, or applications of the EHR itself and/or relate to patient safety issues that can arise from the use of the EHR. The 9 high-risk areas are:

- Computerized provider order entry (CPOE) and e-prescribing
- Clinical decision support
- Test result reporting
- Communication between providers
- Patient identification
- EHR downtime events
- EHR customization and configuration
- System-system interface data transfer
- Health IT safety-related human skills

5.2.1 PROCEDURES FOR DEVELOPING THE SELF-ASSESSMENT GUIDES

We will develop guides in each of the 9 high-risk areas identified above through a stepwise process of consultation and data gathering with subject matter experts as well as stakeholders with specific interests in quality improvement and patient safety. These steps are listed below.

5.2.1.1 STEP 1: CONVENING AN EXPERT PANEL AND WORKGROUP

We will identify and recruit an expert panel and a technical workgroup, each consisting of approximately 4–6 experts, to provide guidance and advice on several aspects of the project including design of the self-assessment guides, facilitating access to key stakeholders and organizations that might ultimately use these guides, and facilitating access to institutions that would be willing to serve as test sites. The areas of expertise to be represented on the expert panel will include EHRs, patient safety, quality improvement, risk management, human factors engineering and usability, knowledge and experience with small physician office practices, and accreditation and certification practices for patient safety. The technical workgroup will include clinical and technical experts who can help develop and prioritize specific components of each of the guides to be developed.

The expert panel will meet in person once in the early phase of the project. The goal of this meeting will be to inform members about this project, help select study sites, identify which individuals will need to be interviewed during site visits, and identify the best strategies to develop the guides and implement them in the field. Over the course of the project, the expert panel will convene twice more through two teleconferences, which will be used to review and refine the tools and to work further on strategies to get the guides into the field.

The technical workgroup will hold one teleconference and will be available to review the guides as needed. The goal of the teleconference will be to discuss and prioritize initial lists of items in the guides, and the reviews will be focused on recommendations for modifying the items after site visits. For expert panel and technical workgroup meetings, both modified Delphi and normative group techniques will be used for assessing content validity for the guides. Between meetings we expect to obtain additional feedback on guide items via e-mail. The following is a preliminary list of questions that would be asked:

1. Question 1-Content: Do you consider the items of high enough importance to be included in this guide, and are there themes or items that we have not considered which must be included?
2. Question 2-Wording/comprehension: Are the items appropriately worded so that a multidisciplinary team assembled by the organization conducting this as a "self-assessment" (including representatives from groups such as Chief Medical Informatics Officers (CMIOs), Chief Information Officers (CIOs), Chief Nursing Informatics Officers, risk management/quality and safety personnel, ambulatory clinic practice managers, and specialized personnel in pharmacy/lab etc.) is able to answer them and find them useful for improvement?

3. Question 3-Prioritization: Are there some items on the checklist that you would consider "essential" for all organizations regardless of size, EHR vendor, type of organization, etc.?

5.2.1.2 STEP 2: LITERATURE SEARCH

We will conduct an updated literature search to inform the development of the self-assessment guides. Because this field is fairly nascent, sources from the web and press reports on EHR safety events will be considered in addition to the peer-reviewed scientific literature.

5.2.1.3 STEP 3: STAKEHOLDER ENGAGEMENT FOR DEVELOPMENT AND FUTURE DISSEMINATION

It will be important to engage with, educate, and solicit feedback from key stakeholders other than the expert panel and technical workgroup members to ensure that the final guides are supported and their use encouraged for the target audiences many of these stakeholders represent. Given the small membership of the expert panel and technical workgroup, stakeholder engagement will also serve to broaden the awareness level of many more influential individuals and organizations about issues related to HIT safety. The key stakeholders invited to participate will vary depending on the topic area for each guide. We will involve some in developing the guides, some in gaining their endorsement of the guides, and others in actually using the guides. We will not reimburse stakeholders for participation in these activities, except for individuals who represent stakeholder groups who might also serve on expert panels.

We will make initial contacts by e-mail and phone. Some discussions will require conference calls. We will prepare explanatory materials for written and oral presentation that include a description of the problem we are trying to address and how this project will address it. Members of our team will also attend conferences where we can give presentations about the objectives of the project and, later, to describe results of our work and solicit suggestions.

5.2.1.4 STEP 4: DEVELOPMENT OF THE SELF-ASSESSMENT GUIDES

We will develop the guides as self-contained chapters or modules, each devoted to one of the 9 different high-risk areas. The guides themselves will be created in the form of question lists with accompanying instructions for completion and an introduction explaining the methodology for developing each guide.

Our multi-disciplinary team of clinicians, health services researchers, human factors engineers, computer scientists, and informaticians will develop and continuously refine on an iterative basis, an itemized list of characteristics of reliability of safety and effectiveness within each of the 8 dimensions. For example, within the hardware/software dimension, we have identified the need for redundant hardware for key system components (e.g., database servers, and distributed clinical workstations) as an important safety feature. Likewise, within the workflow and communication dimension, we have identified methods of promptly notifying clinicians that new, abnormal test results are available as an important characteristic. We expect to initially identify about 7–10 items within each of the 8-dimensions of our sociotechnical model for each of the guides. The initial list will be narrowed down using the steps above.

This development process will incorporate the input of key stakeholders, our expert panel, and the technical workgroup as described above. Lists of items for portions of each tool will be reviewed and prioritized in the most appropriate manner for that module. Once the guides are drafted, we will pilot test them using interviews with intended end users at local facilities.

5.2.1.5 STEP 5: FACE VALIDATION AND REFINEMENT OF THE GUIDES

This step will involve 5 site visits to healthcare organizations for two purposes. First, we will establish the content validity of the checklists with a group of likely users. For example, we will ask individuals to answer the questions posed by the guides and solicit feedback about the questions themselves and the perceived usefulness of the guides. The second purpose of these visits will be to assess the context within which the tools could be used. Data collection will include semi-structured and informal interviews, naturalistic observation of various processes and people, and document analysis. Interview subjects will be selected for their expertise and roles in HIT safety. Observation subjects will represent a cross-section of system users including both "EHR skeptics" and champions. Our team has previously used these methods to understand operations of EHR work systems [30-32]. Our team members from multiple disciplines will be present at the site visits.

The site visits will be essential to understanding the local contexts and constraints in which EHRs are being implemented, as these are expected to be highly variable even within the same organization. The data we obtain will shed light on who would most likely use the guides and how. We will also use the site visits to identify high-risk areas that site leaders believe to be associated with safety and effectiveness problems. Site selection for the 5 sites will be based on geography, size, and type (e.g., academic medical center, privately-owned out-patient clinics, EHR service providers) of organization. We will visit sites with variable types of EHR-related characteristics such as 1) experience with using clinical information systems, 2) comprehensiveness of approach and/or resources to address EHR safety, and 3) size and type of institution (ambulatory vs. inpatient). The tools will be modified based on the information gathered, and later they will be beta tested at other sites to establish the integrity of our findings.

5.2.2 PROCEDURES FOR BETA TESTING THE SELF-ASSESSMENT GUIDES

Each of the 9 guides will be beta tested at five additional sites. Two site visitors will interview key informants at each site for feedback and recommendations about further improvement. On-site evaluation will also help us understand how local site assessors will interpret the items and determine what is considered feasible and useful for the organizations to accomplish. We will work with IT professionals, clinicians, and other personnel to compare each item within our self-assessment guides to acknowledged industry-standard best practices. For instance, items that are not found to be easily operationalized in terms of observations and/or measurements will be edited or removed, and new items will be added as needed. One representative from each site will assist our team and help us compare the organization's responses to the items with our own assessments. This process will help ensure that the assessment guides are valid and that multiple respondents interpret items in the same way.

Although we anticipate that our initial development process will result in preliminary guides that have strong content and content validity, we will leave open the potential for substantial further refinement during the beta testing period. We will approach the refinement of the guides iteratively such that the guides will be modified after each visit and new versions tested at the next site.

The anticipated output of this work will be a set of highly generalizable self-administered EHR assessment guides that transform the final set of items into clear checklist-type items with a set of discrete responses. Each item will include an explanation of its relevance and clear descriptions of response choices. Our future goal is for these guides to automatically generate a narrative report on findings and conclusions and, ultimately, provide a comparison to benchmark results. At the end of the beta testing and refinement phase, the guides will be ready for field testing at additional sites. However, true validation testing is outside the scope of this project.

5.3 DISCUSSION

We propose to develop self-assessment guides that health care organizations can use to help prevent, detect, mitigate, and ameliorate hazards associated with the use of EHRs. Our project is grounded conceptually in a multifaceted sociotechnical model of safe and effective use of health IT. In building upon the work proposed here, future initiatives can help create best practices that can be used by key stakeholders to oversee the successful transformation of their health care system into a highly reliable EHR-enabled clinical work system [29].

The development of proactive risk assessment guides is consistent with the WHO conceptual model for patient safety (see Figure 5.2).

SAFER Guide Development Methods

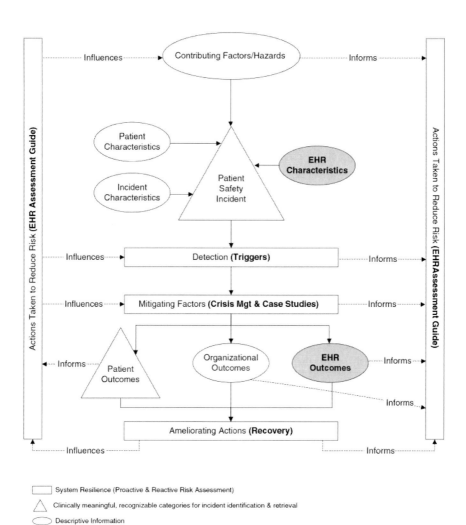

FIGURE 5.2: Modified conceptual model showing relationships between patient safety and resilience within an EHR-enabled work system.

The WHO model focuses on several aspects of system resilience. For example, we will aim to develop guides that help users not only detect risks but also develop "actions or [create] circumstances which prevent or moderate the progression of an incident toward harming the patient" [33]. Users can in turn learn from the ameliorating actions for the event, i.e., identify the changes that were made, and assess the outcomes associated with the patient, organization, and the EHR. Based on this knowledge, organizations can identify and further develop specific actions that will reduce the risk that the error will occur in the future.

Any broad strategy to address patient safety activities within EHR-enabled health care systems must account for variation in stages of EHR implementation and levels of complexity across clinical practice settings. For instance, organizations that have recently deployed or are still planning to implement EHRs may face different challenges to ensuring patient safety (e.g., intraoperability between the EHR and other systems) than their counterparts with established EHRs (e.g., making full use of the EHR to monitor safety events). Thus, implementing proactive assessment of EHR risks could be conceptualized in three phases. The first phase would address safety concerns unique to technology (i.e., ensuring that basic EHR functions are safe and reliable). The second phase would address mitigation of safety concerns from failure to use technology appropriately (i.e. safer application and use of EHRs). In the third phase, organizations will leverage EHR capabilities to monitor and improve patient safety [29]. The assessment is expected to provide input to help organizations or practices to identify and prioritize patient safety issues related to most aspects of EHR-enabled health care delivery. On the basis of this information, they could pursue additional strategies and risk management tools to address the EHR-related patient safety risks identified and engage their leadership in this process.

Our assessment strategies also pose some limitations and challenges. This assessment will require input and time from a number of different individuals, which could include IT managers (e.g., CIO or CMIO), risk managers, practice managers, patient safety and quality personnel, as well as other key stakeholders that are involved in ensuring the safety of an EHR-enabled health care system (nursing, pharmacy, laboratory personnel, and others). Although we will recommend a multidisciplinary team to work together to complete this assessment, this might not always be achievable with internal personnel in every setting. For instance, in small organizations, outside IT expertise will often be required. In some instances, the involvement of the IT vendor/developer may be required. Furthermore, when evaluating specific safety issues that may involve several interacting factors within the sociotechnical model, even large organizations may choose to involve outside organizations or personnel with specific expertise in understanding and addressing health-IT related concerns.

Risk assessment based upon these guides will also be highly context dependent. Thus, the risks associated with not fully implementing practices identified in the health IT self-assessment guides will vary and should be considered within the context of each individual setting. For example, an EHR downtime poses different risks for a small ambulatory practice than it does for a 500-bed level one trauma hospital with an active emergency department and several intensive care units. Settings will need to consider both the severity and the probability (i.e., in terms of frequency) of a safety event that might result from not implementing practices identified in these guides.

EHRs are changing the way we deliver health care. Taken together, error detection, mitigation, and amelioration are the three most important concepts in building system resiliency to reduce the risk of future safety events within the EHR-enabled work system. Lessons learned from the development of the assessment guides will be helpful for both organizations that are beginning the EHR selection and implementation process as well as those that have already implemented systems. The health IT patient safety assessment guides might lead institutions to better leverage the benefits of EHRs. Using a multi-faceted, cross-disciplinary approach to develop evidence-based guidance for the EHR-enabled work system might be a useful step in improving patient safety in technology-enabled health care.

REFERENCES

1. Kilbridge PM, Classen DC: The informatics opportunities at the intersection of patient safety and clinical informatics. J Am Med Inform Assoc 2008, 15:397-407.

2. Metzger J, Welebob E, Bates DW, Lipsitz S, Classen DC: Mixed results in the safety performance of computerized physician order entry. Health Aff (Millwood) 2010, 29:655-663.
3. Sittig DF, Singh H: Eight rights of safe electronic health record use. JAMA 2009, 302:1111-1113.
4. Sittig DF, Ash JS: Clinical information Systems: Overcoming adverse consequences. Sudbury, MA: Jones and Bartlett Publishers, LLC; 2009.
5. Chaudhry B, Wang J, Wu S, Maglione M, Mojica W, Roth E: Systematic review: impact of health information technology on quality, efficiency, and costs of medical care. Ann Intern Med 2006, 144:742-752.
6. Protti D: Comparison of information technology in general practice in 10 countries. Healthc Q 2007, 10:107-116.
7. Blumenthal D, Tavenner M: The "meaningful use" regulation for electronic health records. N Engl J Med 2010, 363:501-504.
8. Sittig DF, Ash JS, Zhang J, Osheroff JA, Shabot MM: Lessons from "Unexpected increased mortality after implementation of a commercially sold computerized physician order entry system". Pediatrics 2006, 118:797-801.
9. Campbell EM, Sittig DF, Ash JS, Guappone KP, Dykstra RH: Types of unintended consequences related to computerized provider order entry. J Am Med Inform Assoc 2006, 13:547-556.
10. Harrington L, Kennerly D, Johnson C: Safety issues related to the electronic medical record (EMR): synthesis of the literature from the last decade, 2000–2009. J Healthc Manag 2011, 56:31-43.
11. Horsky J, Kuperman GJ, Patel VL: Comprehensive analysis of a medication dosing error related to CPOE. J Am Med Inf Assoc: JAMIA 2005, 12:377-382.
12. Koppel R, Metlay JP, Cohen A, Abaluck B, Localio AR, Kimmel SE: Role of computerized physician order entry systems in facilitating medication errors. JAMA 2005, 293:1197-1203.
13. Leviss J: H.I.T. Or Miss: Lessons Learned from Health Information Technology Implementation. Chicago, IL: American Health Information Management Association; 2010.
14. Magrabi F, Ong MS, Runciman W, Coiera E: An analysis of computer-related patient safety incidents to inform the development of a classification. J Am Med Inform Assoc 2010, 17:663-670.
15. Magrabi F, Ong MS, Runciman W, Coiera E: Using FDA reports to inform a classification for health information technology safety problems. J Am Med Inform Assoc 2012, 19:45-53.
16. McDonald CJ: Computerization can create safety hazards: a bar-coding near miss. Ann Intern Med 2006, 144:510-516.
17. Nerich V, Limat S, Demarchi M, Borg C, Rohrlich PS, Deconinck E: Computerized physician order entry of injectable antineoplastic drugs: an epidemiologic study of prescribing medication errors. Int J Med Inform 2010, 79:699-706.
18. Schulte F, Schwartz E: As Doctors Shift to Electronic Health Systems, Signs of Harm Emerge. The Huffington Post. 2010. Ref Type: Newspaper

19. Singh H, Wilson L, Petersen L, Sawhney MK, Reis B, Espadas D: Improving follow-up of abnormal cancer screens using electronic health records: trust but verify test result communication. BMC Med Inf Decis Making 2009, 9:1-7.
20. Sittig DF, Singh H: Defining health information technology-related errors: new developments since to err is human. Arch Intern Med 2011, 171:1281-1284.
21. Committee on Patient Safety and Health Information Technology: Health IT and Patient Safety: Building Safer Systems for Better Care. 11-8-2011. Institute of Medicine; Ref Type: Report
22. Sherman H, Castro G, Fletcher M, Hatlie M, Hibbert P, Jakob R: Towards an International Classification for Patient Safety: the conceptual framework. Int J Qual Health Care 2009, 21:2-8.
23. Sittig DF, Singh H: A new sociotechnical model for studying health information technology in complex adaptive healthcare systems. Qual Saf Health Care 2010, 19:i68-i74.
24. Ash JS, Sittig DF, Poon EG, Guappone K, Campbell E, Dykstra RH: The extent and importance of unintended consequences related to computerized provider order entry. J Am Med Inform Assoc 2007, 14:415-423.
25. Bates DW, Kuperman G, Teich JM: Computerized physician order entry and quality of care. Qual Manag Health Care 1994, 2:18-27.
26. Bates DW, Kuperman GJ, Wang S, Gandhi T, Kittler A, Volk L: Ten commandments for effective clinical decision support: making the practice of evidence-based medicine a reality. J Am Med Inform Assoc 2003, 10:523-530.
27. Singh H, Vij MS: Eight recommendations for policies for communicating abnormal test results. Jt Comm J Qual Patient Saf 2010, 36:226-232.
28. Singh H, Wilson L, Reis B, Sawhney MK, Espadas D, Sittig DF: Ten strategies to improve management of abnormal test result alerts in the electronic health record. J Patient Saf 2010, 6:121-123.
29. Sittig DF, Singh H: Electronic health records and national patient-safety goals. N Engl J Med 2012, 367:1854-1860.
30. Hysong SJ, Sawhney MK, Wilson L, Sittig DF, Espadas D, Davis T: Provider management strategies of abnormal test result alerts: a cognitive task analysis. J Am Med Inform Assoc 2010, 17:71-77.
31. Hysong SJ, Sawhney MK, Wilson L, Sittig DF, Esquivel A, Singh S: Understanding the management of electronic test result notifications in the outpatient setting. BMC Med Inform Decis Mak 2011, 11:22.
32. McMullen CK, Ash JS, Sittig DF, Bunce A, Guappone K, Dykstra R: Rapid assessment of clinical information systems in the healthcare setting. An efficient method for time-pressed evaluation. Methods Inf Med 2011, 50:299-307.
33. Conceptual framework for the international classification for patient safety: World Health Organization. Report: Technical Report. Ref Type; 2009.

CHAPTER 6

OVERVIEW OF SAFER GUIDES

THE SAFER GUIDES: EMPOWERING ORGANIZATIONS TO IMPROVE THE SAFETY AND EFFECTIVENESS OF ELECTRONIC HEALTH RECORDS

Dean F. Sittig, Joan S. Ash, and Hardeep Singh

Electronic health records (EHRs) have the potential to improve the quality and safety of healthcare. [1] Since the enactment of the Health Information Technology for Economic and Clinical Health Act (HITECH), [2] organizations have been adopting EHRs at an unprecedented rate. [3] While the challenges of rapid EHR implementation can be numerous and disruptive, EHRs have clear potential to improve quality and safety with better access to information, [4,5] clinical decision support, [6] and more reliable provider-to-provider communication. [7] Nevertheless, in the early stages of an EHR-enabled healthcare system, benefits thus far have been difficult to achieve and unintended consequences have emerged. [8] Clinicians have experienced safety concerns from EHR design and usability features that are not optimal for complex work flows in real-world practice settings. [9-11] To respond to these challenges, the Office of the National

---

Reprinted with permission. The SAFER Guides: Empowering Organizations to Improve the Safety and Effectiveness of Electronic Health Records. Sittig DF, Ash JS, and Singh H. The American Journal of Managed Care **20**,5 *(2014).*

Coordinator for Health Information Technology (ONC) commissioned the 2012 Institute of Medicine Report, "Health IT and Patient Safety: Building Safer Systems for Better Care" [12] and recently released the Health Information Technology Patient Safety Action and Surveillance Plan that lays out their response to these issues. [13]

National initiatives to improve the safety of EHRs must be accompanied by practical and helpful strategies for those on the front lines of EHR-enabled care delivery. Strategies to address unintended consequences borne from EHR implementation are nonetheless scarce, and frontline clinicians and healthcare organizations (HCOs) are often unaware of best practices for safe EHR implementation and use. For example, they often have minimal guidance to handle problems such as too many alerts, [14,15] a slow EHR, or an EHR that requires an excessive number of "clicks" to complete tasks. These are not skills routinely expected of healthcare providers in the past. [16] Clinicians are also not privy to safety concerns embedded in flawed interfaces between various components of EHRs and in the way EHRs are configured. Solutions to these problems are often multifaceted, requiring analysis and redesign of work flows and organizational processes and procedures that cannot be addressed through improvements in technology alone.

Addressing EHR-related safety concerns is inherently complex and requires a comprehensive and multifaceted systems-based approach. We propose that HCOs equipped with EHRs should consider the strategy of "proactive risk assessment" of their EHR-enabled healthcare to identify and address EHR-related safety concerns. [17] Herein, we describe the conceptual underpinnings of an EHR-related self-assessment strategy to provide clinicians and HCOs a foundation upon which they could build their safety efforts.

6.1.1 CONCEPTUAL FOUNDATION OF SELF-ASSESSMENT

With support from the ONC, we used rigorous, iterative methodologies to develop 9 self-assessment tools to optimize the safety and safe use of EHRs. [18] These tools, the Safety Assurance Factors for EHR Resilience (SAFER) guides, are designed to help clinicians and HCOs self-assess the safety and effectiveness of their EHR implementations, identify specific areas of vulnerability, and create solutions and culture change to mitigate risks.

The goal of SAFER guide-based proactive risk assessment is to eliminate or minimize EHR-related safety hazards to build system resilience, defined as "degree to which a system continuously prevents, detects, mitigates, or ameliorates hazards or incidents so that an organization can bounce back to its original ability to provide care." [19]

Each SAFER guide consists of between 10 and 25 "recommended practices," which can be assessed as fully implemented, partially implemented, or not implemented. Recommended practices help the clinic or organization know "what" to do to optimize the safety and safe use of the EHR. The recommended practices address principles that represent "why" the recommended practices are needed, although any given recommended practice may support several principles that support health IT safety.

The methods used to identify risk areas and associated practices have been described elsewhere, but are briefly summarized here. [18] To develop the content of the guides, we consulted subject matter experts in informatics, patient safety, quality improvement, risk management, and human factors engineering and usability. To ensure generalizability, we conducted site visits at both small and large practices and hospitals. We also reviewed literature to identify existing EHR-related assessment items, which we validated and refined during site visits. To ensure that the guides would be useful to our intended audiences (eg, clinicians, EHR developers, IT professionals, and quality improvement leaders), we engaged with a broad range of stakeholders, such as professional organizations representing diverse groups of intended users. We undertook multiple revisions of items in the guides to increase their applicability and interpretability by individuals with differing degrees of expertise. We also considered the perspectives of those working within organizations at different points in their EHR adoption journey. Thus, we ensured that the SAFER guides are based on best evidence and expertise currently available, as well as on field research and iterative testing. [20] To facilitate wide implementation and use, the SAFER guides have been made available free of charge from ONC's website (http://www.healthit.gov/safer/).

Content of the SAFER guides is organized around 2 conceptual frameworks that account for the complex sociotechnical system in which EHRs are implemented and the risks specific to various phases of implementation, respectively. The first model describes the 8 contextual dimensions of EHR-enabled healthcare systems: (1) hardware and software; (2) clinical content; (3) human-computer interface; (4) people; (5) work flow and communication; (6) internal organization policies, procedures, physical environment, and culture; (7) external rules, regulations, and pressures; and (8) system measurement and monitoring. [21] Along with this sociotechnical framework, we used a 3-phase framework of EHR safety that describes risks along different points of the EHR implementation life cycle. [22] The overall goal, as described by Blumenthal and Tavenner, is for healthcare organizations to move from a paper-based medical record system "up the escalator" to become fully EHR-enabled healthcare systems. [23] Within each phase of the 3-phase framework, all 8 dimensions of the sociotechnical model come into play. Phases remind organizations "which" aspect of safety is being addressed as they adopt EHRs and build safety programs.

The first step on the "escalator" framework (safe health IT) accounts for safety events unique and specific to EHRs and often emerge early in implementation (eg, safety problems owing to unavailable or malfunctioning hardware/software). The second step (using health IT safely) addresses unsafe or inappropriate technology use, including unsafe changes in work flows that emerge due to technology use. The third step (monitoring safety) addresses use of technology to monitor healthcare processes and patient outcomes and to identify potential safety concerns before harm occurs. Together, the sociotechnical framework and 3-phase implementation framework form the conceptual foundation of the self-assessment process.

6.1.2 OPERATIONALIZING SAFER GUIDES

Organizations should consider using the "high priority practices SAFER guide" to identify their most pressing needs and decide which of the more specific SAFER guide(s) to use for a more in-depth self-assessment. In pilot testing at smaller practices (ie, 1-5 physicians with 1-10 clinical and administrative support staff) with experienced EHR users, it took approximately 30 minutes to work through the high priority guide practices. It may take longer for the other SAFER guides.

The SAFER guides include planning worksheets to help organizations set goals and track progress. The worksheets offer a rationale that explains "why" each recommended practice is important. For example, 1 recommended practice within the "Test Results Reporting and Follow-up SAFER guide" is to create "back-up procedures (including use of surrogates) and fail-safe escalation systems" to communicate test results to responsible providers. [24] The rationale is based on known risks introduced at handoffs between providers, especially those involving trainees and part-time providers. [25] Finally, to help operationalize each recommended practice, the worksheets include examples that illustrate "how" recommended practices can be implemented. For example, one way to implement the recommendation above is to have unacknowledged test result notifications forwarded to alternate providers or escalated to supervisors after a certain number of days.

The SAFER guides will need to be integrated within each HCO or clinic's current patient safety programs. However, the existing patient safety structures and processes will likely need to be modified as well to incorporate the unique skill mix needed. For example, larger HCOs need informaticians or clinicians trained in the newly created sub-specialty of clinical informatics [26,27] in addition to a multidisciplinary oversight committee to help identify risks, prioritize interventions, and review EHR-related solutions. A thorough review of the literature combined with our field research and survey suggest that currently most HCOs do not have the tools and expertise to optimize the safety and safe use of EHRs and have not built EHR safety into their patient safety programs. [28,29]

The SAFER guides are intended to be used by stakeholders from several administrative and clinical departments within a large hospital. For example, in larger organizations where major departments have differences in implementation and use of EHRs, it might be beneficial for each of these departments to conduct the assessment. Additionally, departments external to organizations such as EHR vendors, clinical knowledge suppliers, or information technology infrastructure providers might also need to take part in such assessments. These include chief information officers, chief medical informatics officers, risk managers, pharmacists, diagnostic services such as laboratory and radiology services, practice managers, patient safety and quality personnel, key clinical stakeholders, and—very importantly—EHR developers. In smaller practices that rely on remotely hosted EHRs and external IT consultants, representatives of external organizations will be key participants in implementing best practices from the SAFER guides.

Self-assessment should be integrated with ongoing patient safety, quality improvement, and risk management programs both in smaller ambulatory practices and in the different departments within larger hospital-based settings. Any of these parties could leverage the SAFER guides to start a meaningful conversation on optimizing safety and safe use of EHRs. In most settings, SAFER guides could be completed by multidisciplinary teams. Similar to studies of checklists in other high-risk environments such as the operating room, [29] achieving a shared understanding of safe practices would facilitate better teamwork and collaboration and perhaps stimulate implementation of solutions. However, when too many stakeholders are involved, there is a risk that certain team members may lack role clarity or assume that others are primarily responsible for addressing key tasks. [31] We suggest that frontline users must engage in multidisciplinary teams to conduct self-assessments because they interact with all aspects of the EHR-enabled healthcare delivery system. To lead, they must have authority and resources from organizational leadership committed to the safety and safe use of EHRs. Organizational leaders should then work with individuals or organizations (sometimes external ones, such as EHR developers) and assign responsibility for the work to implement recommended practices based on the assessment.

Although a self-assessment is generally intended to stimulate and sustain recommended practices, it can also establish a baseline for measuring effects of interventions designed to improve safety over time. If used for this purpose, the SAFER guides could be used to reassess safety at periodic intervals (eg, annually) or when changes are made to the EHR.

We found in our preliminary field research that the recommended practices in each SAFER guide apply to all types and sizes of EHR-enabled practice settings. Implementation of the recommended practices will vary by type of practice and type of EHR (eg, remotely hosted vs local server). The number and type of people involved in conducting SAFER guide assessments will also vary by practice setting. To address these variations, the SAFER guides have accompanying worksheets with specific examples of different methods to implement the recommended practices.

We also expect variation in the importance or urgency of implementing some of the recommended practices based upon the type of practice setting. Some recommended practices, such as contingency planning to avoid EHR downtimes, may be critical for tertiary care hospitals and less urgent for outpatient clinics that do not perform complex surgical procedures or provide overnight care. [32] In addition, it is important for assessment results to be discussed with EHR vendors and open the dialogue about solutions and patches that might have been created for other clients. For example, EHR vendors may have developed a solution to the specific problem identified but not yet released this fix to all their customers. Furthermore, it is important for parties conducting these assessments to carefully weigh the risks that apply to their settings as well as the difficulty in implementing the required changes in hardware, software, or clinical work flow. [33] The SAFER guides are designed to help all types and sizes of EHR-enabled practice settings target areas of concern specific to their unique circumstances and resources.

6.1.3 CONCLUSIONS

To fulfill the promise of EHRs, the SAFER guides can empower organizations, and practitioners that work within them, to address their EHR-related problems along front lines of care. Furthermore, they could help frontline users and other stakeholders have meaningful conversations on how to optimize EHR functionality and how to use EHRs to drive substantial improvements in the quality and safety of healthcare. These conversations, though likely a first step in a long process for most organizations, could go a long way toward helping leverage the potential of EHRs to improve health and healthcare.

REFERENCES

1. Blumenthal D, Glaser JP. Information technology comes to medicine. N Engl J Med. 2007;356(24):2527-2534.
2. Blumenthal D. Launching HITECH. N Engl J Med. 2010;362(5): 382-385.

3. Wright A, Henkin S, Feblowitz J, McCoy AB, Bates DW, Sittig DF. Early results of the meaningful use program for electronic health records. N Engl J Med. 2013;368(8):779-780.
4. Propp DA. Successful introduction of an emergency department electronic health record. West J Emerg Med. 2012;13(4):358-361.
5. Powsner SM, Wyatt JC, Wright P. Opportunities for and challenges of computerisation. Lancet. 1998;352(9140):1617-1622.
6. Bates DW, Leape LL, Cullen DJ, et al. Effect of computerized physician order entry and a team intervention on prevention of serious medication errors. JAMA. 1998;280(15):1311-1316.
7. Sittig DF, Singh H. Improving test result follow-up through electronic health records requires more than just an alert. J Gen Intern Med. 2012; 27(10):1235-1237.
8. Campbell EM, Sittig DF, Ash JS, Guappone KP, Dykstra RH. Types of unintended consequences related to computerized provider order entry. J Am Med Inform Assoc. 2006;13(5):547-556.
9. Myers RB, Jones SL, Sittig DF. Review of Reported Clinical Information System Adverse Events in US Food and Drug Administration Databases. Appl Clin Inform. 2011;2(1):63-74.
10. ECRI Institute PSO Deep Dive: Health Information Technology. ECRI Institute website. https://www.ecri.org/EmailResources/PSRQ/ECRI_ Institute_PSO_Deep%20 Dive_HIT_TOC.pdf. Published December 2012.
11. Farley HL, Baumlin KM, Hamedani AG, et al. Quality and safety implications of emergency department information systems. Ann Emerg Med. 2013;62(4):399-407.
12. Institute of Medicine. Health IT and Patient Safety: Building Safer Systems for Better Care. Washington DC: The National Academies Press; 2012.
13. Health Information Technology Patient Safety Action & Surveillance Plan. Office of the National Coordinator for Health Information Technology website. http://www.healthit.gov/sites/default/files/safety_plan_ master.pdf. Published July 2, 2013.
14. Weingart SN, Seger AC, Feola N, Heffernan J, Schiff G, Isaac T. Electronic drug interaction alerts in ambulatory care: the value and acceptance of high-value alerts in US medical practices as assessed by an expert clinical panel. Drug Saf. 2011;34(7):587-593.
15. Carspecken CW, Sharek PJ, Longhurst C, Pageler NM. A clinical case of electronic health record drug alert fatigue: consequences for patient outcome. Pediatrics. 2013;131(6):e1970-e1973.
16. Shortliffe EH. Biomedical informatics in the education of physicians. JAMA. 2010;304(11):1227-1228.
17. Marx DA, Slonim AD. Assessing patient safety risk before the injury occurs: an introduction to sociotechnical probabilistic risk modelling in healthcare. Qual Saf Healthcare. 2003;12(suppl 2):ii33-ii38.
18. Singh H, Ash JS, Sittig DF. Safety Assurance Factors for Electronic Health Record Resilience (SAFER): study protocol. BMC Med Inform Decis Mak. 2013;13:46.

19. Sherman H, Castro G, Fletcher M, Hatlie M, Hibbert P, Jakob R. Towards an International Classification for Patient Safety: the conceptual framework. Int J Qual Healthcare. 2009;21:2-8.
20. Vartian CV, Singh H, Sittig DF. Development and field testing of a self-assessment guide for computer-based provider order entry. J Healthc Manag. In press.
21. Sittig DF, Singh H. A new sociotechnical model for studying health information technology in complex adaptive healthcare systems. Qual Saf Healthcare. 2010;19(suppl 3):i68-i74.
22. Sittig DF, Singh H. Electronic health records and national patientsafety goals. N Engl J Med. 2012;367(19):1854-1860.
23. Blumenthal D, Tavenner M. The "meaningful use" regulation for electronic health records. N Engl J Med. 2010;363(6):501-504.
24. Sittig DF, Singh H. Improving test result follow-up through electronic health records requires more than just an alert. J Gen Intern Med. 2012; 27(10):1235-1237.
25. Singh H, Spitzmueller C, Petersen NJ, Sawhney MK, Sittig DF. Information overload and missed test results in electronic health recordbased settings. JAMA Intern Med. 2013;173(8):702-704.
26. Gardner RM, Overhage JM, Steen EB, et al. Core content for the subspecialty of clinical informatics. J Am Med Inform Assoc. 2009;16(2):153-157.
27. Safran C, Shabot MM, Munger BS, et al. Program requirements for fellowship education in the subspecialty of clinical informatics. J Am Med Inform Assoc. 2009;16(2):158-166 [published correction appears in J Am Med Inform Assoc. 2009;16(4):605].
28. Walker JM, Carayon P, Leveson N, et al. EHR safety: the way forward to safe and effective systems. J Am Med Inform Assoc. 2008;15(3):272-277.
29. Menon S, Singh H, Meyer AN, Belmont E, Sittig DF. Electronic health record-related safety concerns: a cross-sectional survey. J Healthc Risk Manag. In press.
30. de Vries EN, Prins HA, Crolla RM, et al. Effect of a comprehensive surgical safety system on patient outcomes. N Engl J Med. 2010;363(20):1928-1937.
31. Singh H, Thomas EJ, Mani S, et al. Timely follow-up of abnormal diagnostic imaging test results in an outpatient setting: are electronic medical records achieving their potential? Arch Intern Med. 2009; 169(17):1578-1586.
32. Sittig DF, Gonzalez D, Singh H. Contingency planning for electronic health record-based care continuity: a survey of recommended practices. UT-Memorial Hermann Center for Healthcare Quality & Safety; Technical Report 2014:1.
33. Campbell EM1, Guappone KP, Sittig DF, Dykstra RH, Ash JS. Computerized provider order entry adoption: implications for clinical work flow. J Gen Intern Med. 2009;24(1):21-26.

A RED-FLAG-BASED APPROACH TO RISK MANAGEMENT OF EHR-RELATED SAFETY CONCERNS

Dean F. Sittig and Hardeep Singh

6.2.1 INTRODUCTION

Although electronic health records (EHRs) have a significant potential to improve patient safety, EHR-related safety concerns have begun to emerge. For instance, some unique risks of EHRs are inherent to the technologies themselves, whereas others are related to how these technologies are applied and used [1].

We previously conducted a web-based survey of the memberships of the American Society for Healthcare Risk Management and the American Health Lawyers Association between August and September 2012. A 17-item survey was developed to capture information about four content areas: (1) extent of EHR use at the primary facility of practice; (2) frequency of EHR-related serious safety events; (3) variables affecting EHR-related serious safety events; and (4) tracking of EHR-related safety measurements. Of 15,400 member e-mail invitations, the survey was completed by 369 respondents (2.4%), a majority of whom worked for large hospitals and healthcare systems. Based on this survey and supplemented by our previous work in EHR-related patient safety, we identified the following common EHR-related safety concerns:

1. Incorrect patient identification
2. Extended EHR unavailability (either planned or unplanned)
3. Failure to heed a computer-generated warning or alert

Reprinted with permission from Wiley. A Red-Flag-Based Approach to Risk Management of EHR-Related Safety Concerns. Sittig DF and Singh H. Journal of Healthcare Risk Management **33**,2 *(2013), DOI: 10.1002/jhrm.21123.*

Overview of SAFER Guides

4. System-to-system interface errors
5. Failure to identify, find, or use the most recent patient data
6. Misunderstandings about time
7. Incorrect item selected from a list of items
8. Open or incomplete orders

Guidance for risk managers on how they should approach these safety concerns is limited. Many EHR-related safety concerns are not visible or apparent to end users. Others are distributed such that one user is often unaware of the broader significance of the safety concern (e.g., errors in system interfaces between the EHR and ancillary systems may not be visible since the person entering the order rarely sees what the ancillary system receives). Thus, voluntary detection and reporting of EHR-related safety problems may be an inadequate strategy. In this chapter, we present a "red flag"-based approach that can be used by risk managers to identify potential EHR safety concerns in their institution. Red flags are indications that something may be wrong and should be given additional consideration or evaluation. In medicine, clinicians commonly look for red flags indicating that a seemingly minor problem may be more serious. For example, a 60-year-old otherwise healthy patient who complains of a cough for a week may be given a diagnosis of upper respiratory infection and a prescription for cough syrup. However, if the same patient indicated he is coughing up blood, that would warrant special attention, perhaps a chest x-ray [2], because the blood in sputum at that age is a "red flag" that could suggest a serious problem such as a lung cancer.

Risk managers routinely collect quality and safety data from multiple sources and are often privy to data from sources unavailable to information technology (IT) specialists or clinicians. Thus, risk managers are in a unique position to conduct a red-flag based analysis. In order to develop these red flags, we conducted an extensive literature search and relied heavily on our extensive experience in EHR-related patient safety research. In the following sections, we define each error type and list several "red flags" that risk managers or other interested parties can use to identify potential EHR-related safety issues within their organizations.

6.2.2 INCORRECT PATIENT IDENTIFICATION

There are two types of patient identification errors. A duplicate record exists when a single patient has more than one medical record. A co-mingled record exists when a single medical record contains information about two or more patients. Duplicate records are created when users create a new record for a patient with an existing record or when patient records from disparate systems are combined without checking for matching records. Co-mingled records result from incorrect patient selection and subsequent use [3]. The likelihood of such events is greatly increased when a) looking up patients in large, multi-institutional healthcare systems that may have over a million patient records; b) looking for patients with "common" names (e.g., Smith, Williams, Jones, Garcia, Rodriguez, etc.); c) attempting to merge patient records from two or more disparate systems without using state-of-the-art patient record matching algorithms; and d) allowing clinicians to open two or more patient records (i.e., either multiple tabs or windows) on the same device.

6.2.2.1 RED FLAGS FOR INCORRECT PATIENT IDENTIFICATION

1. Key patient identifying information (i.e., first and last name, date of birth, gender, medical record number, inpatient location or home address, picture) is missing from EHR screens or printouts [5].
2. Absence of documented processes and procedures for checking patient ID at essential stages of a patient visit (e.g., when patients are called back for rooming, at entering of vital signs, prior to labs, procedures or medication administration, at checkout, etc.).
3. A large number of clinician calls (e.g., > 1/1000 orders or notes entered) request "help desk" or IT support to move their erroneous EHR entries from patient A (the wrong patient) to patient B (the right patient). Incorrect entries could include orders, order sets, clinical notes, or test results.
4. Nurses use copies of one or more patient barcode identification bands taped to their clipboard as a work-around when performing barcoded medication administration [6].

5. A wrong site or wrong patient surgery or procedure was traced back to an order entered on the wrong patient.
6. Greater than expected number of "erroneous notes" in the EHR (i.e., notes entered incorrectly on another patient) as identified by an automated scan of all notes in the system [7].

6.2.3 EXTENDED EHR UNAVAILABILITY (EITHER PLANNED OR UNPLANNED)

Extended (i.e., > 4 hours) EHR unavailability means that some portion, or more likely, all of the patient's medical records are unavailable for review. It results from total or partial failure or planned downtime in any part of the EHR computing infrastructure (e.g., electrical power, network connections, database servers, computer-to-computer interfaces, computer terminals on patient care units, software upgrades, etc.) [8]. These problems can lead to temporary, or even permanent, loss of data or inability to send or receive information from others [9]. The organization must do everything it can to reduce the likelihood of these events as well as prepare to continue providing care in the event a system failure does occur [10].

6.2.3.1 RED FLAGS FOR EXTENDED EHR UNAVAILABILITY

1. Absence of documented EHR downtime and reactivation procedures.
2. Absence of notification procedures for scheduled downtimes, suggesting poor preparedness for downtimes lasting longer than anticipated.
3. No regular off-site backup of all data required to continue caring for patients (i.e., demographic, clinical, and financial).
4. No pre-printed paper order sheets or clinical documentation forms in clinical care areas.
5. Critical clinical computing hardware devices are not configured in a redundant manner (i.e., if one device fails, a backup device does not take over the work).

6. Read-only summaries of recent (< 1 hour old) patient data are not available on a standalone computer connected to the "red" electrical plug in clinical areas.

6.2.4 FAILURE TO HEED A COMPUTER-GENERATED WARNING OR ALERT

Critical information, even if sent to the correct person at the right time and displayed prominently on the computer screen, can be overlooked amidst an overabundance of other false positive information (i.e., items that indicate a given condition exists, when it actually does not). Warnings or alerts can occur either synchronously (i.e., during the activity that the alert pertains to, such as a drug-drug interaction alert during order entry) or asynchronously (i.e., while the user is not engaged in the activity that generated the alert, such as an alert for an abnormal laboratory test result). These missed data can lead to erroneous or delayed diagnoses or treatments.

6.2.4.1 RED FLAGS FOR FAILURE TO HEED A COMPUTER-GENERATED WARNING OR ALERT

1. Reports show widespread non-adherence to computer-generated alerts that are based on recommended guidelines [11].
2. Clinicians report receiving too many irrelevant alerts during order entry or as asynchronous messages in their inboxes [12].
3. Clinicians report intrusive alerts used to present information that is not critical (e.g., a pop-up message reading "Are you sure you want to send this prescription as an e-script?").
4. Clinicians report working at home, staying late after work, or working on weekends to complete all the work in their inboxes (e.g., abnormal laboratory test results, prescription refills, orders to cosign).

Overview of SAFER Guides   167

5. EHR usage logs show that more than 20% of orders entered generate an alert.

6.2.5 SYSTEM-TO-SYSTEM INTERFACE ERRORS

Errors caused by miscommunication (or non-communication) between applications can result in data from one application (e.g., a laboratory system) failing to reach or being corrupted before reaching another application (e.g., the EHR). These errors can occur due to mistakes in the data translation tables (i.e., used to encode and decode orders and results) that are used to transmit information between components of an EHR or between disparate clinical systems. [14] Mismatched data fields may affect orders or results by introducing inadvertent changes (or outright data loss) that are virtually undetectable by the computer, or by the people not privy to the original sender's intentions.

6.2.5.1 RED FLAGS FOR SYSTEM-TO-SYSTEM INTERFACE ERRORS

1. Orders or test results are reported to be missing for certain patients.
2. The "error log" of the interface between components of an EHR contains orders or results that were not able to be transmitted automatically between different components of the EHR system.
3. Laboratory reports in the EHR are reported to be incomplete (e.g., missing measurement units, reference ranges, date and time of result, or comments).
4. Any report of patient receiving incorrect or unnecessary medications.
5. Clinicians report errors or inconsistencies between the structured data fields and free-text comment fields or comments that fail to transfer from system-to-system [15].
6. The organization does not have a method for sending or receiving laboratory tests performed by an outside laboratory through a direct interface to the EHR, (i.e., requiring orders or results to be transmitted via mail or fax) [16].

6.2.6 FAILURE TO IDENTIFY, FIND, OR USE THE MOST RECENT PATIENT DATA

Failure to find or use the most recent patient data (e.g., medication orders, laboratory or radiology results) can cause clinicians to make erroneous clinical decisions and lead to incorrect, unnecessary, or delayed tests, procedures, or therapies. These failures often result from difficulties navigating, seeing, understanding or interacting with user interfaces.

6.2.6.1 RED FLAGS FOR FAILURE TO IDENTIFY, FIND, OR USE THE MOST RECENT PATIENT DATA

1. EHR displays require either horizontal or vertical scrolling to see the most recent orders or results (i.e., data sorted chronologically [earliest to latest] rather than in reverse chronological order [most recent results first]).
2. EHR displays require users to widen data display fields, or columns, to see the complete text of the order or result.
3. Clinicians repeatedly order new diagnostic tests within a short time of the previous result [17].
4. Clinicians take inappropriate therapeutic actions due to missing recent test results (e.g., administering potassium when most recent potassium levels are high) [18].
5. Diagnostic test results are displayed in multiple locations (i.e., different screens or tabs) in the EHR.

6.2.7 EHR TIME MEASUREMENT TRANSLATIONAL CHALLENGES

Translational challenges as a result of the inability of computers to properly translate time measurements as they are conceived and entered by users can lead to many different kinds of errors. For example, users may fail to understand how much time has passed since a displayed date (e.g., patient born 11/23/62 is now 50-years-old and due for a colonoscopy). Users often think and enter information in time relative to the current time. This can be difficult for others to interpret, especially at a later date (e.g., surgery scheduled for tomorrow), when historical information is entered with current time stamp, or when the computer is instructed to carry out a specific action at a future date and time without notifying the clinicians or patients affected.

6.2.7.1 RED FLAGS FOR EHR TIME MEASUREMENT TRANSLATIONAL CHALLENGES

1. Routine tests, medications, or procedures ordered "daily" continue long after they are clinically indicated (i.e., no stop date is documented). Examples include daily chest X-rays for previously intubated patient and prophylactic antibiotics continued after 10 days with no sign of infection.
2. Repeated delays in administration of time-sensitive medications (e.g., antibiotics) ordered as "next routine administration time," or double doses are given when a multi-day course of medication is ordered to begin "now" and then inadvertently repeated when the "next routine administration time" occurs soon after the order time [19].
3. Clinicians report that critical medications have been cancelled automatically with no notice to clinicians [20].
4. Clinicians are unable to create reminders for future important actions within the EHR [21].
5. "Urgent" or "STAT" flags on orders are overused (e.g., more than 50% of orders placed as STAT on acute care hospital units) or any other evidence that clinicians are not confident in the EHR's ability to communicate their routine instructions in a timely manner.

6.2.8 INCORRECT ITEM SELECTED FROM A LIST OF ITEMS

Juxtaposition errors occur when an EHR user inadvertently selects a listed item that is directly adjacent to the item he or she intended to select. These errors can occur if the user does not notice or understand the difference between items or simply selects the incorrect item.

6.2.8.1 RED FLAGS FOR INCORRECT ITEM SELECTED FROM A LIST OF ITEMS

1. Drop-down or static selection lists are too narrow to display the complete text of all items, include too many items, or items are too close together [24].
2. Drop-down or static selection lists are sorted alphabetically (i.e., rather than grouped by similarity of concept) or consist of all CAPITAL LETTERS.
3. Clinicians or members of ancillary services report patient orders for wrong medications, diagnostic tests, or therapeutic procedures.
4. A large number of orders are discontinued soon after they are entered.
5. The EHR user interface has multiple cascading or fly-out submenus (i.e., secondary and tertiary menus displayed on demand from within the primary menu) [25].

6.2.9 OPEN OR INCOMPLETE ORDERS

Open or incomplete orders can result from failure to complete the order entry process including signing and submitting the order(s), or from failure of supervising physicians to co-sign orders that require co-signatures before becoming active. Although these errors may result from clinician oversight, they can also result from user interfaces that make it difficult to understand the current state of user actions.

6.2.9.1 RED FLAGS FOR OPEN OR INCOMPLETE ORDERS

1. Orders requiring co-signature in a queue that are overdue according to the organization's policy (e.g., 24-48 hours old).

2. A large number of incomplete tasks appear in various computer logs (e.g., unsigned orders, discharge summaries, dictated notes, etc.)
3. Clinicians complain that the system is "losing" their orders (i.e., orders that they have entered are not carried out).
4. Some providers use a high percentage of "verbal" orders, rather than entering them into the computer themselves.
5. Referring providers do not receive notification back from specialists about consultations that are completed.

6.2.10 DISCUSSION

We have provided a list of red flags that risk managers can consider using in their ongoing activities to improve patient safety within the context of EHR-enabled healthcare delivery. Identifying that one or more common EHR-related safety concerns has occurred is only the first step in resolving a problem. In most cases, the risk manager or other responsible party should convene a multi-disciplinary group, including members of the IT department and affected clinicians, to investigate the causes of the problem. It may be necessary to work with the EHR vendor to identify the cause of the problem and potential solutions.

While we discussed many types of EHR-related safety concerns, these concerns likely represent only the tip of the iceberg. There might be other concerns we have missed. For instance, adoption of EHRs is still less than 50% of physicians [26] and currently, comprehensive closed claims analysis of EHR-related safety concerns is not available. Thus, the red flags listed could represent only a fraction of the possible factors that can be used to detect EHR-related problems. An organization that routinely conducts EHR-related surveillance activities, such as the ones proposed here, can significantly reduce the risks associated with EHR implementations.

6.2.11 CONCLUSION

EHRs represent one of the most important tools available to improve patient safety in healthcare organizations. Nevertheless, without careful and continuous surveillance of these systems as implemented, safety concerns can arise. Organizations can dramatically reduce both the number and severity of EHR-related serious safety events by addressing these red flags.

REFERENCES

1. Sittig DF, Singh H. Electronic health records and national patient-safety goals. N Engl J Med. 2012 Nov 8;367(19):1854-60. doi: 10.1056/NEJMsb1205420.
2. MedLinePlus. Coughing up blood. Available at: http://www.nlm.nih.gov/medlineplus/ency/article/003073.htm
3. Henneman PL, Fisher DL, Henneman EA, Pham TA, Campbell MM, Nathanson BH. Patient identification errors are common in a simulated setting. Ann Emerg Med. 2010 Jun;55(6):503-9. doi: 10.1016/j.annemergmed.2009.11.017.
4. Hyman D, Laire M, Redmond D, Kaplan DW.The use of patient pictures and verification screens to reduce computerized provider order entry errors. Pediatrics. 2012 Jul;130(1):e211-9. doi: 10.1542/peds.2011-2984.
5. NHS CUI Programme Team, National Health Service Common User Interface (CUI) Design Guide Workstream – Design Guide Entry – Patient Banner v4.0.0.0 Baseline. Last modified on 25 June 2009 Available at: http://www.cuisecure.nhs.uk/CAPS/Patient%20Identification1/Patient%20Banner.pdf
6. Koppel R, Wetterneck T, Telles JL, Karsh BT. Workarounds to barcode medication administration systems: their occurrences, causes, and threats to patient safety. J Am Med Inform Assoc. 2008 Jul-Aug;15(4):408-23. doi: 10.1197/jamia.M2616. Epub 2008 Apr 24.
7. Wilcox AB, Chen YH, Hripcsak G. Minimizing electronic health record patient-note mismatches. J Am Med Inform Assoc. 2011 Jul-Aug;18(4):511-4. doi: 10.1136/amiajnl-2010-000068.
8. Hanuscak TL, Szeinbach SL, Seoane-Vazquez E, Reichert BJ, McCluskey CF. Evaluation of causes and frequency of medication errors during information technology downtime. Am J Health Syst Pharm. 2009 Jun 15;66(12):1119-24. doi: 10.2146/ajhp080389.
9. Sittig DF, Singh H. Defining health information technology-related errors: new developments since to err is human. Arch Intern Med. 2011 Jul 25;171(14):1281-4.
10. Nelson NC. Downtime procedures for a clinical information system: a critical issue. J Crit Care. 2007 Mar;22(1):45-50.
11. McCoy AB, Waitman LR, Lewis JB, Wright JA, Choma DP, Miller RA, Peterson JF. A framework for evaluating the appropriateness of clinical decision support alerts and responses. J Am Med Inform Assoc. 2012 May-Jun;19(3):346-52. doi: 10.1136/amiajnl-2011-000185.
12. Murphy DR, Reis B, Kadiyala H, Hirani K, Sittig DF, Khan MM, Singh H. Electronic health record-based messages to primary care providers: valuable information or just noise? Arch Intern Med. 2012 Feb 13;172(3):283-5.

Overview of SAFER Guides 173

13. Murphy DR, Reis B, Sittig DF, Singh H. Notifications received by primary care practitioners in electronic health records: a taxonomy and time analysis. Am J Med. 2012 Feb;125(2):209.e1-7.
14. Hamblin JF, Bwitit PT, Moriarty HT. Pathology results in the electronic health record. Electronic Journal of Health Informatics 2010;5(2):2010;5(2)e15. Available at http://www.ejhi.net/ojs/index.php/ejhi/article/view/131
15. Singh H, Mani S, Espadas D, Petersen N, Franklin V, Petersen LA.
16. Prescription errors and outcomes related to inconsistent information transmitted through computerized order entry: a prospective study. Arch Intern Med. 2009 May 25;169(10):982-9. doi: 10.1001/archinternmed.2009.102.
17. Sittig DF, Singh H. Improving test result follow-up through electronic health records requires more than just an alert. J Gen Intern Med. 2012 Oct;27(10):1235-7.
18. Bates DW, Kuperman GJ, Rittenberg E, Teich JM, Fiskio J, Ma'luf N, Onderdonk A, Wybenga D, Winkelman J, Brennan TA, Komaroff AL, Tanasijevic M.A randomized trial of a computer-based intervention to reduce utilization of redundant laboratory tests. Am J Med. 1999 Feb;106(2):144-50.
19. Horsky J, Kuperman GJ, Patel VL.Comprehensive analysis of a medication dosing error related to CPOE. J Am Med Inform Assoc. 2005 Jul-Aug;12(4):377-82. Epub 2005 Mar 31.
20. FitzHenry F, Peterson JF, Arrieta M, Waitman LR, Schildcrout JS, Miller RA. Medication administration discrepancies persist despite electronic ordering. J Am Med Inform Assoc. 2007 Nov-Dec;14(6):756-64.
21. Institute for Safe Medication Practices (ISMP) Alert. Let's put a stop to problem-prone automatic stop order policies. August 9, 2000. Available at: http://www.ismp.org/newsletters/acutecare/articles/20000809_2.asp
22. Poon EG, Kuperman GJ, Fiskio J, Bates DW. Real-time notification of laboratory data requested by users through alphanumeric pagers. J Am Med Inform Assoc. 2002 May-Jun;9(3):217-22.
23. Walsh KE, Adams WG, Bauchner H, Vinci RJ, Chessare JB, Cooper MR, Hebert PM, Schainker EG, Landrigan CP. Medication errors related to computerized order entry for children. Pediatrics. 2006 Nov;118(5):1872-9.
24. Zhan C, Hicks RW, Blanchette CM, Keyes MA, Cousins DD. Potential benefits and problems with computerized prescriber order entry: analysis of a voluntary medication error-reporting database. Am J Health Syst Pharm. 2006 Feb 15;63(4):353-8.
25. Khajouei R, Jaspers MW. The impact of CPOE medication systems' design aspects on usability, workflow and medication orders: a systematic review. Methods Inf Med. 2010;49(1):3-19. doi: 10.3414/ME0630. Epub 2009 Jul 6. Review.
26. Tullis TS, Connor E, LeDoux L, Chadwick-Dias A, True M, Catani M. A Study of Website Navigation Methods. Usability Professionals Association (UPA) 2005 Conference in Montreal, Quebec. Available at: http://www.eastonmass.net/tullis/WebsiteNavigation/WebsiteNavigationPaper.htm
27. Wright A, Henkin S, Feblowitz J, McCoy AB, Bates DW, Sittig DF. Early results of the meaningful use program for electronic health records. N Engl J Med. 2013 Feb 21;368(8):779-80. doi: 10.1056/NEJMc1213481.

HIGH PRIORITY PRACTICES FOR EHR SAFETY

SAFER Guides

Recommended Practices	Rationale for Practice or Risk Addressed	Examples of Potentially Useful Practices/Scenarios
Phase 1 – Make Health IT Safer		
Principle: Data Availability (EHRs and the data contained within them are available to authorized individuals where and when required to support healthcare delivery and business operations.)		
1. Data and application configurations are backed up and hardware systems are redundant. [8-10] C, IT, H	Hardware and software failures are inevitable. Without redundant backup hardware, delays in restoring system operation can affect business continuity. Without data backups, key clinical and administrative information can be lost.	• Mission-critical hardware systems (e.g., database servers, network routers, connections to the internet) are duplicated. • Data are encrypted and backed up frequently, and transferred to an off-site storage location at least weekly. • System backups are tested (e.g., restored to the test environment) on a monthly basis.
2. EHR downtime and re-activation policies and procedures are complete, available, and reviewed regularly. [11] C, IT, H	Failure to prepare for the inevitability of EHR downtimes greatly increases the potential for errors in patient care during these difficult times.	Policies describe: • When a downtime should be called (including when the EHR is functionally unavailable (e.g., very slow response time), • Who will be in-charge during the downtime, • How everyone will be notified, and • Who is responsible for entering data collected during the downtime. • Hospital personnel are trained (and tested annually) in these procedures. • The organization regularly conducts tabletop downtime and reactivation simulations or "drills."

Overview of SAFER Guides 175

Recommended Practices	Rationale for Practice or Risk Addressed	Examples of Potentially Useful Practices/Scenarios
3. Allergies, problem list entries, and diagnostic test results (including interpretations of those results such as "normal" and "high"), are entered/stored using standard, coded data elements in the EHR. [7,12-21] C, Ev, MU	Free text data cannot be used by clinical decision support logic [22] to check for data entry errors or notify clinicians about important new information.	• RxNorm is used for coding medications and NDF-RT for medication classes. • SNOMED-CT is used for coding allergens, reactions, and severity. • SNOMED-CT, ICD-9, or ICD-10 (after 10/2015) is used for coding clinical problems and diagnoses. • LOINC and SNOMED CT are used for coding clinical laboratory results. • Abnormal laboratory results are coded as such.
4. Evidence-based order sets and charting templates are available for common clinical conditions, procedures, and services. [7,23] C, Ev, IT	Requiring clinicians to enter individual orders for routine clinical practices increases risk of overlooking one or more items. Allowing individual clinicians to create order sets runs the risk of institutionalizing poor practice.	• Clinical content is developed or modified based on evidence through consensus by experts relying, where available, on nationally recognized, consensus-based clinical decision support (CDS) recommendations. See AHRQ's Clinical Decision Support Initiative • Institute for Safe Medication Practices (ISMP) order set guidelines [24] are used to create order sets. • Order sets exist for top 10 most common clinical conditions (e.g., management of chest pain), procedures (e.g., insulin administration and monitoring), and clinical services (e.g., admission to labor & delivery).
5. Interactive clinical decision support features and functions (e.g., interruptive warnings, passive suggestions, or info buttons) are available and functioning.. [25-30] (MU) C, Ev, IT	Interactive clinical decision support interventions help reduce the risks associated with ordering inappropriate, contraindicated, and non-therapeutic doses (i.e., under or overdoses), and provide just-in-time clinical knowledge to clinicians.	• Each practice identifies a minimum number of highly specific CDS features and functions and monitors their availability and use. • Appropriate CDS features and functions include: o Alerts for abnormal laboratory test results. [5] o Tiered drug-drug interactions. [26] o Drug-allergy interactions. [31] o "Reverse allergy" checking occurs when a new allergen is entered on a patient. o Drug-food interactions.

Recommended Practices	Rationale for Practice or Risk Addressed	Examples of Potentially Useful Practices/Scenarios
		o Drug-condition interactions (e.g., Accutane or tetracycline prescribed for a pregnant woman).
		o Drug-patient age interactions (e.g., medications contraindicated in the elderly).
		o Drug dosing support for maximum (dose, daily, and lifetime), minimum, renal [32], weight-based, and age-appropriateness.
Phase 1 – Make Health IT Safer		
Principle: Data Integrity (Data are accurate, consistent and not lost, altered or created inappropriately.)		
6. Hardware and software modifications and system-system interfaces are tested (pre- and post- go-live) to ensure data are not lost, incorrectly entered, displayed, or transmitted within or between EHR system components. [33-36] C, Ev, IT, H	Failure to test new or modified hardware and software functionality along with system-system interfaces, both pre- and post-go-live, increases the risk of inadvertent errors and patient harm. Routine changes can result in unexpected side-effects leading to incomplete or unreliable functionality.	• Hardware and software should be tested both pre- and post-go-live. Include tests using clearly named "test" patients (e.g., ZZtest345 with patient ID 999999999) in the "live" environment.
		• High priority clinical processes should be simulated using real clinicians.
		• Use Leapfrog's "Evaluation Tool for Computerized Physician Order Entry" (or similar automated tool) to assess point-of-care clinical decision support (CDS) intervention completeness and reliability on a regular basis. [33]]
		• Applications and system-system interfaces are tested to ensure that data are not neither lost nor incorrectly entered, displayed, or transmitted.
		• Interfaces (e.g., HL-7) capable of sending, receiving, acknowledging, and cancelling orders and results exist and are tested between ADT – Laboratory, pharmacy, radiology; CPOE – pharmacy, Laboratory, Radiology.
		• Error logs are regularly inspected and errors fixed.

Overview of SAFER Guides 177

Recommended Practices	Rationale for Practice or Risk Addressed	Examples of Potentially Useful Practices/Scenarios
7. Clinical knowledge, rules, and logic embedded in the EHR are reviewed and addressed on a regularly and whenever changes are made in related systems.30,37-40 C, IT	Medical knowledge is constantly evolving. Failure to review and update clinical content can result in outdated practices continuing long after they should be discontinued or updated.	• Clinical content (e.g., order sets, default values, charting templates, patient education materials, and health maintenance reminders) are reviewed at least bi-annually or as needed (e.g., following user feedback or manufacturer alert) against recent evidence and best-practices.

Phase 1 – Make Health IT Safer

8. Policies and procedures ensure accurate patient identification at each step in the clinical workflow. Ev, IT, H	Wrong patient charting is one of the more common safety problems in EHRs and can result in both data integrity and data confidentiality issues when protected health information is disclosed in the wrong chart and is missing from the right chart. Accurate and consistent patient identification is one of the most important patient safety measures in an EHR enabled healthcare system.	• Information required to facilitate positive patient ID is visible on all screens and printouts and includes: Last name, First name, date of birth (with calculated age in appropriate units), gender, medical record number, in-patient location (or home address), recent photograph (recommended), responsible physician (optional). • The master patient index uses a probabilistic matching algorithm that uses patient's first and last names, date of birth, gender, and zip code or telephone number or social security number. [41] • System generates a pop-up alert when a user attempts to create a record for a new patient or looks up an existing patient with the same first and last name as an existing patient. • Before allowing the user to change the current patient (and display data for another patient), the system checks that all entered data has been saved (i.e., signed). [42]

Principle: Data Confidentiality (Patient data is only available to those authorized to see it)

Phase 2 – Safer Application and Use of IT

Principle: Complete/Correct EHR Use (Correct system usage [i.e., features and functions used as designed, implemented, and tested] is required for mission-critical clinical and administrative processes throughout the organization.)

Recommended Practices	Rationale for Practice or Risk Addressed	Examples of Potentially Useful Practices/Scenarios
9. Information required to accurately identify the patient is clearly displayed on screens and printouts. [42,43]	If clinicians cannot clearly identify the patient they are working on, they are at increased risk making EHR entries in the wrong record or relying on information on the wrong patient, resulting in patient care and treatment errors, which are among the most common types of errors in the modern EHR-enabled healthcare system.	• Information required for patient ID includes: • Last name • First name • Date of birth (with calculated age) • Gender • Medical record number • In-patient location (or home address) • Recent photograph (optional) • Responsible physician (e.g., attending, PCP, or admitting). • The duplicate patient ID rate, number of patient records with the same first name, last name, and date of birth in the EHR database, is monitored.
10. The human-computer interface is easy to use and designed to ensure that required information is visible, readable, and understandable. [43-46] Ev, IT	Clinicians are constantly under time pressure. User interfaces that are difficult to see, comprehend, and use significantly increase the risk of error and patient harm.	• Visible: columns are wide enough to view critical data. [45] • Readable: appropriate font sizes and contrast are used. • Understandable: only standardized abbreviations are used; the most recent orders and results are clearly marked. [43] • Consistent: similar functions have similar labels; different functions have different labels. • When possible, items that are related, or have similar functions, are grouped and displayed together rather than alphabetically. • System response time is adequate (e.g., mean under 3 seconds; max under 10 seconds). • User input data fields are large enough to enter required information, and selection options are easy to select.

Overview of SAFER Guides   179

Recommended Practices	Rationale for Practice or Risk Addressed	Examples of Potentially Useful Practices/Scenarios
11. The status of orders can be tracked in the system. [7] (MU) Ev, IT	Errors often occur when users assume that orders entered into the computer will be done as specified. To facilitate closed loop communication and tracking of tasks and orders, the EHR should provide users with information regarding their status.	• Users are notified of key actions (or inactions) relating to their orders, such as when ordered medications are discontinued (manual or auto), antibiotic renewals are not processed, and when orders placed at later times of the day and will not be acted upon till the next day. • Users are able to track the status of orders (e.g., specimen collected, specimen received, resulted). • There is clear distinction (e.g., different font or color) between newly entered and copied data.45
12. Clinicians are able to override computer-generated clinical interventions when they deem necessary. [47,48] C, Ev, IT	Computers cannot practice medicine. Disallowing clinician overrides of computer-generated interventions implies that computers have access to more accurate data and greater medical knowledge and expertise than clinicians. This is rarely true.	Hard stop alerts (i.e., the user must take an action before proceeding) are used only for the most egregious potential errors. Hard stop alert overrides are closely monitored and reviewed often. [47] The alert override rate (i.e., the number of point-of-care alerts that clinicians ignore divided by the total number of point-of-care alerts generated) is monitored, and alerts with high override rates are reviewed.

Phase 2 – Safer Application and Use of IT

Principle: System Usability (All EHR features and functions required to manage the treatment, payment, and operations of the healthcare system are designed, developed, and implemented in such a way to minimize the potential for errors. In addition all information in the system must be clearly visible, understandable, and actionable to authorized users.)

Recommended Practices	Rationale for Practice or Risk Addressed	Examples of Potentially Useful Practices/Scenarios
13. EHR is used for ordering medications, diagnostic tests, and procedures. [7] (MU) C, Dx, IT, Rx	Partial EHR use means that clinicians must look in two separate places to find the most recent orders which increases the potential to miss or delay filling critical orders. Hybrid systems, part electronic and part paper, are particularly hazardous. [53]	• The CPOE rate (i.e., the number of orders electronically entered by clinicians divided by the total number of orders entered) is monitored. • The percentage of verbal or paper orders that are entered by ancillary personnel is less than 10 percent • Free text and "miscellaneous" orders are discouraged.. • Policies and procedures are in place that clearly identify and manage hazards associated with ordering that continues to occur outside of the EHR.
14. Knowledgeable people are available to train, test, and provide continuous support for clinical EHR users. [49] C, IT	Clinicians cannot use EHRs safely if they have not been trained and do not have access to assistance when needed. EHRs are complex tools. In order to maximize patient safety, clinicians must not be expected to "learn the basics on the job."	• All clinicians receive training appropriate to their expected use of the EHR. An assessment is made of the need for such specialized training beyond system-wide, generic training. • Trainers have advanced EHR and/or informatics training. • Trainers are available before and after go-live as well as to provide on-going support for users during EHR optimization. [49] • All clinicians are trained and tested on basic EHR and CPOE operations before being issued login credentials. • The clinician training rate (i.e., the number of clinicians trained to use the EHR who have passed a basic competency test divided by the total number of clinicians with EHR user privileges) is monitored. • When any category of clinician users of EHRs requests training, especially when they also indicate that they are not adequately trained to safely do their jobs, such training is promptly provided. Organizations have processes to identify training opportunities that would optimize the safe use of EHRs

Overview of SAFER Guides 181

Recommended Practices	Rationale for Practice or Risk Addressed	Examples of Potentially Useful Practices/Scenarios
15. Pre-defined orders have been established for common medications, diagnostic (laboratory/radiology) testing. [50] C, Dx, IT, Rx	Unnecessary clinical practice variation should be minimized. Forcing clinicians to enter specific values that are then matched to a list of allowable values or to select from a set of possible values increases variability and can result in errors.	Complete medication order sentences exist for the most commonly ordered medications, laboratory and radiology tests.51 Complete medication order sentences exist for the most commonly ordered medications, laboratory and radiology tests. [51]

Phase 3 – Leverage IT to Facilitate Oversight and Improvement of Patient Safety

Principle: Safety Surveillance and Optimization (Monitor, detect and report on safety-critical clinical and administrative aspects of EHRs and healthcare processes and make iterative refinements to optimize safety.)

16. Key EHR safety metrics related to the practice/organization are monitored.52 C, Ev, IT	Measurement and monitoring of key performance indicators is essential for improvements in safety.	• EHR uptime rate—minutes the EHR was available to clinicians divided by number of minutes in the reporting period. [52] • System response time—mean time to display a recent CBC result on a test patient, measured every minute of every day in the reporting period. • Serious EHR-related adverse events—list of reported EHR-related adverse events (whether they resulted in patient harm or not, includes any reported breaches of patient confidentiality). • Potential wrong patient error rate—Requests to "change" orders that result in cancellation of 1st order and creation of an order for the same item on a different patient by the same user.

Recommended Practices	Rationale for Practice or Risk Addressed	Examples of Potentially Useful Practices/Scenarios
17. EHR-related patient safety hazards are reported to all responsible parties, and steps are taken to address them. C, Dx, Ev, IT, Rx	Ensuring that EHR-related patient safety hazards are systematically identified, reported, and addressed is essential to improving the safety of EHRs.	• The organization clearly identifies through policies and procedures how to address reports of EHR safety hazards. • The organization ensures that reports of hazards and adverse events are reported, as appropriate, to EHR developers as well as senior leadership and boards. • The organization has a relationship with a patient safety organization experienced in investigating and addressing EHR-related patient safety incidents. • The number of EHR-related software errors (i.e., bugs) reported is monitored. • The serious EHR error fix rate (i.e., the number of errors with potential for causing direct patient harm fixed within 3 months divided by the total number of errors reported) is monitored.
18. Activities to optimize the safety and safe use of EHRs include clinician engagement. C, Ev, Dx, IT, Rx	Unless clinicians are included in decisions that affect their use of the EHR, they may not understand or accept changes, which increases risks. Clinicians should be engaged in identifying opportunities for the EHR to support safe and effective clinical use.	Representatives from the following groups are involved in decision making about EHR safety: clinicians, administrators, patients, IT/informatics, board and CEO, quality, and legal staff.

REFERENCES

1. Ash JS, Berg M, Coiera E. Some unintended consequences of information technology in health care: the nature of patient care information system-related errors. J Am Med Inform Assoc. 2004;11:104-112.
2. Harrington L, Kennerly D, Johnson C. Safety issues related to the electronic medical record (EMR): synthesis of the literature from the last decade, 2000-2009. J Healthc Manag. 2011;56:31-43.
3. Singh H, Wilson L, Petersen LA et al. Improving follow-up of abnormal cancer screens using electronic health records: trust but verify test result communication. BMC Med Inform Decis Mak. 2009;9:49.
4. Singh H, Thomas EJ, Mani S et al. Timely follow-up of abnormal diagnostic imaging test results in an outpatient setting: are electronic medical records achieving their potential? Arch Intern Med. 2009;169:1578-1586.
5. Singh H, Thomas EJ, Sittig DF et al. Notification of abnormal lab test results in an electronic medical record: do any safety concerns remain? Am J Med. 2010;123:238-244.
6. Sittig DF, Classen DC. Safe electronic health record use requires a comprehensive monitoring and evaluation framework. JAMA. 2010;303:450-451.
7. Sittig DF, Singh H. Electronic health records and national patient-safety goals. N Engl J Med. 2012;367:1854-1860.
8. Lee OF, Guster DC. Virtualized disaster recovery model for large scale hospital and healthcare systems. International Journal of Healthcare Information Systems and Informatics. 2010;5.
9. Hogan B. Backing up every byte, every night. Del Med J. 2005;77:415-418.
10. Schackow TE, Palmer T, Epperly T. EHR meltdown: how to protect your patient data. Fam Pract Manag. 2008;15:A3-A8.
11. Scholl, M., Stine, K., Hash, J., Bowen, P., Johnson, A., Smith, C. D., and Steinberg, D. I. An introductory resource guide for implementing the Health Insurance Portability and Accountability Act (HIPAA) security rule. Revision 1, 800-866. 2008. NIST Special Publications.
12. Carvalho CJ, Borycki EM, Kushniruk A. Ensuring the safety of health information systems: using heuristics for patient safety. Healthc Q. 2009;12 Spec No Patient:49-54.
13. Kuperman GJ, Bobb A, Payne TH et al. Medication-related clinical decision support in computerized provider order entry systems: a review. J Am Med Inform Assoc. 2007;14:29-40.
14. Sittig DF, Singh H. Eight rights of safe electronic health record use. JAMA. 2009;302:1111-1113.
15. Callen JL, Westbrook JI, Georgiou A, Li J. Failure to follow-up test results for ambulatory patients: a systematic review. J Gen Intern Med. 2012;27:1334-1348.
16. Dalal AK, Poon EG, Karson AS, Gandhi TK, Roy CL. Lessons learned from implementation of a computerized application for pending tests at hospital discharge. J Hosp Med. 2011;6:16-21.

17. El-Kareh R, Roy C, Williams DH, Poon EG. Impact of automated alerts on follow-up of post-discharge microbiology results: a cluster randomized controlled trial. J Gen Intern Med. 2012;27:1243-1250.
18. Elder NC, McEwen TR, Flach J, Gallimore J, Pallerla H. The management of test results in primary care: does an electronic medical record make a difference? Fam Med. 2010;42:327-333.
19. Murphy, D. R., Laxmisan, A., Reis, B, Thomas, E. J., Esquivel, A., Forjuoh, S. N., Parikh, R., Khan, M. M., and Singh, H. Electronic Health Record-Based Triggers to Detect Potential Delays in Cancer Diagnosis. 2012.
20. Singh H, Wilson L, Reis B, Sawhney MK, Espadas D, Sittig DF. Ten strategies to improve management of abnormal test result alerts in the electronic health record. J
21. Sittig DF, Singh H. Improving Test Result Follow-up through Electronic Health Records Requires More than Just an Alert. J Gen Intern Med. 2012;27:1235-1237.
22. Wright A, Goldberg H, Hongsermeier T, Middleton B. A description and functional taxonomy of rule-based decision support content at a large integrated delivery network. J Am Med Inform Assoc. 2007;14:489-496.
23. Wright A, Feblowitz JC, Pang JE et al. Use of Order Sets in Inpatient Computerized Provider Order Entry Systems: A Comparative Analysis of Usage Patterns at Seven Sites. J Am Med Inform Assoc. In press.
24. ISMP's Guidelines for Standard Order Sets. http://www.ismp.org/Tools/guidelines/StandardOrderSets.asp . 2012. Institute for Safe Medication Practices.
25. Hoffman S, Podgurski A. Drug-Drug interaction alerts: emphasizing the evidence. St Louis University Journal of Health Law and Policy. 2012;5.
26. Paterno MD, Maviglia SM, Gorman PN et al. Tiering drug-drug interaction alerts by severity increases compliance rates. J Am Med Inform Assoc. 2009;16:40-46.
27. Phansalkar S, van der SH, Tucker AD et al. Drug-drug interactions that should be non-interruptive in order to reduce alert fatigue in electronic health records. J Am Med Inform Assoc. 2012.
28. Ridgley M, Greenberg M. Too many alerts, too much liability: sorting through the malpractice implications of drug-drug interaction clinical decision support. St Louis University Journal of Health Law and Policy. 2012;5:257-296.
29. Strom BL, Schinnar R, Aberra F et al. Unintended effects of a computerized physician order entry nearly hard-stop alert to prevent a drug interaction: a randomized controlled trial. Arch Intern Med. 2010;170:1578-1583.
30. Wright A, Phansalkar S, Bloomrosen M et al. Best Practices in Clinical Decision Support: the Case of Preventive Care Reminders. Appl Clin Inform. 2010;1:331-345.
31. Isaac T, Weissman JS, Davis RB et al. Overrides of medication alerts in ambulatory care. Arch Intern Med. 2009;169:305-311.
32. Chertow GM, Lee J, Kuperman GJ et al. Guided medication dosing for inpatients with renal insufficiency. JAMA. 2001;286:2839-2844.
33. Overview of the Leapfrog Group Evaluation Tool for Computerized Physician Order Entry. http://www.leapfroggroup.org/media/fil/Leapfrog-CPOE_Evaluation2.pdf . 2001. 11-1-2012.
34. Birkmeyer, JD and Dimick, JB. Leapfrog safety standards: potential benefits of universal adoption. 2004. Washington, DC, The Leapfrog Group.

35. Kilbridge PM, Welebob EM, Classen DC. Development of the Leapfrog methodology for evaluating hospital implemented inpatient computerized physician order entry systems. Qual Saf Health Care. 2006;15:81-84.
36. Metzger JB, Welebob E, Turisco F, Classen DC. The Leapfrog Group's CPOE Standard and Evaluation Tool. Patient Safety & Quality Healthcare. 2008.
37. Horsky J, Schiff GD, Johnston D, Mercincavage L, Bell D, Middleton B. Interface design principles for usable decision support: A targeted review of best practices for clinical prescribing interventions. J Biomed Inform. 2012.
38. Osheroff J, Teich J, Levick D et al. Improving Outcomes with Clinical Decision Support: An Implementer's Guide. Second Edition ed. Healthcare Information and Management Systems Society; 2012.
39. Sittig DF, Wright A, Ash JS, Middleton B. A set of preliminary standards recommended for achieving a national repository of clinical decision support interventions. AMIA Annu Symp Proc. 2009;2009:614-618.
40. Wright A, Sittig DF, Ash JS et al. Governance for clinical decision support: case studies and recommended practices from leading institutions. J Am Med Inform Assoc. 2011;18:187-194.
41. Smith J. Fundamentals for Building a Master Patient Index/Enterprise Master Patient Index (Updated). Journal of AHIMA. 2010.
42. Sittig DF, Teich JM, Yungton JA, Chueh HC. Preserving context in a multi-tasking clinical environment: a pilot implementation. Proc AMIA Annu Fall Symp. 1997;784-788.
43. Horsky J, Kuperman GJ, Patel VL. Comprehensive analysis of a medication dosing error related to CPOE. J Am Med Inform Assoc. 2005;12:377-382.
44. Khajouei R, Jaspers MW. CPOE system design aspects and their qualitative effect on usability. Stud Health Technol Inform. 2008;136:309-314.
45. Lowry, S. Z., Quinn, M. T., Ramaiah, M., Brick, D., Patterson, E. S., Zhang, J., Abbott, P., and Gibbons, M. C. A Human Factors Guide to Enhance EHR Usability of Critical User Interactions when Supporting Pediatric Patient Care. http://www.nist.gov/customcf/get_pdf.cfm?pub_id=911520 . 6-28-2012. 11-1-2012.
46. Sengstack P. CPOE configuration to reduce medical errors. Journal of Health Care Information Management. 2010;24:26-32.
47. Sittig DF, Singh H. Rights and responsibilities of users of electronic health records. CMAJ. 2012;184:1479-1483.
48. van der SH, Aarts J, Vulto A, Berg M. Overriding of drug safety alerts in computerized physician order entry. J Am Med Inform Assoc. 2006;13:138-147.
49. Ash JS, Stavri PZ, Dykstra R, Fournier L. Implementing computerized physician order entry: the importance of special people. Int J Med Inform. 2003;69:235-250.
50. Teich JM, Merchia PR, Schmiz JL, Kuperman GJ, Spurr CD, Bates DW. Effects of computerized physician order entry on prescribing practices. Arch Intern Med. 2000;160:2741-2747.
51. Payne TH, Hoey PJ, Nichol P, Lovis C. Preparation and use of preconstructed orders, order sets, and order menus in a computerized provider order entry system. J Am Med Inform Assoc. 2003;10:322-329.

52. Sittig DF, Campbell E, Guappone K, Dykstra R, Ash JS. Recommendations for monitoring and evaluation of in-patient Computer-based Provider Order Entry systems: results of a Delphi survey. AMIA Annu Symp Proc. 2007;671-675.
53. Sparnon E. Spotlight on Electronic Health Record Errors: Paper or Electronic Hybrid Workflows. Pennsylvania Patient Safety Advisory, 2013 Jun;10(2):55-58. Available at: http://patientsafetyauthority.org/ADVISORIES/AdvisoryLibrary/2013/Jun;10(2)/Pages/55.aspx.

CHAPTER 7

MITIGATING EHR DOWNTIMES

CONTINGENCY PLANNING FOR ELECTRONIC HEALTH RECORD-BASED CARE CONTINUITY: A SURVEY OF RECOMMENDED PRACTICES

Dean F. Sittig, Daniel Gonzalez, and Hardeep Singh

7.1.1 INTRODUCTION

The United States of America's (USA) Health Information Technology for Economic and Clinical Health (HITECH) Act of 2009 [1] has led to increased adoption and use of health information technologies (HIT), particularly use of electronic health record systems (EHRs) [2] in previously paper-based healthcare systems. As such, healthcare processes are increasingly dependent on availability of HIT. However, HIT is not infallible and is subject to disruptions and downtimes that may threaten the continuity of operations [3] and cause adverse patient care outcomes, both of which can lead to financial and operational difficulties for healthcare organizations [4].

Reprinted from International Journal of Medical Informatics, *Sittig DF, Gonzalez D, and Singh H, Contingency Planning for Electronic Health Record-Based Care Continuity: A Survey of Recommended Practices, Copyright 2014, with permission from Elsevier.*

Over the last several years, there have been several highly publicized, widespread (i.e., affecting multiple facilities simultaneously), extended (i.e., lasting greater than 12 h) EHRs downtimes in the USA and Canada [5], [6], [7], [8], [9], [10], [11] and [12]. EHRs downtimes have also been reported in China [13]. However, there is little published description of practices that institutions are using to maintain the safety and effectiveness of continuous healthcare delivery while EHRs are unavailable. Our study goal was to describe EHRs downtime practices across a variety of healthcare institutions and identify practices that could be useful for planning for and dealing with EHRs unavailability. By describing and highlighting important elements of contingency plans across a variety of EHRs-enabled healthcare systems, our goal was to provide healthcare organizations with more comprehensive information to prepare for the risks of potential operational disruptions and avoid harm to patients.

7.1.2 METHODS

7.1.2.1 SURVEY DEVELOPMENT

Before survey development, we reviewed the existing literature and did not find any previous survey that systematically described or assessed EHRs downtime practices within healthcare organizations. Therefore, we developed a survey for the purposes of the present study. The conceptual foundation for the survey was Sittig and Singh's eight-dimension sociotechnical model of safe and effective HIT use. Although not specific to EHRs downtime, this model describes the complex interactions within eight components or "dimensions" of a HIT system and/or process [14]. These include hardware and software; clinical content; user interface; people; workflow and communications; organizational policies, procedures, and the physical environment; external rules, regulations, and pressures; and system measurement and monitoring. By applying these dimensions to downtime processes, we developed survey items that addressed multiple, interrelated aspects of downtime preparedness and processes.

Following review of published articles describing noted EHRs downtimes along with articles describing best practices for contingency planning, we conducted fact-finding interviews (April–September 2011) at three large academic institutions and two community hospitals to elicit policies, procedures, and practices related to EHRs downtimes. Interview participants included IT personnel and hospital administrators. These interviews revealed a large degree of heterogeneity between institutions in policies, procedures, and practices and informed the development of items related to each dimension of our sociotechnical conceptual model. For example, in the "people" dimension, representatives from all institutions mentioned the need to train key personnel on appropriate downtime procedures, although there were significant differences in the type and extent of training offered. In addition to interviews, we observed a planned downtime (November 2011) at one of the academic hospitals to enable a better understanding of practices related to the "workflow and communication" dimension. Thus, a combination of data from interviews, our observations of a planned downtime, and a pre-publication copy of the American Health Lawyers Association (AHLA) Emergency Preparedness Checklist [15], developed by a team of lawyers with extensive experience in managing the aftermath of unexpected EHRs downtimes, provided information required to create items for our EHRs downtime survey. These data also provided us with a conceptual basis to discover potentially useful practices for EHRs contingency planning. Early drafts of the survey were pilot tested with five subjects, not involved in the original survey development, who had extensive experience working in EHRs-enabled healthcare organizations. Several questions and many of the response options were modified in response to their feedback. The final version of the survey consisted of 96 multiple choice and free text items where respondents could describe their institutions' policies, procedures and practices during scheduled or unscheduled downtimes (available at http://goo.gl/fCdc9o).

7.1.2.2 SURVEY ADMINISTRATION

We administered an online version of the downtime survey through a web-based questionnaire hosting service (https://www.SurveyMonkey.com). Following approval by our local institutional review board (December 2011), the survey was distributed to the Scottsdale Institute's member email distribution list in February 2012 [16]. At the time of the survey, the Scottsdale Institute consisted of 59 member organizations focused on improving their organization's HIT practices. The Scottsdale Institute reported that their members have a mean, Health Information and Management Systems Society (HIMSS) Electronic Medical Record Adoption Model (EMRAM) score of 4.6. In addition, 75% of these members reported a score of 4 or greater. Healthcare organizations with HIMSS EMRAM scores of 4 or greater are using computerized physician order entry with clinical decision support, have implemented the major ancillary systems (i.e., pharmacy, laboratory, and radiology), have a clinical data repository for results review, and have an electronic medication administration record. These organizations are likely at much greater risk in the event of system unavailability for any reason [17]. Members included institutional leaders (e.g., chief executive officers, chief information officers, and chief financial officers) and HIT experts from large healthcare organizations across the USA. Participants were asked to base their responses on the current EHRs downtime practices of their respective organizations. One email reminder was sent to prospective participants after 2 weeks, and the survey was closed after one month.

7.1.2.3 SURVEY DATA ANALYSIS

Survey responses were downloaded from the web-based survey administration tool and were analyzed using Microsoft Excel (Microsoft Corporation, Redmond, WA). We generated descriptive statistics to summarize the characteristics of respondents and their responses. Free text responses were reviewed to identify common themes and provide context for specific items.

7.1.3 RESULTS

We received survey responses from representatives of 50 of the 59 (84%) institutional members of the Scottsdale Institute (i.e., unit of analysis was the institutional member), although not all respondents answered all ques-

Mitigating EHR Downtimes

tions on the survey. Respondents were either chief information officers or other personnel directly responsible for maintaining the organization's HIT infrastructure. Most (96%) represented non-profit organizations, and 80% were affiliated with large (>600 bed) hospital systems. Nearly all respondents had experience with downtime events, with 95% reporting at least one unplanned downtime (of any length) in the last 3 years and 70% reporting at least one unplanned downtime greater than 8 h in the last 3 years. Three respondents reported that one or more patients were injured as a result of either a planned or unplanned downtime.

TABLE 7.1.1: Overview of infrastructure for backup systems (positive responses calculated based on the percent of Scottsdale Institute members responding positively to each survey item).

Dimension	Item	Positive response
Hardware/software	Have an uninterruptable power supply (UPS)	100%
Workflow	Test UPS at least monthly	50%
Hardware/software	Have a back-up generator dedicated to HIT infrastructure	96%
Workflow	Test generator at least monthly	79%
Hardware/software	Greater than 2 days of fuel	79%
Hardware/software	Greater than 75% of power replaced by generator	68%
Hardware/software	Have a warm site aback up	80%
Workflow	Warm-site available in 8 h or less	78%
Workflow	Test warm site at least quarterly	31%
Workflow	Warm-site has come online	70%
Workflow	Warm-site has come online because of an emergency	33%
Workflow	Switched over to warm site before 4 h	50%
Hardware/software	Redundant path to the internet at organizational level	92%
Internal policy/procedure	Different internet provider as their redundant path	68%
Hardware/software	Have an EHRs interface transaction error log	60%
Internal policy/procedure	Greater than 75% errors investigated and fixed	50%

[a] Remote site with pre-configured hardware and network connectivity on which an organization's application software and data can be quickly loaded.

TABLE 7.1.2: Overview of point-of-care components during downtimes (positive responses calculated based on the percent of Scottsdale Institute members responding positively to each survey item).

Dimension	Item	Positive response
Practices related to downtime, read-only EHR		
Hardware/software	Have a network-accessible, hospital-wide read-only back-up	77%
Workflow	Backup data updated at least every hour	85%
Workflow	Test central read-only back-up system at least monthly	33%
User interface	Downtime, read-only version of EHRs is clearly marked	62%
User interface	Downtime, read-only EHRs disabled during normal operation	45%
Hardware/software	Have a local, clinic-level read-only back-up system	75%
Workflow	Update data in clinic-level read-only back-up system at least hourly	90%
Internal policy/procedure	Clinicians activate clinic-level read only back-up system	50%
External rules and regulations	Read only clinic-level back-up system generic password protected	52%
Workflow	Test clinic-level read-only back-up system at least monthly	33%
Hardware/software	Clinic-level read-only back-up system connected to UPS	94%
Practices related to data backup		
Content	Back up patient data in a secure, off-site location	100%
Content	Data at off-site location is complete and encrypted	48%
Workflow	Back-up their data to an off-site location daily	100%
Workflow	Organization conducts at least quarterly tests to ensure data can be reloaded	15%
Content	Back-up includes complete, up to date copy of all data used to configure system	96%
Workflow	Organization backs up data before every upgrade	100%
Personnel	Organization trains staff on what to do in planned and unplanned downtime	100%
Workflow	Have a yearly unannounced downtime drill	28%
Monitoring and surveillance	Follow up on drills looking for opportunities for improvement	100%

Mitigating EHR Downtimes

TABLE 7.1.2: *Cont.*

Dimension	Item	Positive response
Internal policy/procedure	Have a written downtime policy and procedure	94%
Workflow	Review and update downtime policy at least every 2 years	41%
Workflow	Planned downtime communication strategy via email	81%
Workflow	Unplanned downtime communication strategy via email	59%
Workflow	Out of band downtime communication strategy (pager, overhead, people)	92%
Availability of paper forms before downtime		
User interface	Order and document medications	88%
User interface	Order and document laboratory tests and results	92%
User interface	Order and document radiology tests and results	92%
User interface	Document RN observations and care delivered	83%
User interface	Document MD observations and plans	79%
Workflow	Enough paper on hand to last >48 h	43%
Personnel	Staff trained in use of paper forms	92%

We organized survey responses according to several key hardware infrastructure components (see Table 7.1.1) and point-of-care components (see Table 7.1.2) involved in either preparing for or dealing with a downtime event. Table 7.1.1 and Table 7.1.2 list for each item the percentage of positive responses (i.e., the proportion of institutional respondents who reported having the components in place), along with the corresponding sociotechnical dimension from our conceptual model. Of note, we included follow-up items to determine not only the availability of certain "backup systems" but also more detailed information about how these systems were tested and used (see Table 7.1.1). For example, all respondents reported having uninterruptable power supplies (UPS), but only 50% reported that they tested them on a monthly basis. Similarly, 96% reported having a back-up generator, but only 79% tested their generators on a monthly basis, and only 79% had more than 2 days of fuel available to keep it running.

Table 7.1.2 lists respondents' endorsement of various point-of-care system components to ensure that staff can continue to access critical information and perform needed clinical functions during planned and unplanned downtimes. Overall, about three-fourths of organizations reported having a central or hospital-wide, read-only, back-up system, and a similar proportion reported having clinic-level, read-only, back-up systems. The large majority backed up their data at least hourly. However, far fewer organizations—only a third—tested their read-only backup systems at least monthly. Other practices related to backup systems were endorsed more variably.

All respondents reported that they maintained a daily backup copy of their patient data in a secure off-site location, although less than half reported that their data was complete and encrypted, and relatively few (15%) attempted to restore their backups on a quarterly basis. In fact, two organizations reported that they had never tested their backup. All organizations trained their staff on how to handle either a planned or unplanned downtime, but less than a third (28%) had yearly, unannounced downtime drills on any shift. Sites that performed these drills followed up with clinicians to identify areas for improvement.

7.1.4 DISCUSSION

We surveyed representatives of large integrated healthcare systems that were members of a professional organization created to share EHRs-related practices. The vast majority of these organizations were advanced EHRs users as indicated by their mean HIMSS EMRAM score of 4.6. Almost all of our respondent organizations had experienced an unplanned downtime, and most had experienced an unplanned downtime exceeding 8 h in the last 3 years. Three organizations reported patient injury had resulted from these events.

While many EHRs downtimes could be uneventful, the risk of patient injury remains. Downtime events could result in patient harm due to delays in test performance, delivery of abnormal test results, or in the administration of time-critical medications. In a subsequent on-line survey of the American Society for Healthcare Risk Management (ASHRM) and the American Health Lawyers Association (AHLA) on frequency and types of various EHR-related serious safety events, we also found downtime to be a significant issue [18]. Respondents from these organizations confirmed the findings from the Scottsdale Institute and reported similar rates and severities of events related to downtime. These data sources suggest that the risk of major system outages is likely more common and hazardous than most organizations currently plan for. For example, while all of the organizations surveyed had an off-site backup copy of their data, not all had implemented key care continuity infrastructure such as uninterruptable power supplies, backup generators, redundant paths to the Internet, and paper-based backup ordering and documentation procedures. Only three-quarters had some sort of read-only backup system capability either at the organization or clinical unit level. In light of the high rates of downtimes, potential severity of these events and potential for patient harm, it is incumbent on all organizations to take the necessary precautions required to mitigate the effects of inevitable downtime events.

Furthermore, while the vast majority of institutions had the proper system components in place, far fewer were using them completely or correctly. For example, 80% of organizations had a warm-site backup (i.e., a remote site with pre-configured hardware and network connectivity on which an organization's application software and data can be quickly loaded), but less than a third tested it at least quarterly. Similar findings were seen for other practices such as testing backup generators and encrypting backup databases. Even fewer organizations had routine measurement and monitoring systems in place to enable continuous surveillance of the infrastructure required to maintain the organization's EHRs and ensure continuity of care. For example, only 60% of organizations had a computer-generated EHRs interface error transaction log, and only half of those that did, reported that greater than 75% of the errors identified were investigated and fixed. Our findings thus suggest the need for increasing the resilience of the existing hardware and software infrastructure as well as streamlining policies, procedures, and people required to implement, test, and maintain these systems [19] in order to achieve the promise of transforming our healthcare delivery system through state-of-the-art health information technology.

7.1.4.1 NEED FOR BEST PRACTICES TO AVOID EHR-RELATED DISRUPTIONS AND DOWNTIMES

A recent USA Institute of Medicine report recommends that healthcare organizations have contingency plans in place to avoid if possible and, if not, to mitigate any potential issues related to planned and unplanned EHRs downtimes [20]. While there are several USA federal regulations (e.g., HIPAA Security Rule 45 CFR Sec. 164.x) and compliance standards (e.g., USA Center for Medicare and Medicaid (CMS) Services' Hospital Conditions of Participation [21] and The USA Joint Commission) that cover legal aspects of EHR-related disruptions and downtimes [22], to our knowledge a list of best practices for managing these disruptions that focus on patient safety does not exist in the USA or abroad.

While both CMS and the Joint Commission recognize safety issues posed by EHRs downtimes and the loss of continuous access to patient information, healthcare providers are generally expected to establish their own record maintenance and security systems to ensure that the EHRs meets their requirements and is reliably available when needed. Literature related to downtime preparedness emphasizes the need to have an EHRs disaster plan that is tested, accessible, and regularly audited [23], including provisions for assessing employees' knowledge of downtime practices (e.g., testing an organization's plan in a "fire drill"-type scenario) [24]. Experiences along the USA's gulf coast during hurricanes Katrina (August 2005) [25] and Ike (September 2009) [26] as well as along the New York/New Jersey coast during hurricane Sandy (October 2012) [27] illustrated the necessity for such contingency plans, providing useful insights regarding needs for internal communications and widespread clinician access to patient information [28]. Some guidance is available from the American Health Lawyer's Association's (AHLA) Emergency Preparedness, Response and Recovery Checklist, which provides a few items to help healthcare organizations prepare their EHRs services for emergencies [15] Although many professionals have shown the need for planning and some have suggested a limited number of best practices or tips, there is no comprehensive guidance to inform the development of EHRs-specific downtime best practices. Thus, institutions might have a highly variable spectrum of contingency planning practices, and it is unknown to what

Mitigating EHR Downtimes 197

extent any of the practices are implemented or used. Therefore, while the evidence in this area is still emerging, there is a pressing need to develop appropriate downtime procedures to minimize risks to patients [29].

The USA HIPAA Security Rule, which applies to healthcare providers as "covered entities" and to their "business associates" (which includes EHRs vendors), is the USA federal law that most explicitly addresses contingency planning related to electronic health records. Its general requirements include ensuring the "availability" of all electronic protected health information, as well as its confidentiality and integrity, USA Security Rule 45 CFR Sec. 164.306(a)(1), and protecting against reasonably anticipated threats or hazards to the security or integrity of such information, USA Security Rule 45 CFR Sec. 164.306(a)(2). This standard has implementation specifications on data backup plans, disaster recovery plans, emergency mode operation plans, testing and revision procedures, applications and data criticality analysis. These provisions of the HIPAA Security Rule overlap with and reinforce the practices we identified in our survey on avoiding and mitigating downtimes.

7.1.4.2 DEVELOPING POTENTIALLY USEFUL PRACTICES TO AVOID EHR-RELATED DISRUPTIONS AND DOWNTIMES

Our survey findings from the work with the Scottsdale Institute suggested the need for best practices to avoid EHRs downtimes. It also suggested what some of those practices might be. We used the findings from our survey to inform work on a separate project, funded by the Department of Health and Human Services (DHHS) Office of the National Coordinator for Health Information Technology (ONC) to develop self-assessment guides in 9 areas, including contingency planning, to optimize the safety and safe use of EHR. These guides are called the Safety Assurance Factors for EHRs Resilience (SAFER) Guides (Available at: http://www.healthit.gov/safer/) and the recommended practices in these guides are centered on six key principles: data availability, data quality and integrity, data confidentiality, complete and correct EHRs system use, system usability, and system surveillance and monitoring [30]. The Contingency Planning SAFER Guide ("Downtime Guide") was developed based upon prior research, including the survey described above, expert input, and field research, including site visits to 9 healthcare organizations ranging in size from large, multi-site, integrated health care organizations to single physician practices [31]. Interview data from key representatives provided additional context to develop our list of recommended practices on contingency planning (see Downtime SAFER guide pg 202).

7.1.4.3 USE OF CONTINGENCY PLANNING SAFER GUIDE AS A RESOURCE TO PLAN FOR DOWNTIMES

In the SAFER Guide, we provide examples of activities that institutions could perform in order to operationalize good clinical practices and, where possible, cited additional literature to support our recommendations. As highlighted in some of the open-ended comments from our survey, we acknowledge that many of the recommended practices are not entirely under the control of the healthcare organization. Often EHRs developers must create new, or modify existing functionality, to enable organizations to configure their products to fully implement a practice. For example, when the EHRs is down and EHRs users must rely on the "read only" back-up, EHRs developers often control whether the user interface of the read-only system is clearly distinguishable from that of the "live" system (recommendation #9). This is not a feature that is easily "configurable" by the EHRs implementation team within the healthcare organization.

The Contingency Planning SAFER Guide acknowledges overlap with the HIPAA Security Rule requirements on contingency planning, and encourages healthcare providers and their business associates to use the SAFER Guides in conjunction with required compliance with the Security Rule. The Contingency Planning SAFER Guide includes references to provisions of HIPAA that are implicated by the recommended practices. Following the recommended practices in the SAFER Guide could help with compliance with the HIPAA Security Rule, but the HIPAA Security Rule is broader than the recommendations in the Contingency Planning SAFER Guide.

In the future, healthcare organizations of all sizes could use contingency planning guides, such as the ones we developed, to help them assess their readiness for the inevitable system outages and identify areas for improvement. Over time and in organizations at different points in their development of an EHR-enabled healthcare system, we anticipate that these guides will need to be modified. Finally, as we learn more about the impact of EHRs on patient safety, we expect that the standards that these recommendations reflect, will progressively become more stringent.

7.1.5 STUDY LIMITATIONS

The major limitation of the survey was that respondents were from a relatively small number of USA-based, large, integrated, hospital-centric, healthcare delivery systems with significant experience in implementation, use, and ongoing optimization of their HIT and EHRs infrastructures. There were no small, self-contained ambulatory medical practices involved. We do not know how the survey findings on compliance with the practices identified in the survey would differ if the respondents were ambulatory practices. In addition, the goal of the survey was to explore the breadth of the current status of organizations with regards to contingency planning and suggest a minimum set of requirements. However, it may be difficult to get people responsible for purchasing, installing, and maintaining these systems to agree to anything more stringent [19]. Finally, with regard to the development of the "best practices," it is possible that new technologies, for example, tapeless backup systems that enable backups to be completed much faster and promise faster restore times [32], could make specific recommendations obsolete, although the underlying principles on which they are based will remain valid.

7.1.6 CONCLUSION

Extended EHR-related downtimes occurred in the majority of organizations surveyed. Most institutions had only partially implemented comprehensive contingency plans to maintain safe and effective healthcare during unexpected EHRs downtimes. Preparing for these unexpected downtimes should be a part of every EHR-enabled healthcare organization's overall patient safety strategy. The best practices identified in this survey and in the SAFER Guide on Contingency Planning could help the EHR-enabled healthcare system prepare for continuity of operations to ensure safe and effective healthcare.

REFERENCES

1. Health and Human Services Federal Register: HIPAA Administrative Simplification: Interim Final Rule Department of Health and Human Services (October 2009) http://www.hhs.gov/ocr/privacy/hipaa/administrative/enforcementrule/enfifr.pdf (retrieved 14.12.11)
2. A. Wright, S. Henkin, J. Feblowitz, A.B. McCoy, D.W. Bates, D.F. Sittig Early results of the meaningful use program for electronic health records N. Engl. J. Med., 368 (February (8)) (2013), pp. 779–780 http://dx.doi.org/10.1056/NEJMc1213481
3. P. Kilbridge. Computer crash – lessons from a system failure. N. Engl. J. Med., 348 (March (10)) (2003), pp. 881–882
4. D.F. Sittig, H. Singh. Legal ethical and financial dilemmas in electronic health record adoption and use. Pediatrics, 127 (April (4)) (2011), pp. e1042–e1047 http://dx.doi.org/10.1542/peds.2010-2184
5. J. Merrick. 'Serious' computer crash hits hospital trusts. Daily Mail (July 31, 2006), p. 3984 Available at: http://www.dailymail.co.uk/news/article-77/Serious-crash-hits-hospital-trusts.html
6. L. Rosencrance. Problems abound for Kaiser e-health records management system: an internal report details hundreds of technical issues and outages. Computer World (November 13, 2006), p. 9005 Available at: http://www.computerworld.com/s/article/004/Problems_abound_for_Kaiser_e_health_records_management_system_
7. B. Brewin. August VA systems outage crippled western hospitals, clinics. Government Executive (October 5, 2007) Available at: http://www.govexec.com/defense/2007/10/august-va-systems-outage-crippled-western-hospitals-clinics/9/2546
8. NPR Staff. Anti-virus program update wreaks havoc with PCs. National Public Radio (April 21, 2010) Available at: http://www.npr.org/templates/story/story.php?.storyId=126168997&sc=17&f=1001
9. C. Terhune. Patient data outage exposes risks of electronic medical records. Los Angeles Times (August 3, 2012) Available at: http://articles.latimes.com/2012/aug/03/business/la-fi-hospital-data-outage-20120803
10. K. Robertson. Sutter electronic records system crashed. Sacram. Bus. J., 27 (August 2013) Available at: http://www.bizjournals.com/sacramento/news//08/27/sutter-electronic-records-system-down.html?.page=all 2013
11. E. McCann. Network glitch brings down Epic EMR. Healthcare IT News (January 28, 2014) Available at: http://www.healthcareitnews.com/news/network-glitch-brings-down-epic-emr

12. G. Slade. System failure has docs, patients upset. Medicine Hat News (June 10, 2014) Available at: http://medicinehatnews.com/news/local-news//06/10/system-failure-has-docs-patients-upset/2014
13. J. Lei, P. Guan, K. Gao, X. Lu, Y. Chen, Y. Li, Q. Meng, J. Zhang, D.F. Sittig, K. Zheng. Characteristics of health IT outage and suggested risk management strategies: an analysis of historical incident reports in China. Int. J. Med. Inform., 83 (2) (February 2014), pp. 122–130 http://dx.doi.org/10.1016/j.ijmedinf.2013.10.006
14. D.F. Sittig, H. Singh. A new sociotechnical model for studying health information technology in complex adaptive healthcare systems. Qual. Saf. Health Care (2010), pp. i68–i74
15. E. Belmont, S. Chao, A.L. Chestler, S.J. Fox, M. Lamar, K.B. Rosati, E.F. Shay, D.F. Sittig, A.J. Valenti. Emergency Preparedness Checklist for Information Technology Infrastructure and Software Applications. American Health Lawyer's Association, Washington, DC (2013) Available at: http://www.healthlawyers.org/hlresources/PI/InfoSeries/Documents/For%20the%20Healthcare%20Executive/EHR/Emergency%20Preparedness%20Checklist%20For%20Information%20Technology%20Infrastructure%20and%20Software%20Applications.pdf
16. The Scottsdale Institute – The healthcare executive resource for information management. Available at: http://www.scottsdaleinstitute.org/
17. D.F. Sittig, D.C. Classen. Monitoring and evaluating the use of electronic health records—Reply. J. Am. Med. Assoc., 303 (19) (2010), pp. 1918–1919 http://dx.doi.org/10.1001/jama.2010 591
18. S. Menon, H. Singh, A.N.D. Meyer, E. Belmont, D.F. Sittig. Electronic health record-related safety concerns: a cross-sectional survey. J. Healthc. Risk Manag., 34 (1) (2014) http://dx.doi.org/10.1002/jhrm.21146
19. M.W. Smith, J.S. Ash, D.F. Sittig, H. Singh. Resilient practices in maintaining safety of health information technologies. J. Cogn. Eng. Decis. Mak., 8 (September (3)) (2014), pp. 265–282 http://dx.doi.org/10.1177/1555343414534242
20. Institute of Medicine. Health IT and Patient Safety: Building Safer Systems for Better Care. Institute of Medicine (November 2011) http://www.iom.edu/Reports/2011/Health-IT-and-Patient-Safety-Building-Safer-Systems-for-Better-Care.aspx (retrieved 14.12.11)
21. Center for Medicare and Medicaid Services. Conditions of Participation. U.S. Government Printing Office (June 1986) http://edocket.access.gpo.gov/cfr_2004/octqtr/pdf/42cfr482.24.pdf (retrieved 14.12.11)
22. The Joint Commission. Comprehensive Accreditation Manual Oakbrook Terrace. The Joint Commission, Oak Brook, IL (2011)
23. P. Spath. Health information disaster planning 101. Hosp. Peer Rev. (2002), pp. 112–114
24. N.C. Nelson. Downtime procedures for a clinical information system: a critical issue. J. Crit. Care (2007), pp. 45–50
25. S. Fink. The deadly choices at memorial. The New York Times (August 25, 2009) Available at: http://www.nytimes.com/2009/08/30/magazine/30doctors.html?pagewanted=all
26. Anonymous. Texas hospital leaders say Katrina's lessons helped them better prepare for Hurricane Ike. Health Facil. Manage., 22 (1) (2009), pp. 5–7

27. F. Mogul. Four NYC hospitals still closed by hurricane sandy. Kaiser Health News (November 18, 2012) Available: http://www.kaiserhealthnews.org/stories//november/19/nyc-hospitals-still-closed-hurricane-sandy.aspx 2012
28. K.H. Gamble. Weathering the Storm. Healthcare Informatics (November 2008) http://www.healthcare-informatics.com/ME2/dirmod.asp?sid=9B6FFC446FF7486981EA3C0C3CCE4943&nm=Articles%2FNews&type=Publishing&mod=Publications%3A%3AArticle&mid=8F3A7027421841978F18BE895F87F791&tier=4&id=96F1EDDC8CB24008870DBF8E7A0A5775 (retrieved 14.12.11)
29. T.L. Hanuscak, S.L. Szeinbach, E. Seoane-Vazquez, B.J. Reichert, C.F. McCluskey. Evaluation of casues and frequency of medication errors during information technology downtime. Am. J. Health. Syst. Pharm. (2009), pp. 1119–1124
30. D.F. Sittig, J.S. Ash, H. Singh. The SAFER guides: empowering organizations to improve the safety and effectiveness of electronic health records. Am J Managed Care, 20 (5) (2014), pp. 418–423
31. H. Singh, J.S. Ash, D.F. Sittig. Safety assurance factors for electronic health record resilience (SAFER): study protocol. BMC Med. Inform. Decis. Mak., 13 (April (1)) (2013), p. 46
32. J. Boucher. Ochsner health system transforms its backup and recovery with EMC. EMC Press Release (September 25, 2012) Available at: http://www.emc.com/about/news/press/2012/-04.htm 20120925

DOWNTIME

SAFER Guides

Recommended Practices	Rationale for Practice or Risk Addressed	Examples of Potentially Useful Practices/Scenarios
Phase 1 – Make Health IT Safer		
Principle: Data Availability (EHRs and the data contained within them are available to authorized individuals where and when required to support healthcare delivery and business operations.)		
1. Hardware that runs applications critical to the organization's operation is duplicated. C, Ev, IT	Organizations should take steps to prevent and minimize impact of technology failures. A single point of failure greatly increases risk.	• The organization has a remotely located (i.e., > 50 miles away and > 20 miles from the coastline) "warm-site" (i.e., a site that can be activated in less than 8 hours) backup facility that can run the entire EHR.

Mitigating EHR Downtimes

Recommended Practices	Rationale for Practice or Risk Addressed	Examples of Potentially Useful Practices/Scenarios
		• The warm-site is tested at least quarterly.
		• The organization maintains a redundant path to the internet consisting of two different cables, in different trenches (a microwave or other form of wireless connection is also acceptable), provided by two different internet providers.5
2. An electric generator and sufficient fuel are available to support the EHR during an extended power outage. C, IT	Most health care organizations must be able to continue running their health IT infrastructure and preserve data and communication capabilities in cases of sustained power outage.	• Organizations evaluate the consequences to patient safety and to business operations due to loss of power that shuts down the EHR, and implement concrete plans to keep the EHR running to the extent needed to avoid unacceptable consequences
		• In the event of a power failure, there is an uninterruptible power supply (UPS)), either batteries or a "flywheel," capable of providing instantaneous power to maintain the EHR for at least 10 minutes.
		• The UPS is tested regularly (optimally on at least a monthly basis).
		• The on-site, backup electrical generator is capable of maintaining EHR functionality critical to the organization's operation (e.g., results review, order entry, clinical documentation).
		• The organization maintains 2 days of fuel for the generator on-site.
		• The generator is tested regularly (optimally at least on a monthly basis).
		• The UPS and the generator are housed in secure locations not likely to flood.

Recommended Practices	Rationale for Practice or Risk Addressed	Examples of Potentially Useful Practices/Scenarios
3. Paper forms are available to replace key EHR functions during downtimes. C	Clinical and administrative operations need to continue in the event of a downtime.	• The organization maintains enough paper forms to care for patients on the unit for at least 8 hours. Paper forms could include those required to enter orders and document the administration of medications, labs, radiology on each unit.10
		• There is a process in place to ensure that the information recorded on paper during the downtime gets entered and reconciled into the EHR following its reactivation (e.g., could be entering in data as coded data or scanning of documents).10

Phase 1 – Make Health IT Safer

Principle: Data Integrity (Data are accurate, consistent and not lost, altered or created inappropriately.)

4. Patient data and software application configurations critical to the organization's operations are backed-up.* C, Ev, IT	Backup of mission-critical patient data and EHR system configuration allows system restoration to a "pre-failure" state with minimal data loss.	• The organization has a daily, off-site, complete, encrypted backup of patient data.*
		• The offsite backup is tested regularly (optimally on at least a monthly basis, i.e., complete restore).*
		• The content required to configure the system is backed up on a regular basis (optimally on a monthly basis and before every system upgrade).
		• The organization maintains multiple backups, created at different times.
		• The organization maintains multiple backups, created at different times.
		• Backup media are physically secured.*
		• Backup media are rendered unreadable (i.e., use software to scramble media contents or better yet, physically destroy/shred media) before disposal.*

Mitigating EHR Downtimes

Recommended Practices	Rationale for Practice or Risk Addressed	Examples of Potentially Useful Practices/Scenarios
		• The organization has a "read-only" backup EHR system that is updated frequently (optimally at least hourly).
		• The read-only EHR system is tested regularly (optimally at least a weekly basis).
		• Users can print from the read-only EHR system.
		• If there is a "unit-level" read-only backup EHR system, it is connected to a local UPS or "red plug."
Phase 1 – Make Health IT Safer		
Principle: Data Confidentiality (Patient data is only available to those authorized to see it.)		
5. Processes and procedures are in place to ensure accurate patient identification when preparing for, during, and after downtimes.*C, Ev	Without policies, procedures, and processes in place to manage patient identification during downtimes, mismatches and lost records could compromise patient confidentiality and safety. Patient confidentiality and careful identification should be maintained, to the extent possible, at all times.	• The read-only EHR system should have user-specific passwords (i.e., should not use a generic password for all users).
		• There is a mechanism in place to register new patients during downtime including assignment of unique temporary patient record numbers along with a process for reconciling these new patient IDs once the EHR comes back on-line.*
		• Ensure that paper documents created during downtime are protected using standard HIPAA safeguards and policies.*
Phase 2 – Safer Application and Use of IT		
Principle: Complete/Correct EHR Use (Correct system usage [i.e., features and functions used as designed, implemented, and tested] is required for mission-critical clinical and administrative processes throughout the organization.)		
6. Staff are trained and tested on downtime and recovery procedures.*C	In organizations that have not had a significant downtime in over a year, there is an increased risk of having employees who do not know how to function in a paper-based environment.	• Organizations establish and follow training requirements so that each employee knows what to do to keep the organization operating safely during EHR downtimes.
		• Clinicians are trained in use of the paper-based ordering and charting tools.*

Recommended Practices	Rationale for Practice or Risk Addressed	Examples of Potentially Useful Practices/Scenarios
		• The organization conducts unannounced EHR "downtime drills" at least once a year.
		• Clinicians have been trained on how and when to activate the "read-only" backup EHR system.
7. A communication strategy that does not rely on the computing infrastructure exists for downtime and recovery events.*C, IT	Institutions need to be prepared to communicate with key personnel without use of the computer.	• The organization has methods other than electronic-based (i.e., NOT email, twitter, voice-over-IP, etc.) to notify key organizational administrators and clinicians about times when the EHR is down (either planned or unplanned).*
		• The organization has a mechanism in place to activate the read-only backup EHR system and notify clinicians how to access it.
		• The organization has a mechanism in place to notify clinicians when the EHR is back on-line (either planned or unplanned).
8. Written policies and procedures on EHR downtimes and recovery processes ensure continuity of operations with regard to safe patient care and critical business operations. C, IT	Policies and procedures on EHR downtime and recovery keep everyone "on the same page" so they are able to care for patients and maintain critical business operations during inevitable downtimes, whether planned or unplanned.	• The organization has a written downtime and recovery policy that describes key elements such as when a downtime should be called; how often further communication will be delivered; who will be in-charge during the downtime (both on the clinical and technical side); how everyone will be notified; and how information collected during the downtime is entered into the EHR.
		• The downtime policy is reviewed at least every 2 years.
		• The EHR downtime policy describes when the warm-site backup process should be activated (ideally, before the system has been down for 2 hours).

Mitigating EHR Downtimes

Recommended Practices	Rationale for Practice or Risk Addressed	Examples of Potentially Useful Practices/Scenarios
		• A paper-based copy of the current downtime and recovery policy is available on clinical units. • A paper copy of the current downtime and recovery policy is stored in a safe, off-site location.
Phase 2 – Safer Application and Use of IT Principle: System Usability (All EHR features and functions required to manage the treatment, payment, and operations of the healthcare system are designed, developed, and implemented in such a way to minimize the potential for errors. In addition all information in the system must be clearly visible, understandable, and actionable to authorized users.)		
9. The user interface of the locally-maintained backup, read-only EHR system is clearly differentiated from the live/production EHR system. C, Ev	When the usual system is unavailable, a read-only copy can enable access to patient records, though it can't support adding or editing patient data. If it looks the same to users it could easily result in attempts to enter data that will not be recorded.	• Access to the "read-only" backup EHR is disabled (e.g., icons on the computer screens are "greyed out" or not available) during periods of normal EHR operations. • The user interface of the read-only backup EHR system is visibly different than the fully operational system (e.g., there is a different background color for screens, a watermark across screens, or data entry fields are greyed out). • Clinicians are trained on appropriate use of the read-only backup EHR.
Phase 3 – Leverage IT to Facilitate Oversight and Improvement of Patient Safety Principle: Safety Surveillance and Optimization (Monitor, detect and report on safety-critical clinical and administrative aspects of EHRs and healthcare processes and make iterative refinements to optimize safety.)		
10. There is a comprehensive testing and monitoring strategy in place to prevent and manage downtime events. C, Ev, IT	Comprehensive testing and monitoring strategies can prevent and minimize impact of future technology failures.	• The organization regularly monitors and reports on system downtime events. • The organization regularly monitors and reports on system response time (optimally under 2 seconds). • The organization has a written policy describing the different hardware, software, process, and people-related testing procedures.

Recommended Practices	Rationale for Practice or Risk Addressed	Examples of Potentially Useful Practices/Scenarios
		• The organization maintains a log of all testing activities.
		• Unplanned downtimes and the effectiveness of follow- up to prevent them from recurring are monitored by the top leadership.

REFERENCES

1. Lee OF, Guster DC. Virtualized Disaster Recovery Model for Large Scale Hospital and Healthcare Systems. International Journal of Healthcare Information Systems and Informatics (IJHISI) 5(3); 2010. DOI: 10.4018/jhisi.2010070105
2. Hogan B. Backing up every byte, every night. Del Med J. 2005 Oct;77(10):415-8.
3. Schackow TE, Palmer T, Epperly T. EHR meltdown: how to protect your patient data. Fam Pract Manag. 2008 Jun;15(6):A3-8.
4. McKinney M. Technology. What happens when the IT system goes down? Hosp Health Netw. 2007 Dec;81(12):14.
5. Nelson NC. Downtime procedures for a clinical information system: a critical issue. J Crit Care. 2007 Mar;22(1):45-50.
6. Scholl M, Stine K, Hash J, Bowen P, Johnson A, Smith CD, Steinberg DI. An Introductory Resource Guide for Implementing the Health Insurance Portability and Accountability Act (HIPAA) Security Rule. NIST Special Publications 800-66 Revision 1; October 2008. Available at: http://csrc.nist.gov/publications/nistpubs/800-66-Rev1/SP-800-66-Revision1.pdf
7. Sittig DF, Campbell E, Guappone K, Dykstra R, Ash JS. Recommendations for monitoring and evaluation of in-patient Computer-based Provider Order Entry systems: results of a Delphi survey. AMIA Annu Symp Proc. 2007 Oct 11:671-5.

CHAPTER 8

SAFELY CONFIGURING AND MAINTAINING EHRS AND SYSTEM-TO-SYSTEM INTERFACES

FIELD STUDY OF THE SYSTEM INTERFACES SAFER GUIDE

Rodney E. Howell, Hardeep Singh, and Dean F. Sittig

8.1.1 BACKGROUND AND SIGNIFICANCE

The US Institute of Medicine (IOM) report on medical errors dramatically called attention to dangers inherent in the U.S. health care system when it said that up to 98,000 deaths in hospitals, costing an estimated $38 billion per year, can be attributed to errors [1,2]. The IOM report brings attention to and acknowledges the importance of patient safety and the serious impacts it has on quality of care in the American health care system. Many clinicians, researchers, and government agencies are looking to the computerization of the health care system to be a major factor in the reduction of the number of medical errors seen today and possibly in the future [3]. The introduction of technology into the health care system was in part to reduce errors that occur in paper-based systems. However, it appears that

---

Adapted with permission from: Howell RE. Field Study of the System-to-System Interfaces and Patient Identification ONC SAFER Assessment Guides. Masters Thesis, School of Biomedical Informatics, University of Texas Health Science Center at Houston, TX 2014.

we may be trading one set of problems for another by switching to an electronic health record (EHR) enabled-system [4].

To address the initially slow EHR adoption rate, the US Federal Government introduced the Health Information Technology for Economic and Clinical Health (HITECH) Act, a part of the American Recovery and Reinvestment Act of 2009 (ARRA) [5]. This legislation dedicated $27 billion to the promotion of health information technology, with part of the funds used as incentive payments to physicians and health care organizations, encouraging them to, in a meaningful and significant way, adopt and use electronic health records [6]. The incentive is scheduled to be paid over a six year period, 2009 to 2015, and is awarded to health care organizations meeting certain EHR usage criteria. The push to meet HITECH requirements is drastically increasing the adoption rate of EHRs and associated technologies, and as a consequence, has the potential to significantly influence the amount of errors encountered by and affecting all those who are part of the system [7]. The concern about these errors is centered on the potential for unintended consequences to arise from the introduction of these new tools, placing patient safety and quality of care at greater risk [8]. The Committee on Quality of Health Care in America under the Institute of Medicine notes that health care organizations "should expect any new technology to introduce new sources of error and should adopt the custom of automating cautiously, alert to the possibility of unintended harm" [9].

Many of the technology driven errors and associated risks occur due to the limitations of poorly designed information systems, as EHRs are only one component of a complex system of interfacing components that must work together in order to ensure safe and effective health care delivery [10]. Technology-related errors have the potential to be generated in any one of the cluster of patient care systems being used in the health care environment, which include the organization's EHR (usually with computerized provider order entry and clinical decision support systems built-in), any laboratory or imaging systems, administrative systems, pharmacy systems (e-prescribing), outside health system or hospital EHRs, etc. that may interact with the original organization's EHR "behind the scenes." These integrated systems are continuously sending patient data to and receiving it from other systems and in some cases this exchange occurs without any validation that important data was not lost. Ash et al. refer to these errors as "silent errors" [2]. According to Ash et al., there are "two major kinds of silent errors caused by health care information systems: those related to entering and retrieving information and those related to communication and coordination" [2]. For example, a registrar creates a new patient record for an already existing patient, resulting in clinicians missing important information due to duplicate patient records, or providers retrieving incorrect data from the EHR because of the non-integrated or poorly integrated IT infrastructure, leading to the administration of inappropriate medications [1,2,11].

Recently, national attention has been directed at the need to improve these electronic health record systems and to reduce the patient safety events that occur as a result of them. Following the movement towards rapid adoption of EHR systems, the IOM authored a report addressing methods to improve their safety and quality, Health IT and Patient Safety: Building Safer Systems for Better Care [21a]. The recently created Office of the National Coordinator for Health Information Technology (ONC) sponsored this IOM report and has convened expert panels and supported research efforts in the area of safety improvement for EHR systems, especially concerning unintended consequences of these new technologies. One of the research efforts that came out of this ONC interest was the Safety Assurance Factors for EHR Resilience (SAFER) project, the goal of which was to develop self-assessment guides for use by health care organizations. These guides each contain a set of practices that organizations can use to assess their level of safety as it pertains to the topic of the guide. In addition to the checklist-like recommended practice section, each guide contains a worksheet with examples and advice for assessing the organization's level of implementation of the recommended practices. With these guides, organizations will be able to proactively assess potential EHR-related safety risks in 9 areas, such as contingency planning, computerized provider order entry, and of course, system-system interfaces, with the help of these informative guides, now freely available on the ONC's website, http://www.healthit.gov/policy-researchers-implementers/safer.

In order to improve the safety of system-to-system interfaces, it is essential to determine ways to monitor and prevent these aforementioned "silent errors," which are difficult to detect and often noticed only after patient safety events have transpired, from occurring. Before that can happen, it will be important to understand and develop methods to promote safe use of these interfacing systems. Prospective assessment of these risks using a sociotechnical approach that takes into account the complex context of the health care system and all its varied components will put us on a path towards mitigation and eventually prevention of these errors. Thus, the System Interfaces SAFER guide is an essential starting point for the assessment of the safety of an organization's electronic health system.

The objective of the research described in this chapter is two-fold. We compared and contrasted current system interface processes using the ONC SAFER system interface guide and then identified gaps and made recommendations for incorporation of SAFER practices moving forward.

8.1.2 MATERIALS AND METHODS

8.1.2.1 ASSESSMENT GUIDE SELECTION

The System Interfaces SAFER assessment guide was chosen for evaluation because of its focus on mitigating errors that occur during and result from the transmission of data between systems. If a system interface is not managed correctly, the result could be patient harm or poor quality of care. The System Interfaces SAFER guide enumerates 18 recommended practices for ensuring the safety of system interfaces in an EHR-enabled health care system. For instance, testing system interfaces is critical to ensuring transmission of clean data. Untested data interfaces that are out of sync can result in the creation of erroneous patient records [12]. This SAFER guide was developed to address the complex nature of system interfaces, which include physical equipment, software containing data and concepts, and social and organizational factors that interact with the hardware and software components. It is meant to help organizations prioritize, understand, and improve on interface-related safety issues.

8.1.2.2 HOSPITAL AND SKILLSET REQUEST

The guide is meant to be answered by a multidisciplinary team, so to identify the most appropriate people to complete the guide and be interviewed at each site, we created a skillset selection request. This request was designed to identify the respondents possessing the knowledge needed to answer a majority of questions contained in the guide. Interviewing multiple personnel at all three hospitals was difficult to arrange, so we focused on the specific knowledge required to answer the questions on the SAFER assessment. For the System Interface assessment guide, the key informant choice was a person with detailed knowledge of the interoperability of system interfaces, such as an Integration Specialist or Health Information System Analyst.

The selection process started with identifying key contacts at three major hospitals in the Texas Medical Center in Houston, Texas (see Table 8.1.1). The hospitals have been labeled "Hospital A" to "Hospital C" in all results, and the order listed below is random to maintain their anonymity.

TABLE 8.1.1: Participating Hospitals

Hospital	Specialty	# of Beds	EHR
MD Anderson	Cancer	656	Clinic Station (Custom)
The Women's Hospital of Texas	Gynecology / Obstetrical	397	Meditech
Houston Methodist Hospital	Multiple Specialties	1633	Allscripts / Eclipsys

The contacts were asked if they could assist with supplying key personnel to answer questions on the assessment guide. Four participants were selected and interviewed. One hospital provided 2 people.

8.1.2.3 KEY INFORMANT INTERVIEW

We scheduled a date and time for the interviews with each participant. All interviews were done face to face in order to build rapport and to have the ability to probe deeper into areas of interest. Prior to the interview, we described the research project to the informants and gave them the SAFER assessment protocol paper, which discusses the development of the guides, for review [13]. A total of four participants from 3 hospitals were interviewed, and the interviews ranged in time from 30 minutes to a little over an hour. During the interview session, we verbally went through each item on the assessment guide with the key informant, obtaining their answer to the practice item (response choices are: 1. Not Implemented 2. Partially implemented in some areas 3. Fully implemented in all areas, or 4. N/A) and listening to any feedback, concerns, or comments they had about the items. Impromptu questions regarding comments informants made about their system or workflow were asked during the interview session if time allowed and the participant was willing. Each participant answered all 18 recommended practice items on the System Interface guide according to their own knowledge of their organization's systems. If they felt that they did not have the answer, a (N/A) was assigned. [14].A set of 7 follow-up questions was developed and sent to each participant via email after the interview (see Table 8.1.3). Of the seven questions, three of them (questions 1, 2 and 3), requested a comment from the participant.

8.1.2.4 ANALYSIS APPROACH AND METHODS

Once the four respondents had answered all 18 questions on the self-assessment guide, the data was summarized across the three institutions. The aggregate percentage of each level of implementation was calculated to determine the percentage of practices that were fully, partially, or not implemented at all across the 3 sites. Responses were also summarized per each practice. A content analysis was performed on the interview data, derived from field notes, and the follow-up question responses.

8.1.3 RESULTS

8.1.3.1 SYSTEM INTERFACE ASSESSMENT

The System Interface self-assessment guide consists of 18 suggested practices that focus on mechanisms, procedures, and features that should be in place in any EHR-enabled healthcare environment that relies on interconnected systems. Four participants with specific skillsets and job functions—Senior Integration Analyst (2), HIS programmer (1) and Chief Information Officer CIO (1) from 3 hospitals, representing large academic medical facilities with state-of-the-art EHR implementations, answered the 18 self-assessment items (Table 8.1.2) .

After aggregating the 4 participant responses, the percentage of the levels of implementation was calculated. Within our sample, 47.2% of the checklist items were fully implemented, 52.3% partially implemented, and 0% not implemented at all. No single item garnered all "fully implemented" responses; however, five items, which included topics such as using up-to-date versions of software, testing hardware/software, and having adequate security procedures, came close (Table 8.1.2).

Hospitals A and B showed a mix of fully implemented and partially implemented practices. The participant from Hospital A, a Sr. Integration Analyst, responded that 67% of practices were partially implemented, and the remaining 33% were fully implemented. The two participants from Hospital B, a Sr. Integration Analyst and HIS Manager, did not agree on the level of implementation of 44% of the practices, though they did agree that all practices were either partially or fully implemented. The respondent from Hospital C, a CIO, reported that his institution is fully compliant with all questions in the System Interface assessment guide.

Analyzing the system interface data by practice yields a possible value (i.e., where level of implementation is scored from 1-3 and this summed across all respondents) ranging from a minimum of 4 up to a maximum of 12 (Table 8.1.2). Because all sites indicated that they had at least partially implemented all practices, the analysis of the data by item revealed very little deviation across hospitals, with cumulative scores ranging from 9 to 11. The participants took great care in formulating a response and visited the worksheet portion of the guide several times to understand the meaning and significance of the question as it related to their environment. The analysis of the interview data showed that these respondents felt that the lack of resources and vendor willingness to add certain features to the systems are preventing their organizations from implementing some of the guide's recommendations. Some of the specific comments regarding this topic included:

- "If we had resources to implement every single recommendation in the guide, it probably would make the interfaces more reliable"
- "I think using the guides would improve quality of data being delivered since it would be in a standard format and tested as such. Some of the recommendations on the guides are items IT can't control and vendor functionality and inability to implement custom items limit IT ability"

TABLE 8.1.2: System to System Interface Assessment Question Responses

	System Interface Assessment Questions	Hospital A	Hospital B	Hospital E
1.	The EHR supports and uses standardized protocols for exchanging data with other systems.	2	3 / 2	3
2.	Established and up-to-date versions of operating systems, virus and malware protection software, application software, and interface protocols are used.	2	3 / 3	3
3.	System-to-system interfaces support the standard clinical vocabularies used by the connected applications.	2	2 / 2	3
4.	System-to-system interfaces are properly configured and tested to ensure that both coded and free-text data elements are transmitted without loss of or changes to information content.	2	2 / 3	3
5.	The intensity and the extent of interface testing is consistent with its complexity and with the importance of the accuracy, timeliness, and reliability of the data that traverses the interface.	2	2 / 3	3
6.	At the time of any major system change or upgrade that affects an interface, the organization implements procedures to evaluate whether users (clinicians or administrators) on both sides of the interface correctly understand and use information that moves over the interface.	2	2 / 2	3
7.	Changes to hardware or software on either side of the interface are tested before and monitored after go-live.	3	2 / 3	3
8.	There is a hardware and software environment for interface testing that is physically separate from the live environment.	3	2 / 3	3
9.	Policies and procedures describe how to stop and restart the exchange of data across the interface in an orderly manner.	2	2 / 2	3
10.	Security procedures, including role-based access, are established for managing and monitoring key designated aspects of interfaces and data exchange.	3	2 / 3	3

TABLE 8.1.2: *Cont.*

	System Interface Assessment Questions	Hospital A	Hospital B	Hospital E
11.	The organization has access to personnel with the skills required to configure, test, and manage all operational system-to-system interfaces.	3	2 / 2	3
12.	Administrative, financial, and clinical data exchange needs are clearly documented and include how data will be used and who is responsible for maintaining the interface and the systems connected to it.	2	2 / 2	3
13.	The organization notifies people involved in maintenance or use of system interfaces when changes are made that affect the content of the standard data files or allowable values transmitted via the interface (e.g., the orderable catalog or charge master).	3	2 / 3	3
14.	The operational status of the system interface is clear to its users with regard to clinical use, such as knowing when the interface cannot transmit or receive messages, alerts, or crucial information.	3	2 / 2	3
15.	The interface is able to transmit contextual information, such as units for measures or sources of information, to enable clinicians to properly interpret information.	2	2 / 2	3
16.	Interface problems associated with known system interface risks and data field size limits are managed to avoid readily preventable errors.	2	2 / 2	3
17.	The organization monitors the performance and use of system interfaces regularly, including monitoring the interface error log and the volume of transactions over the interface.	2	2 / 3	3
18.	When interface errors are detected, they are reported, fixed, and used to construct new test cases to improve the interface testing.	2	2 / 2	3

Assessment Question Legend:
1 = Not Implemented
2 = Partially Implemented
3 = Fully Implemented
N/A = No Answer
X / Y indicate the score of two participants at the same hospital

8.1.3.2 FOLLOW-UP QUESTIONNAIRE

The follow-up questionnaire consisted of 7 structured questions. The questions were developed to elicit the participant's thoughts concerning the assessment guide they completed. The first three questions asked the participant to comment on the usefulness of the guide. The questionnaire was sent to each participant within two weeks after the interview. All 4 participants completed and returned the questionnaire. The questionnaire offered the following scale for the participant to select from (Table 8.1.2).

- 1 = 0%
- 2 = 25% / once a month
- 3 = 50% / twice a year
- 4 = 75% / 3 times a year
- 5 = 100% / yearly

The scale was setup to give an increasingly positive response.

Analyzing the follow up question data across hospitals by adding the responses for each question (i.e., Total column - using the scale of 0 to 20 with 5 being the most positive and 4 participants; 5x4 = 20), (Table 8.1.3). The cumulative results across hospitals for the follow up questions reflected a positive attitude of acceptance for the assessment guides as a viable resource. According to the respondents, questions 1, 5, and 6 received the highest rating of (17), and questions 3 and 4 received the lowest rating of (13). This could be attributed to the participant's sense of their work priorities and their ability to present the necessary changes to upper management with positive results. This type of apprehension from some of the participants was observed during the interview process and speaks to the resource comment above. However, only one participant acknowledged that they did not have any plans to implement the recommendations from assessment guide that they weren't already doing.

TABLE 8.1.3: Answers to Follow Up Questions

Follow up Questions		Hospital A	Hospital B	Hospital C	Total
1.	By following the assessment guide would it improve patient safety at your facility?	5	4 / 3	5	17
2.	In your opinion how complete are the guides?	4	4 / 3	4	15
3.	In your opinion how usefulness are the guides?	3	4 / 2	4	13

TABLE 8.1.3: *Cont.*

Follow up Questions		Hospital A	Hospital B	Hospital C	Total
4.	Do you or your team plan on following some or all of the assessment guides?	3	3 / 3	4	13
5.	At what frequency do you feel reassessment should be done using the guides?	5	4 / 3	5	17
6.	How helpful was the work sheet in the appendix in answering the questions?	5	4 / 5	3	17
7.	How much of information in the guides raised your attention concerning your organization?	2	5 / 5	3	15

Follow Up Question Legend:
1 = 0%
2 = 25% / once a month
3 = 50% / twice a year
4 = 75% / 3 times a year
5 = 100% / yearly
X / Y indicate the score of two participants at the same hospital

8.1.4 DISCUSSION

The SAFER Assessment guides are tools freely available online for health care organizations to conduct self-assessments for optimizing the safety and safe use of EHRs. We evaluated the system interface guide.

8.1.4.1 CHALLENGES

One of the anticipated challenges going into the research was the apprehension of interviewees to reveal potential holes in their systems where silent errors could develop. This apprehension was not founded; participants were open to answering the questions honestly. Several of the participants agonized over their response to certain questions and visited the worksheet to get more information before selecting a final answer. This response was mainly seen at hospitals where the selected individual was given the clearance from upper management to participate in the interview process. It can be presumed that clearance from upper management was a signal to the individual that this type of research is important to the organization and should be taken seriously.

Another challenge was to insure that the interviewee had the requisite knowledge about the EHR, other interfacing systems, such as laboratory or pharmacy, and how they interact with each other. By design the SAFER Assessment guides are multidisciplinary tools, so finding a key person with the overall system knowledge was important. All but one of the interviewees was currently in a technical role, and most had the requisite understanding of their system to answer the questions.

8.1.4.2 ADDITIONAL PRACTICE RECOMMENDATIONS

This research validated that there are a number of areas for organizations to allocate time and resources for safety improvement regarding system interfaces. Below is a framework for addressing system interface concerns and furthering research in this understudied area. Until organizations start changing and monitoring systems and processes for potential silent threats, vulnerabilities in system interfaces will remain, and errors will occur. It is important to:

1. Recognize the seriousness of the problem.

- Acquire full support from senior management; clinical, administrative, and IT for implementing a technology driven safety program
- Hire an informaticist, or someone with graduate-level training in informatics to head a technology safety group

2. Take a snapshot of the system for the purpose of measuring, evaluating, and reporting.

- Assemble the appropriate stakeholders to engage in and take ownership of their area of concern. Use external resources as needed (internal stakeholders, technology vendors, accrediting bodies and the ONC) [15].
- Capture the system as much as possible in its current state (i.e., often organizations want to discuss the "next generation" of their plans which may never come to fruition).

- Review methods to plug holes and refine processes in your system (SAFER Assessment and other tools)

3. Work to resolve existing and potential threats.

- Prioritize threats to the system.
- Work to eliminate the threat based on the largest return to patient safety.
- Monitor, report, and adjust continuously

Medical errors, many of which are the result of unintended consequences of health information technology (HIT), are seriously affecting patient safety and quality of care as well as costing billions of dollars to the health care industry and patients alike. Everyone has a stake in reducing these errors, and it is not until organizations pool their resources and work across disciplines that systems safety will be improved. Technology has been called to lead the charge for patient safety and cost reduction in the form of automation and efficiency resulting in improved clinical and administrative processes. The EHR is one tool that offers great hope for improving patient safety and workflow efficiency; however, it also increases the potential for new and unforeseen risk to patients. Data entry to and retrieval from the EHR system is dependent upon how well the HIT system functions as a whole.

The report put together by the Office of the National Coordinator for Health Information Technology on July 1, 2013, "Health Information Technology Patient Safety Action & Surveillance Plan," states that Health IT can only fulfill its enormous potential to improve patient safety if the risks associated with its use are identified, if there is a coordinated effort to mitigate those risks, and if it is used to make care safer [16].

8.1.5 CONCLUSION

In this research project, three major hospitals in the Texas Medical Center took an important first step toward improving patient safety by opening their doors and allowing key staff members to assess their systems and processes using one of the ONC SAFER guides. We identified and prioritized patient safety concerns by introducing the participants to the

System Interface assessment guide that focuses on the interactions of the entire EHR-enabled health care system. The evaluations of these guides were compared and contrasted across and within these three hospitals, highlighting opinions from real users of the assessments. Many of the responses indicated that the guide was complete, useful, and would improve patient safety at their facility. At the same time, respondents indicated that they were not planning on following some or all of the guidelines. These responses convey the guide's effectiveness in revealing concerns surrounding system interfaces and related processes, but also uncovered recognition of a lack of resources and the belief that processes were already in place in many of the organizations to manage the potential threats. As we move forward, it will be increasingly important to heighten awareness of system interface issues and promote methods to monitor systems safety in order to improve patient safety overall.

REFERENCES

1. Khoumbati, K. , Brunel University, UK, Themistocleous, M. ; Irani, Z.: Integration Technology Adoption in Healthcare Organizations: A Case for Enterprise Application Integration. System Sciences, 2005. HICSS '05. 03-06 Jan. 2005. p 149a. issn: 1530-1605. Doi: 10.1109/HICSS.2005.331
2. Ash JS, Berg M, Coiera E. Some unintended consequences of information technology in health care: the nature of patient care information system-related errors. J Am Med Inform Assoc. 2004 Mar-Apr;11(2):104-12. Epub 2003 Nov 21. PubMed PMID: 14633936; PubMed Central PMCID: PMC353015.
3. Middleton B, Bloomrosen M, Dente MA, Hashmat B, Koppel R, Overhage JM, Payne TH, Rosenbloom ST, Weaver C, Zhang J; American Medical Informatics Association. Enhancing patient safety and quality of care by improving the usability of electronic health record systems: recommendations from AMIA. J Am Med Inform Assoc. 2013 Jun;20(e1):e2-8. doi: 10.1136/amiajnl-2012-001458. Epub 2013 Jan 25. PubMed PMID: 23355463; PubMed Central PMCID: PMC3715367.

4. Sittig, Dean F., Ash, Joan S.,: Clinical Information Systems: Overcoming Adverse Consequences. Jones and Bartlett Publishers.: ISBN-13: 978-0763757649, ISBN-10: 0763757640
5. http://www.gpo.gov/fdsys/pkg/BILLS-111hr1enr/pdf/BILLS-111hr1enr.pdf
6. Hoffman S, Podgurski A. Meaningful use and certification of health information technology: what about safety? J Law Med Ethics. 2011 Mar;39 Suppl 1:77-80. doi: 10.1111/j.1748 720X.2011.00572.x. PubMed PMID: 21309903.
7. Weaver E. Will the rush to EHRs harm patients? Small-, medium-size groups most at risk. MGMA Connex. 2011 Aug;11(7):21-2. PubMed PMID: 21913607.
8. Weiner JP, Fowles JB, Chan KS. New paradigms for measuring clinical performance using electronic health records. Int J Qual Health Care. 2012 Jun;24(3):200-5. doi: 10.1093/intqhc/mzs011. Epub 2012 Apr 6. PubMed PMID: 22490301.
9. Walker JM, Carayon P, Leveson N, Paulus RA, Tooker J, Chin H, Bothe A Jr, Stewart WF. EHR safety: the way forward to safe and effective systems. J Am Med Inform Assoc. 2008 May-Jun;15(3):272-7. doi: 10.1197/jamia.M2618. Epub 2008 Feb PubMed PMID: 18308981; PubMed Central PMCID: PMC2409999.
10. Singh H, Graber ML, Kissam SM, Sorensen AV, Lenfestey NF, Tant EM, Henriksen K, LaBresh KA.: System-related interventions to reduce diagnostic errors: a narrative review.: BMJ Qual Saf. 2012 Feb;21(2):160-70. doi: 10.1136/bmjqs-2011-000150. Epub 2011 Nov 30. PMID: 22129930. PMCID: PMC3677060
11. McCoy AB, Wright A, Kahn MG, Shapiro JS, Bernstam EV, Sittig DF. Matching identifiers in electronic health records: implications for duplicate records and patient safety. BMJ Qual Saf. 2013 Mar;22(3):219-24. doi: 10.1136/bmjqs-2012-001419. Epub 2013 Jan 29. PubMed PMID: 23362505.
12. Just BH, Proffitt K. Do you know who's who in your EHR? Healthc Financ Manage. 2009 Aug;63(8):68-73. PubMed PMID: 19658327.
13. Singh H, Ash JS, Sittig DF. Safety Assurance Factors for Electronic Health Record Resilience (SAFER): study protocol. BMC Med Inform Decis Mak. 2013 Apr 12;13:46. doi: 10.1186/1472-6947-13-46. PubMed PMID: 23587208; PubMed Central PMCID: PMC3639028.
14. SAFER Self-Assessment, System-System Interfaces: Available at: http://www.healthit.gov/safer/guide/sg005
15. Health Information Technology Patient Safety Action & Surveillance Plan: Available at: http://www.healthit.gov/sites/default/files/safety_plan_master.pdf

SYSTEM-SYSTEM INTERFACES

SAFER Guides

Recommended Practices	Rationale for Practice or Risk Addressed	Examples of Potentially Useful Practices/Scenarios
Phase 1 – Make Health IT Safer		
Principle: Data Availability (EHRs and the data contained within them are available to authorized individuals where and when required to support healthcare delivery and business operations.)		
1. The EHR supports and uses standardized protocols for exchanging data with other systems . C, Dx, Ev, IT, Rx	Standards such as those developed by HL-7 greatly simplify the establishment and maintenance of safe and effective interfaces between external systems (e.g., ancillaries like Laboratory, Radiology, or Pharmacy) thereby reducing errors of misscommunication.	• At a minimum, the EHR satisfies ONC's certification requirements related to electronic exchange of information. • The EHR is capable of sending and receiving clinical and administrative data using HL7 2.x messages where the sending and receiving systems use the same version. • The EHR has 2-way, HL7 v 2.x compatible interfaces to mission critical ancillary systems (at a minimum: Pharmacy, Laboratory, Blood Bank, Radiology). • The EHR is capable of generating, exporting, importing, and decoding clinical patient summary documents encoded in the Continuity of Care Document (CCD) standard . This includes procedures such as placing the correctly decoded clinical data into the proper location in the EHR, rather than just adding a human-readable version of the document to the patient's list of free text reports. • If the organization has an "interface engine", the hardware running this application is duplicated (i.e., operational backup hardware is installed . • Both the sending and receiving side of the interfaces are documented in sufficient detail to allow both sides to validate the adequacy of the interface for use. • The EHR has links to external clinical information reference resources using the HL7 InfoButton standard.

Configuring and Maintaining EHRs and System-to-System Interfaces

Recommended Practices	Rationale for Practice or Risk Addressed	Examples of Potentially Useful Practices/Scenarios
2. Established and up-to-date versions of operating systems, virus and malware protection software, application software, and interface protocols are used. Dx, Ev, IT, Rx	Failure to stay up-to-date with the latest versions of software and interface protocols places the organization at risk of clinical and administrative data loss, corruption, or theft.	• The organization has policies and procedures to determine how soon version testing and implementation will occur after the release of new software. • The organization has employees or service providers responsible for monitoring and upgrading software and communication protocols as needed. • Operating systems, virus and malware protection software, application software, and interface protocols in use are supported by their suppliers.

Phase 1 – Make Health IT Safer

Principle: Data Integrity (Data are accurate, consistent and not lost, altered or created inappropriately.)

Recommended Practices	Rationale for Practice or Risk Addressed	Examples of Potentially Useful Practices/Scenarios
3. System-to-system interfaces support the standard clinical vocabularies used by the connected applications. Ev, IT	Use of standard clinical vocabularies is essential to ensure semantic interoperability (i.e., consistent interpretation of the meaning of terms) between systems.	• The interface supports and encourages use of clinical vocabularies from ONC's certification requirements, for example: RxNorm for medication names, SNOMED-CT for clinical problems, and LOINC for laboratory tests. • A process is in place to ensure that standard clinical vocabularies are updated and consistent in all interfaced software applications. • Organizations evaluate interfaced software prior to purchase to ensure that it uses compatible versions of standard clinical vocabularies.
4. System-to-system interfaces are properly configured and tested to ensure that both coded and free-text data elements are transmitted without loss of or changes to information content. Ev, IT	Maintaining a system-to-system interface within a rapidly evolving clinical information system is challenging in part because many changes are required. Without the ability to implement and test these changes prior to go-live, patients would be placed at significantly increased risk of data loss, corruption or theft. Failure to test system interface components is one of the leading causes of EHR-related patient safety events.	• System-to-system interfaces are tested going into production and after changes to hardware, software, or content (i.e., the allowable list of data elements to be exchanged) on either side of the interface. • Free text data fields accessible to clinical end users of one system are transferred intact (i.e., no changes or truncation of characters) to the secondary system. • The organization (or interface developer) should develop a reference or validation data set that includes boundary cases (i.e., data that are slightly below, at, and slightly above key thresholds). These test data are run through the interface repeatedly after any change to the hardware or software on either end of the interface to document that the interface is working appropriately.

226 SAFER Electronic Health Records

Recommended Practices	Rationale for Practice or Risk Addressed	Examples of Potentially Useful Practices/Scenarios
5. The intensity and the extent of interface testing is consistent its complexity and with the importance of the accuracy, timeliness, and reliability of the data that traverses the interface . C, Dx, Ev, IT, Rx	While ideally everything should be carefully tested, the demands of testing must also be reasonable. The more important the data is to patient safety the more interface testing that should be conducted.	• When testing an interface, both anticipated and unanticipated types of data (e.g., text characters in a numeric field) and amounts of data should be used to ensure that the interface does not respond incorrectly in either case. • Organizations, through policies and/or job descriptions, address responsibility for evaluation of the intensity and extent of interface testing for all new software purchases or upgrades of systems that must be interfaced. • Organizations address the role of EHR technology developers in the testing of interfaces, and incorporate expectations in contractual and service obligations
6. At the time of any major system change or upgrade that affects an interface, the organization implements procedures to evaluate whether users (clinicians or administrators) on both sides of the interface correctly understand and use information that moves over the interface . C, Dx, Ev, IT, Rx	At the time of major system changes, social factors can interact with technical factors to create new risks. Information, even when correctly encoded and transmitted, can be misinterpreted because of differences in how users conceptualize their work.	• Testing uses a wide range of cases and scenarios including those where users of the external application or users in the external facility or service may interpret things different (e.g., Check to see if "day," means the same thing to a 24/7 facility and a 9-5 facility; if "home phone" means the same thing for a college campus clinic, a nursing home, an urban "safety net" community clinic, and a private physician practice). • When a new system is connected or integrated, testing includes looking for ways that correctly transmitted and coded information could nevertheless be misinterpreted. For example, in the first few weeks of using a newly integrated system, staff is designated to observe use of the software or to talk to users (in person or by phone) to confirm the receipt and intended interpretation and use information and messages sent via the interface . • Testing should include real-world, clinical scenarios of information exchange, such as: schedule an appointment; admit a patient; place an order; process order in ancillary lab; report results; record medication administration.
7. Changes to hardware or software on either side of the interface are tested before and monitored after go-live. Dx, Ev, IT, Rx	Hardware and software updates are inevitable. If the new hard- ware or software is unable to handle the load of transactions or otherwise work as intended in the actual workplace, it may shut down or compromise data integrity.	• Upgrades to EHR and ancillary systems are supported by additional testing of the system-to-system interfaces involved. • The organization carries out "load testing" (e.g., run a large number of transactions through the interface in a short period of time and "stress testing" (e.g., send erroneous random data through the interface to induce unexpected outputs) to ensure that the system can handle the required load at peak times and when confronted with erroneous data .

Configuring and Maintaining EHRs and System-to-System Interfaces 227

Recommended Practices	Rationale for Practice or Risk Addressed	Examples of Potentially Useful Practices/Scenarios
8. There is a hardware and software environment for interface testing that is physically separate from the live environment. Ev, IT	EHRs and the many applications they must interface with are continually changing. System administrators and application developers need a "safe" place to develop and test their changes without fear of causing harm to patients.	• Changes to applications (or the content to be exchanged) on either side of the interface, or to the interface itself, are implemented and tested in the test/development environment before being put into production. • Develop and test batch processing jobs for applications and interfaces. • Regression testing (i.e., to ensure that all previous functionality is still working appropriately) is conducted in the test environment before the changes are moved to production.
9. Policies and procedures describe how to stop and re-start the exchange of data across the interface in an orderly manner. Dx, Ev, IT, Rx	Failure to stop and re-start an interface properly can result in "in transit" data being lost or corrupted without any warning to users.	• Ensure that all system interface buffers are empty prior to stopping or re-starting the system. • If the interface must be disconnected while the sending system continues to produce data for transmission, e.g., lab tests ordered through CPOE, the buffers are of adequate size and behavior to prevent any loss of data. • The organization has a method of communicating to users when a clinical interface is not functioning properly (e.g., alert on the login page, or alert in the EHR whenever data retrieval or transmission is attempted but not completed). • Ensure reliable procedures are in place and used for stopping and starting system interfaces. The procedures are available and consulted during hardware/software upgrades.

Phase 1 – Make Health IT Safer
Principle: Data Confidentiality (Patient data is only available to those authorized to see it.)

Recommended Practices	Rationale for Practice or Risk Addressed	Examples of Potentially Useful Practices/Scenarios
10. Physical and logical security procedures are established based on user roles for managing different aspects of the interface or data exchange (e.g., content mapping applications, error logs, and clinical data). Ev, IT	The integrity and confidentiality of data within applications are well-protected. When data moves between systems there is an increased risk of data loss, corruption, or theft. Both physical and logical security controls are required over this exchange of data are required to prevent unintended changes.	• The server hosting the "interface engine" is maintained in a physically secure (i.e., locked room) location. • The server hosting the interface hardware and software is maintained in a physically secure (i.e., locked room) location. • The server hosting the "interface engine" has a secure administrator login to prevent unauthorized changes to the interface configuration or access to the data as it crosses the interface. • System security is tested to ensure that unauthorized individuals or applications cannot gain access to protected health information. • The security procedures identify and protect key designated aspects of the interfaces, including content mapping applications, the content maps themselves, error logs, and clinical data.

Phase 2 – Safer Application and Use of IT

Principle: Complete/Correct EHR Use (Correct system usage [i.e., features and functions used as designed, implemented, and tested] is required for mission-critical clinical and administrative processes throughout the organization.)

11. The organization has access to personnel with the skills required to configure, test, and manage system-to-system interfaces. IT	Configuring, testing, and managing system-to-system interfaces are complex tasks. The organization must ensure that staff are adequately trained and afforded the opportunity to learn to configure, test, and manage the system prior to go-live.	• Help desk operator manuals for quick reference are developed, readily available, and up-to-date. • Assigned personnel are trained on all system-to-system interface maintenance and monitoring activities, or have appropriate access to qualified personnel. • The organization identifies who is able to access help from the EHR developer and other external experts. • The organization has a plan for getting access to key individuals during off-hours (i.e., after routine business hours and on weekends and holidays).

Configuring and Maintaining EHRs and System-to-System Interfaces 229

Recommended Practices	Rationale for Practice or Risk Addressed	Examples of Potentially Useful Practices/Scenarios
12. Administrative, financial, and clinical data exchange needs are clearly documented and include how data will be used and who will be responsible for maintaining the interface and the systems connected to it. Dx, Ev, IT, Rx	Failure to document the business needs and responsibilities for the interface can result in miscommunication regarding the meaning and timing of the exchange of various data items and lead to patient harm.	• All types of data to be exchanged via the interface are clearly specified including: allowable values (e.g., text, numeric, length or size of fields); clinical vocabularies used; and how associated values (i.e., metadata) will be communicated (e.g., representation of units on measurements, sources of data, etc.). • The interface is designed to handle the estimated mean and maximum amounts of data expected to cross the interface with acceptable performance and errors generated. • The organization maintains a comprehensive data dictionary that includes for each data element: o Data type (e.g., coded, freetext, numeric) o Data definition o Metadata – creator, date created, users • The organization maintains a comprehensive interface data map that includes data recodes or conversions, as required. • The organization maintains a set of interface system performance requirements including the expected throughput of the system, uptime requirements, and protocols supported.
13. The organization notifies people involved in maintenance or use of system interfaces when changes are made that affect the content of the standard data files or allowable values transmitted via the interface (e.g., the orderable catalog or charge master).Dx, Ev, IT, Rx	EHR-related hardware and software change frequently. Failure to notify all parties involved in the maintenance or use of the system interfaces often results in interface errors. Some of these errors may be subtle and difficult to identify. Failure to account for and manage these changes can lead to serious patient safety events.	• Changes are clearly communicated and tested prior to go-live, including changes to: conversion programs, interfaces, databases, screens (e.g., length of data entry or display fields), tables (e.g., data interpretation, numeric values, times, dates, or text-based data fields), and vocabularies • Documentation that appropriate testing has occurred after all system modifications is available. • There is a policy describing configuration control procedures that includes: who must be notified before any change is made, who can make the changes, who is responsible for testing the changes, who is responsible for approving the changes, and when can the changes be implemented in the live system.

Recommended Practices	Rationale for Practice or Risk Addressed	Examples of Potentially Useful Practices/Scenarios
Phase 2 – Safer Application and Use of IT		
Principle: System Usability (All EHR features and functions required to manage the treatment, payment, and operations of the healthcare system are designed, developed, and implemented in such a way to minimize the potential for errors. In addition, pertinent information should be easily accessible by authorized users, visible, understandable, prioritized and organized by relevance to the specific user.		
14. The operational status of the system interface is clear to its users with regard to clinical use, such as knowing when the interface cannot transmit or receive messages, alerts, or crucial information. Ev, IT	Users must be notified when the interface between clinical systems is not functioning properly. Failure to distinguish between "there are no results" and "the interface to the system containing the results is not functioning" could lead to diagnostic or therapeutic delays.	• The user is informed when the interface cannot transmit a message. • The user is informed when the remote system from where they are requesting information is unavailable either due to errors in the interface or the remote system itself. • The user is notified when drug-allergy testing is performed on local medications only, not those identified by remote pharmacy or health information exchanges. • EHR applications that depend on system interfaces should report the interface status when in use (e.g., while reviewing imaging studies, the EHR shows last update time or current connection with PACS system).
15. The interface is able to transmit the relevant contextual information, such as units for measures or sources of information, to enable clinicians to properly interpret information. Dx, Ev, IT, Rx	Failure to transmit the relevant metadata (i.e., context or details) related to the data, and necessary for its interpretation, can lead to misunderstandings and erroneous decisions.	• The interface can transmit the "units" for measurements along with the measurements, and the units are stored in structured data fields (e.g., 175 lbs. or 500 mg). • The interface can transmit information associated with a particular measure (e.g., fraction of inspired oxygen along with the arterial blood gas results to allow clinicians to interpret the blood gas values in the proper context).

Configuring and Maintaining EHRs and System-to-System Interfaces

Recommended Practices	Rationale for Practice or Risk Addressed	Examples of Potentially Useful Practices/Scenarios
16. Interface problems associated with known system interface risks and data field size limits are managed to avoid readily preventable errors. Dx, Ev, IT, Rx	Physical and logical interfaces have limitations. Failure to acknowledge and plan for these limitations often results in patient safety events.	• The sending system identifies and restricts messages that are not transmittable (e.g., incorrect data type). • The user is notified if what they are typing exceeds the buffer size for either the storage location or the system-to-system interface. • The organization has a process for managing and minimizing known risks associated with interface problems, such as two systems with different field size limits. The system with the smaller limit can cause data to be truncated unless the risk is addressed properly

Phase 3 – Leverage IT to Facilitate Oversight and Improvement of Patient Safety

Principle: Safety Surveillance and Optimization (Monitor, detect and report on safety-critical clinical and administrative aspects of EHRs and healthcare processes and make iterative refinements to optimize safety.)

17. The organization monitors the performance and use of system interfaces regularly, including monitoring the interface error log and the volume of transactions over the interface. Dx, Ev, IT, Rx	System-to-system interfaces are complex and many of their actions are not directly visible. Extensive system monitoring is required to help identify and track hidden errors before they affect patients.	• The system-to-system interface error log is automatically monitored and all failed transactions are brought to the attention of the appropriate supervisor, investigated and fixed within one week. • The number of transactions crossing the interface is monitored to ensure that the number of transactions is "normal" (e.g., displayed in control chart showing the mean and reasonable upper and lower bounds (e.g., 2 or 3 standard deviations from the mean).
18. When interface errors are detected, they are are reported, fixed, and used to construct new test cases to improve the interface testing. Dx, Ev, IT, Rx	Failure to fix interface errors in a timely manner can lead to patient harm or to loss of clinicians' confidence in the data.	• After any interface error is detected and fixed, additional test data are added to the standard set of tests to check for the same error in future releases.

231

REFERENCES

1. Sittig DF, Wright A, Ash JS, Middleton B. A set of preliminary standards recommended for achieving a national repository of clinical decision support interventions. AMIA Annu Symp Proc. 2009 Nov 14;2009:614-8.
2. Dolin RH, Alschuler L, Boyer S, Beebe C, Behlen FM, Biron PV, Shabo Shvo A. HL7 Clinical Document Architecture, Release 2. J Am Med Inform Assoc. 2006 Jan-Feb;13(1):30-9.
3. Kim HS, Cho H, Lee IK. The Development of a Graphical User Interface Engine for the Convenient Use of the HL7 Version 2.x Interface Engine. Healthc Inform Res. 2011 Dec;17(4):214-23. doi: 10.4258/hir.2011.17.4.214.
4. Wichmann B, Barker R, Cox M, Harris P. Software Support for Metrology: Best Practice Guide No. 1 Measurement System Validation: Validation of Measurement Software. National Physical Laboratory, Middlesex, United Kingdom, April 2000 pp 86.
5. Cimino JJ, Jing X, Del Fiol G. Meeting the electronic health record "meaningful use" criterion for the HL7 infobutton standard using OpenInfobutton and the Librarian Infobutton Tailoring Environment (LITE). AMIA Annu Symp Proc. 2012;2012:112-20
6. RxNorm Files. Available at: http://www.nlm.nih.gov/research/umls/rxnorm/docs/rxnormfiles.html
7. The International Health Terminology Standards Development Organisation. SNOMED Clinical Terms® User Guide. Available at: http://ihtsdo.org/fileadmin/user_upload/doc/
8. Forrey AW, McDonald CJ, DeMoor G: Logical observation identifier names and codes (LOINC) database: a public use set of codes and names for electronic reporting of clinical laboratory test results. Clin Chem 1996, 42:81-90.
9. Sparnon E, Marella WM. The Role of the Electronic Health Record in Patient Safety Events. Pa Patient Saf Advis 2012 Dec;9(4):113-21.
10. Amland, S. (2000). Risk-based testing: Risk analysis fundamentals and metrics for software testing including a financial application case study. Journal of Systems and Software, 53(3), 287-295.
11. Klein CS. LIMS user acceptance testing. Qual Assur. 2003 Apr-Jun;10(2):91-106.
12. Aarts J. Towards safe electronic health records: A socio-technical perspective and the need for incident reporting. Health Policy and Technology 2012; 1(1):8-15.
13. Mostefaoui, GK, Wilson G, Ma X, Simpson A, Power D, Russell D, and Slaymaker M. "The Development, Testing, and Deployment of a Web Services Infrastructure for Distributed Healthcare Delivery, Research, and Training. IGI Global (2009). Available at: http://pdf.aminer.org/000/237/319/daview_a_linux_webdav_client_supporting_effective_distributed_authoring.pdf

14. Nanda, A., Mani, S., Sinha, S., Harrold, M. J., & Orso, A. (2011, March). Regression testing in the presence of non-code changes. In Software Testing, Verification and Validation (ICST), 2011 IEEE Fourth International Conference on (pp. 21-30). IEEE.
15. Smokers prescribed Viagra to quit. Available at: http://newsvote.bbc.co.uk/mpapps/pagetools/print/news.bbc.co.uk/2/hi/uk_news/scotland/glasgow_and_west/6175271.stm
16. Mykkänen J, Porrasmaa J, Rannanheimo J, Korpela M. A process for specifying integration for multi-tier applications in healthcare. Int J Med Inform. 2003 Jul;70(2-3):173-82.
17. Rinard M, Cadar C, Dumitran D, Roy DM, and Leu T. "A dynamic technique for eliminating buffer overflow vulnerabilities (and other memory errors)." In Computer Security Applications Conference, 2004. 20th Annual, pp. 82-90. IEEE, 2004.
18. Um KS, Kwak YS, Cho H, Kim IK. Development of an HL7 interface engine, based on tree structure and streaming algorithm, for large-size messages which include image data. Comput Methods Programs Biomed. 2005 Nov;80(2):126-40.
19. Sittig DF, Campbell E, Guappone K, Dykstra R, Ash JS. Recommendations for monitoring and evaluation of in-patient Computer-based Provider Order Entry systems: results of a Delphi survey. AMIA Annu Symp Proc. 2007 Oct 11:671-5.

HARDWARE/SOFTWARE CONFIGURATION

SAFER Guides

Legend	Key Facilitators of Practice Implementation
C	Clinicians, support staff, and/or clinical administration (e.g., Medical Records and Risk Managers)
Dx	Diagnostic services, such as laboratory or radiology—could be local or remote
Ev	EHR vendor
IT	IT support staff, could be local or contracted. Responsible for maintaining the EHR and infrastructure
Rx	Pharmacy – could be local or remote

Recommended Practices	Rationale for Practice or Risk Addressed	Examples of Potentially Useful Practices/Scenarios
Phase 1 – Make Health IT Safer		
Principle: Data Availability (EHRs and the data contained within them are available to authorized individuals where and when required to support healthcare delivery and business operations.)		
1. There are an adequate number of EHR access points in all clinical areas. C, IT	Rapid, reliable access to the patient's computer-based record is essential for safe and effective care. Such access depends critically on configuring the EHR in clinical care areas such that a computer is always conveniently available.	• Organizational policy sets minimum standards for EHR access by clinicians (e.g., clinicians walk no more than 50 feet to access an EHR and, if there are wait times, they are minimal and ensure that urgent clinical needs can be addressed). • Resources are dedicated to acquiring sufficient computer hardware to ensure appropriate access, in accordance with policy. • Workflows have been mapped to ensure ready and timely access to all needed EHR functionality in clinical areas. • There is at least 1 EHR access point for every clinician and administrative staff member in an out-patient clinic.[4] • Computer terminals used to access the EHR are mapped to the appropriate (e.g., in close physical proximity) printer. • There is at least one printer available for use on all acute care nursing units or within easy reach of each out-patient exam room (e.g., less than 25 feet). • There is a mapping table that shows the physical location of all hard-wired, network-attached devices (end-user workstations and printers) • Critical hardware is connected to uninterrupted power supplies (UPS). • Clinicians should not have to wait for, or walk more than 50 feet on a clinical unit to find an available EHR access point.

Configuring and Maintaining EHRs and System-to-System Interfaces 235

Phase 1 – Make Health IT Safer

Principle: Data Integrity (Data are accurate, consistent and not lost, altered or created inappropriately.)

Recommended Practices	Rationale for Practice or Risk Addressed	Examples of Potentially Useful Practices/Scenarios
2. The EHR is hosted safely in a physically and electronically secure manner. IT, Ev	Whether the EHR is hosted locally or remotely, it can only provide reliable support for safe, effective care if it is available and secure.	• Key data required to take care of patients and run the organization are available 24 hours/7 days per week, are not altered inadvertently or maliciously, and are kept confidential. • All data and operational systems are maintained on at least 2, geographically-distinct hosting sites that are mirrored in real-time ("hot" or "warm" sites). This redundancy reduces the risk of a single natural or man-made disaster to disable operating capacity. • There are at least 2 physically-distinct network connections between the two hosting sites. • Within a data center (i.e., hosting center) all servers are mirrored on physically-separate servers. • The healthcare organization has a contract in place that describes in detail how they will get access to their data in the event that either the EHR vendor or the remote hosting site vendor goes out of business (e.g., EHR and database management software has been placed in escrow and current data backups are independently accessible). • In an EHR's shared, remote hosting facility the data from different healthcare organizations are maintained within their own virtual machine (VM) environments or on separate physical servers.
3. The organization's information assets are protected using strong authentication mechanisms. IT	Failure to implement and manage authentication access to any system or data (e.g., strong passwords, fingerprints, and role-based access) is an avoidable source of erroneous data that can lead to patient harm.	• Organizations have policies and procedures and conduct regular risk assessments to define, implement, and monitor person authentication. • Access to the organization's "backbone network" via wireless devices is password protected.

Recommended Practices	Rationale for Practice or Risk Addressed	Examples of Potentially Useful Practices/Scenarios
		• Two-factor authentication is required for remote access to the servers' "administrative" accounts [e.g., root privileges on Unix] and clinicians remote access to patient data. Two-factor authentication involves using at least 2 means of identification, information one knows [i.e., password], information one has [i.e., electronic ID or random number token], or information unique to a person [e.g., iris or fingerprint scan]).
		• All users have a unique username and "strong" password (i.e., contains letters, numbers, and special characters (e.g., $, %, &).
		• Periodic changes to passwords are required) .
		• Employee login credentials are revoked as soon as their employment ends.
		• The organization has implemented a "single sign-on" solution that allows authorized clinicians to rapidly move between disparate clinical applications without requiring any additional login information.
4. System hardware and software required to run the EHR (e.g., operating system) and their modifications are tested individually and as installed before going live and are closely monitored after go-live. IT	Failure to adequately test system hardware and software can lead to sub-optimal performance as measured by response time, reliability, and error-free operation.	• Critical system infrastructure components, such as database servers, network routers, and end-user terminals, are regularly load tested.
		• All system software updates are installed and tested in the "test" environment before they are moved into the production or "live" environment and re-tested.
		• The organization monitors the system downtime and response time.
		• Organizational policies and procedures address post-installation issues (e.g., 24x7 support, help desk availability, and leadership walk-arounds).10
		• Organizational policies define criteria for testing (e.g., testing in a simulated environment, day of week testing, minimum # of test cases, types of user roles associated with test cases, facility defined vs. developer defined test cases).

Configuring and Maintaining EHRs and System-to-System Interfaces 237

Recommended Practices	Rationale for Practice or Risk Addressed	Examples of Potentially Useful Practices/Scenarios
5. Clinical applications and system interfaces are tested individually and as installed before go-live and are closely monitored after go-live. IT, C	One of the most common sources of adverse events is poor configuration between critical applications, such as between CPOE and pharmacy. Failure to adequately test applications and their interfaces can lead to data integrity issues as well as impede response time, availability, and error-free operation.	• New application software and updates (e.g., both major upgrades and small "patches") are installed and tested in the "test" environment before they are moved into the production or "live" environment, re-tested and closely monitored in the "live" environment for several days. • System-system interfaces between key clinical applications (e.g., CPOE and pharmacy, or laboratory and EHR) are tested and continuously monitored to detect new errors. • Simulations are conducted for clinical processes such as order entry, including Pharmacy review, RN notification, Medication fill, RN administration, RN document administration to ensure that the application works as designed and addresses the organization's needs.
6. Computers and displays in publicly accessible areas are configured to ensure that patient identifiable data are physically and electronically protected. IT, C	Failure to physically protect patient identifiable data to ensure that it is not inadvertently or maliciously viewed, changed, or deleted is vital to ensuring safe and effective use of clinical applications.	• Terminals used to access patient data in publicly accessible locations have an automatic screen locking feature set, appropriate to the clinical setting (e.g., lock after idle for three minutes). • Devices used to access patient data have their screens facing away from publicly accessible locations and/or have "privacy filters" (i.e., filters that restrict screen viewing angles). • Public displays of patient names on EHRs are masked (i.e., only a portion of the patient's name is visible in public areas, e.g., ED and waiting rooms). • The server room has physical security controls in place (e.g., room is locked, there is non-water-based fire suppression, room is above ground to prevent flooding, and backups are kept in a different location). • All portable computing devices used to access EHR data have encrypted hard drives. • Backups containing patient-identifiable data are encrypted.

Recommended Practices	Rationale for Practice or Risk Addressed	Examples of Potentially Useful Practices/Scenarios

7. There are processes in place to ensure data integrity during and after major system changes, such as upgrades to hardware, operating systems, or browsers.	Major system changes create the risk of loss or corruption of patient data. Data persistence must be ensured independent of hardware and software changes to maintain continuity of care. Losing data due to "improvements" in the underlying systems is unacceptable.	• Organizations have change management and internal control policies and procedures in place, designed to ensure data integrity, which apply to all major system changes. Major system changes include, at a minimum, operating system or browser version upgrades, or adding new system software (e.g., virus protection upgrades).
		• There are processes in place to migrate existing data to the new system while ensuring it remains accurate, valid, and accessible to the:
		  o application (e.g., from one EHR vendor to another),
		  o format (e.g., from freetext to structured data),
		  o coding system (e.g., from ICD-9 to ICD-10),
		  o storage mechanism (e.g., from magnetic tapes to solid state hard drives), etc.
		• Standard clinical and administrative reports are generated and reviewed regularly to ensure that the data on which they are based has not changed in a way that renders the report meaningless.
		• If data becomes corrupted, the facility has policies and processes for reverting to a backup version of the data that precedes the corruption.

Phase 2 – Safer Application and Use of IT

Principle: Complete/Correct EHR Use (Correct system usage [i.e., features and functions used as designed, implemented, and tested] is required for mission-critical clinical and administrative processes throughout the organization.)

Recommended Practices	Rationale for Practice or Risk Addressed	Examples of Potentially Useful Practices/Scenarios
8. Clinical content used, for example, to create order sets and clinical charting templates and to generate reminders within the EHR, is up-to-date, complete, available, and tested. C, IT	Clinical content drives significant parts of the user experience. Failure to update, test, and maintain this content can result in significant degradations in performance.	• There are no "broken links" to internet-based clinical information resources. • The organization has a naming convention and unambiguous synonyms for common orders, results, procedures, order sets, charting templates, and macros (e.g., dot phrases or "canned text"). • Default values are available for common orders (e.g., medication order sentences, routine laboratory draw times).

Configuring and Maintaining EHRs and System-to-System Interfaces 239

Recommended Practices	Rationale for Practice or Risk Addressed	Examples of Potentially Useful Practices/Scenarios
		• Items necessary to provide clinical care are available as orderable items within the CPOE system.
		• Clinical content is tested to ensure that items entered in one system are accurately transmitted through the system-to-system interface and received by the remote system unchanged.
		• Clinical content is reviewed by the organization at least annually.
		• The organization has a clinical informatics committee to review content.
9. There is a role-based access system in place to ensure that all applications, features, functions, and patient data are accessible only to users with the appropriate level of authorization. C, Ev, IT	There is a role-based access system in place to ensure that all applications, features, functions, and patient data are accessible only to users with the appropriate level of authorization.	• User roles with different data input and review capabilities are defined for both clinical and non-clinical users. Within each of these groups, sub-categories of users are defined with very specific capabilities (e.g., only credentialed MDs, DO's, or NP's can order schedule 2 (narcotics) medications without a co-signature)
		• There is a multi-disciplinary committee responsible for creating new roles and determining that the appropriate features and functions are assigned to each role.
		• Employees that change jobs are re-assigned to the appropriate roles promptly.
		• Periodically (e.g., yearly) supervisors are prompted to review and re-authorize or revoke their clinical and administrative staff's authorizations to access various clinical systems and functions.

Recommended Practices	Rationale for Practice or Risk Addressed	Examples of Potentially Useful Practices/Scenarios
10. The EHR is configured to ensure EHR users work in the "live" production version, and do not confuse it with training, test, and read-only backup versions. C, IT	Failure to clearly differentiate training, testing and live EHR environments can lead to data review and entry errors.	• There is a dedicated "training" environment for the EHR that includes de-identified patient data to allow high-fidelity testing with real-world data.. • Both the training and test environments are as complete as possible (e.g., within the training and test environments users can enter and sign orders that will display for another user, can review laboratory data, and can see alerts firing). • There is a dedicated "test" environment for the EHR that facilitates the configuration and testing of all new software and hardware updates. • The read-only backup system is password protected and clearly identifiable as read-only. • The EHR is configured to make it difficult to confuse the live version of the EHR with other versions. For example, the screen background color or the color of the patient headers could be different. • The organization has a policy and process for creating and naming test patients. Avoid "cute" names like Dr. Spock, and instead use unmistakable test names like "ZZZ" as a prefix for the name and at least 4 leading zeroes for Medical record number.
11. System configuration settings that limit clinical practice are minimized, carefully implemented following clinician acceptance, and closely monitored. C, IT	Configuration decisions that result in mismatches between institutional policies, routine practices, and EHR settings often result in "workarounds" by clinicians, which increase patient safety risks and lead to suboptimal use of EHRs.	• Organizational policies on EHR change/configuration management that address decisions that limit clinical practice, such as mandatory clinical alert settings (e.g., hard stops that cannot be overridden by clinicians or alerts that cannot be turned off by clinicians), are developed with clinicians, and are judiciously implemented and carefully monitored.7 • Organizational policy minimizes configurations that limit clinicians' ability to continue practicing (e.g., enter new orders) due to incomplete work (e.g., overdue cosignatures or incomplete discharge summaries).

Configuring and Maintaining EHRs and System-to-System Interfaces 241

Recommended Practices	Rationale for Practice or Risk Addressed	Examples of Potentially Useful Practices/Scenarios

Phase 2 – Safer Application and Use of IT

Principle: System Usability (All EHR features and functions required to manage the treatment, payment, and operations of the healthcare system are designed, developed, and implemented in such a way to minimize the potential for errors. In addition all information in the system must be clearly visible, understandable, and actionable to authorized users.)

| 12. The human-computer interface is configured for optimal usability for different users and clinical contexts. Ev, IT | Failure to support differences in user interface requirements for different locations, specialties, and users can lead to suboptimal system safety and effectiveness. | • The EHR user interface (those aspects of an EHR that users see and use) is configured (and configurable) to enable users with different capabilities and requirements to use the system safely and effectively (e.g., fonts large enough for all users to see; reduced screen brightness on night shifts; variable color and contrast schemes to accommodate color-blind users).
• The EHR user interface is monitored for safe use (e.g., user-reported usability hazards) and user satisfaction, and is improved over time.
• Default column widths are set wide enough to see key data.
• The EHR user interface is configured to address clinical specialty requirements. Clinical specialties have their "favorites" or 20 most commonly ordered medications, clinical laboratory, and imaging tests available on a single screen. |

Phase 3 – Leverage IT to Facilitate Oversight and Improvement of Patient Safety

Principle: Safety Surveillance and Optimization (Monitor, detect and report on safety-critical clinical and administrative aspects of EHRs and healthcare processes and make iterative refinements to optimize safety.)

| 13. The organization has processes and methods in place to monitor the effects of key configuration settings to ensure they are working as intended. Ev, IT | Failure to monitor configuration settings associated with key clinical components (e.g., CPOE interface to pharmacy) and processes (e.g., medication reconciliation) can lead to serious safety events that are otherwise difficult to identify. | • Key configuration settings include the number and size of database servers dedicated to the EHR application, password strength, system timeouts, and other similar settings. Organizations have policies and procedures that identify the key configuration settings and the persons responsible for monitoring them.
• The organization has a method of automatically monitoring (e.g., by periodically checking) all Internet-based links presented within the EHR. |

Recommended Practices	Rationale for Practice or Risk Addressed	Examples of Potentially Useful Practices/Scenarios
		• System response time is measured and reported regularly. • The interface error log is regularly reviewed and all errors are identified and fixed promptly. • The alert override rate is monitored and regularly reviewed. Alerts that are ignored 100 percent of the time (or nearly so) are re-evaluated and fixed or disabled.[8] • Clinical decision support is monitored using statistical processes (e.g., control charts) to identify malfunctions.[9]

REFERENCES

1. Haskins M. Legible charts! Experiences in converting to electronic medical records. Can Fam Physician. 2002 Apr;48:768-71.
2. O'Connor KJ. Everything you always wanted to know about software escrow agreements--and then some! J Healthc Inf Manag. 2005 Winter;19(1):10-2.
3. Committee on Maintaining Privacy and Security in Health Care Applications of the National Information Infrastructure. For the Record Protecting Electronic Health Information. NATIONAL ACADEMY PRESS, Washington, D.C. 1997.
4. Berger RG, Baba J. The realities of implementation of Clinical Context Object Workgroup (CCOW) standards for integration of vendor disparate clinical software in a large medical center. Int J Med Inform. 2009 Jun;78(6):386-90. doi: 10.1016/j.ijmedinf.2008.12.002. Epub 2009 Jan 20.
5. Sittig DF, Campbell EM, Guappone KP, Dykstra RH, Ash JS. Recommendations for Monitoring and Evaluation of In-Patient Computer-based Provider Order Entry Systems: Results of a Delphi Survey. Proc. Amer Med Informatics Assoc Fall Symposium (2007) p 671-675.
6. Bobb AM, Payne TH, Gross PA. Viewpoint: controversies surrounding use of order sets for clinical decision support in computerized provider order entry.
7. J Am Med Inform Assoc. 2007 Jan-Feb;14(1):41-7.

CHAPTER 9

ASSESSMENT OF PATIENT IDENTIFICATION RELATED PRACTICES

MATCHING IDENTIFIERS IN ELECTRONIC HEALTH RECORDS: IMPLICATIONS FOR DUPLICATE RECORDS AND PATIENT SAFETY

Allison B. McCoy, Adam Wright, Michael G. Kahn, Jason S. Shapiro, Elmer Victor Bernstam, and Dean F. Sittig

9.1.1 BACKGROUND AND SIGNIFICANCE

Increasing adoption of electronic health records (EHRs) and renewed emphasis on health information exchange (HIE) are leading to growing quantities of electronically available patient data that have the potential to reduce errors in a variety of ways, including improved clinical decision support. [1,2] However, health information technology (HIT) may also be associated with new errors. [3] One scenario that may lead to HIT-related errors is duplicate patient records. Three scenarios exist in relation to du-

---

Matching Identifiers in Electronic Health Records: Implications for Duplicate Records and Patient Safety. Reproduced from BMJ Quality and Safety, *McCoy AB, Wright A, Kahn MG, Shapiro JS, Bernstam EV, and Sittig DF,* **22**, *pp. 219-224, 2013, with permission from BMJ Publishing Group Ltd.*

plicate patient records: a correct registration exists when a single individual has a single medical record, a duplicate record exists when a single individual has more than one medical record, and a commingled record exists when a single medical record contains information about two or more individuals (Figure 9.1.1).

When duplicate records occur, a clinician could easily miss important information that exists in a different record than the one that was initially accessed, and such gaps in information are associated with increased adverse outcomes. [4,5] One recent study found that duplicate records, even when empty, were associated with missed abnormal test results. [6] Prior research describes the occurrence of duplicate patient records in clinical datasets, frequently in the setting of creating master patient indexes (MPIs) for HIEs and data warehouses, and reports a number of methods for identifying and merging duplicate records that range from comparison of basic demographic data to probabilistic algorithms that include various spellings and forms of demographics or clinical data. [7–19]

A number of approaches to managing duplicate records have been described previously. The first approach is prevention, where institutions try to keep duplicate records from occurring. Prevention approaches include effective training, centralised registration, notifications to users when creating a new record that is similar to an existing record, and use of MPI technology. [10,20] When prevention is not implemented or is insufficient, institutions must have methods in place to detect duplicate records. Detection methods may include automated (eg, deterministic or probabilistic) and manual reviews for similarity. [7,11–13,17,21] After duplicate records have been detected, they must be removed. Finally, institutions should adopt error mitigation approaches, to prevent or reduce patient harm when duplicate or potentially duplicate patient records persist within systems. Methods for mitigating errors include notifying users who access a record that is similar to another record, alternating row colours in patient lists, requiring photo or biometric (eg, vein pattern matching) identification during patient registration and including an up-to-date picture of the patient in the record. [8,20,22–26]

One important indicator of a duplicate record is matching identifiers, for example, two records that have the same first and last names and dates of birth. However, two distinct individuals may have the same identifiers, especially when only first and last name are used. [27] These demographic collisions may be more common in settings with less heterogeneity among names, such as an institution with a larger percentage of minority patients. [9] While potentially duplicate records are technically correct, they may leave patients at risk for misidentification, or a wrong patient error. Extensive previous literature describes these errors, which commonly include ordering or administration of medication or care to the wrong patient. [20,22–26,28–31] Duplicate records may also exist when the demographics are different; for example, a woman could have one record under her maiden name and one record under her married name. However, these scenarios are more difficult to detect and may require more complex, automated or manual approaches beyond matching first name, last name and date of birth.

In this paper, we quantify the number of patients at five institutions with identical first and last names and dates of birth as an indicator of duplicate or potentially duplicate records. We then identify solutions that may assist in detection, prevention and removal of duplicate records, reporting the rate of adoption of each method across the five institutions, and describing best practice recommendations for maximising patient safety related to records with matching identifiers.

9.1.2 METHODS

9.1.2.1 DATA COLLECTION

We performed a retrospective evaluation in five healthcare settings utilising EHRs: two large urban teaching hospitals with affiliated ambulatory practices across a variety of specialties using a locally developed EHR and Allscripts Enterprise EHR, a free-standing children's hospital using Epic, a regional HIE, and a large teaching hospital using Epic that is a member of the regional HIE. The study was approved by the institutional review board at each location.

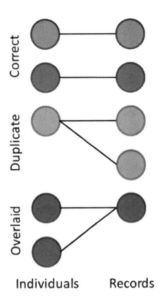

FIGURE 9.1.1: Scenarios for duplicate records.

9.1.2.2 QUANTIFICATION OF MATCHING PATIENT IDENTIFIERS

In each setting, we retrieved record counts, which were reported as the number of distinct medical record numbers (MRNs) having the matching identifier; individual elements of a patient's electronic medical record (eg, result reports, visit notes) were not counted separately, unless the patient had multiple MRNs. We first retrieved deidentified counts for records with exact matching first and last names, determining the occurrence of matching name pairs and the number of records associated with each matching name pair. We then repeated the process to retrieve counts for records with matching first name, last name and date of birth. Middle names or initials were not used, even if present. To depict the worst case for matching identifiers, we also retrieved the top 250 most common name pairs in each setting, determining the number of records occurring for each.

9.1.2.3 REVIEW OF METHODS FOR MANAGING DUPLICATE RECORDS OR RECORDS WITH MATCHING IDENTIFIERS

Through literature review and assessment of practices in place at each site of analysis, we identified methods for preventing, detecting and removing duplicate records in EHRs, in addition to methods for mitigating errors that may result from records with matching identifiers. For each site, we recorded whether the identified methods had been implemented.

9.1.3 RESULTS

9.1.3.1 QUANTIFICATION OF MATCHING PATIENT IDENTIFIERS

Table 9.1.1 depicts the occurrence of matching patient first and last names, and Table 9.2.2 depicts the occurrence of matching patient first name, last name and date of birth. The number of total records and rates of matching identifiers varied across settings; Sites A and B had the highest percentage of patients having a matching identifier when using first and last names (40.07% and 40.66%, respectively) and when including date of birth (13.36% and 15.47%, respectively). Site C had the lowest percentage of patients having a matching identifier when using first and last name (16.49%), while Site D had the lowest percentage of patients having a matching identifier when including date of birth (0.16%). The rate of matches decreased substantially for each site when including the date of birth as an identifier.

TABLE 9.1.1: Patient records with matching first and last name

	Site A	Site B	Site C	Site D	Site E
Total records in the study setting database	4256844	2678033	1014969	1012059	779736
Unique records identified by first and last name (%)	2551050 (59.93)	1589030 (59.34)	847642 (83.51)	767397 (75.83)	598776 (76.79)
Records having one or more matching first and last name (%)	1705794 (40.07)	1089003 (40.66)	167327 (16.49)	244662 (24.17)	180960 (23.39)

TABLE 9.1.2: Patient records with matching first name, last name, and date of birth

	Site A	Site B	Site C	Site D	Site E
Total records in the study setting database	4256844	2678033	1014969	1012059	779736
Unique records identified by first name, last name, and date of birth (%)	3720412 (87.40)	2263616 (84.53)	969753 (95.55)	1010476 (99.84)	747596 (95.88)
Records having one or more matching first name, last name, and date of birth (%)	536432 (13.36)	414417 (15.47)	45216 (4.45)	1583 (0.16)	32140 (4.35)

Figure 9.1.2 illustrates the number of records having the same first and last name for the 250 most commonly occurring names in each setting. The most frequently occurring name pair for Sites A, B, C, D and E had 2552, 744, 41, 53 and 1634 distinct records, respectively.

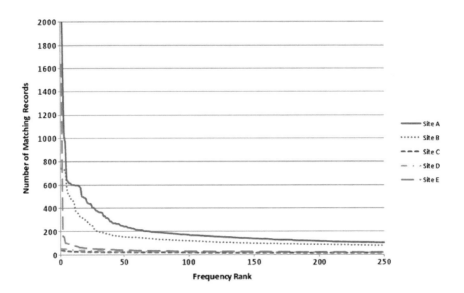

FIGURE 9.1.2: Frequency of matching first and last names.

TABLE 9.1.3: Methods for managing duplicate patient records or records with matching identifiers

	Site				
	A	B	C	D	E
Prevention					
Notifications to users when creating a record similar to an existing record	No	No	Yes	Yes	Yes
Centralised registration	No	No	Yes	No	No
Required training for registrars	Yes	No	Yes	Yes	Yes
Master patient index	Yes	No	Yes	No	Yes
Detection					
Automated demographics comparison	Yes	No	Yes	No	No
Name similarity comparison (eg, Soundex, look-alike, common variations, common misspellings)	No	No	Yes	Yes	No
Automated chart comparison (eg, clinical problems, labs, medications)	No	No	No	Yes	No
Manual reviews for similarity	Yes	No	Yes	Yes	No
Removal					
User ability to report duplicate records	No	Yes	No	No	No
Regularly scheduled institution efforts to merge records	No	No	Yes	Yes	No
Error mitigation					
Notification to users when accessing a record similar to other records	No	No	No	No	Yes
Photo identification during patient registration	No	No	No	Yes	No
Biometric identification during patient registration	No	No	No	No	No
Alternating row colours in patient lists	No	No	No	No	No
Total methods	3	1	8	7	3

9.1.3.2 IMPLEMENTATION OF METHODS FOR MANAGING DUPLICATE RECORDS OR RECORDS WITH MATCHING IDENTIFIERS

Table 9.1.3 depicts the methods for each approach implemented by each institution to manage duplicate records or records with matching identifiers. The most frequently implemented approaches included those for prevention, with four sites implementing more than one method. Of the three sites having implemented MPI technology, two implemented technology from Initiate Systems (acquired by IBM) and one implemented Cerner Enterprise Master Patient Index. Detection and removal approaches were also frequently implemented, while only two sites implemented error mitigation approaches.

9.1.3.3 DETAILED SITE ANALYSIS

Site C had the fewest records with matching patient identifiers and the most advanced methods for preventing, detecting and correcting duplicate patient errors. Thus, we present their methods in detail. Duplicate prevention starts at registration. At Site C, all registration is handled by a central phone registration service; clinics cannot register their own patients. To register a new patient, the registration clerk first searches for existing records using a variety of criteria, including name, insurance information, social security number and address. If an existing record for the patient is found, the registration clerk confirms that the record matches the patient and provides his or her existing MRN. If no record is found, the clerk queries a regional transaction clearinghouse to confirm the patient's insurance information and verify that there are no existing MRNs linked to that patient's identity in the clearinghouse. Only at this point will the clerk create a new record for the patient. The patient is then issued an identification card with their MRN, which they are encouraged to bring to all of their visits, and which is used to identify them; if the patient already has an MRN, they are issued a duplicate card with their previously issued MRN.

Although the initial central registration efforts prevent the creation of most duplicate records, there are still some cases where a duplicate record is created inadvertently. To manage these, a team in the Site C health information management department runs regular queries to find potentially duplicate records in their MPI. The query tool uses a variety of heuristics based on demographic information to identify potentially duplicate records, which are placed onto an action list for the team to investigate and, when necessary, merge records. The team also merges records in cases where a patient has to be temporarily registered without identifying information (eg, if the patient is not conscious and cannot be identified), with the goal of ensuring that there is always one record per patient.

9.1.4 DISCUSSION

We evaluated the number of matching patient identifiers and methods implemented for managing duplicate or potentially duplicate records at five unique sites. The rates varied across settings, from 16.49% to 40.66% of first and last name pairs having two or more records. With inclusion of patient date of birth in the search, the rates decreased, ranging from 0.16% to 15.47%. The number of records for the most frequently occurring name at each site ranged from 41 to 2552. Similarly, institutions varied widely in the implemented methods for managing duplicates or potential duplicates.

Previous work on duplicate patient records has most often focused on describing methods for identifying duplicate records or linking records in HIEs. [7–19] Our study is the first, to our knowledge, to quantify the rate of matching identifiers and the methods for managing duplicate records independently across multiple organisations. The rates we found are consistent with previous findings, [14,15] indicating there are potential patient safety problems across healthcare organisations despite various existing approaches for managing duplicate records.

Several factors may contribute to our observed rates of matching patient identifiers across sites. We anticipated that the number of matching identifiers would positively correlate with the number or type of implemented methods for managing duplicates; however, the data did not indicate any trends. Still, the fact that some institutions were very effective at preventing or removing duplicate records indicates that effective methods for managing duplicate records do exist. Further investigation is necessary to determine how effective these methods will be across institutions.

The large observed decrease in numbers of records with matching identifiers when including the date of birth in the search suggests that it is an important data element when identifying patients. It may also indicate that many of the first and last name pairs with two or more records belong to patients with common names (eg, John Smith, Maria Garcia), and those records with matching first name, last name and date of birth identifiers are more likely to be true duplicate records. The high number of records for the most frequently occurring names at each site confirms this theory, as it is more plausible that tens or thousands of different individuals have the same name than so many different records were created independently.

9.1.5 LIMITATIONS

We reported only the rates of identical patient identifiers and did not distinguish duplicates from potential duplicates, where different patients simply had the same demographic information. However, both these scenarios merit consideration as potential threats to patient safety. We also only used exact text matching, which did not account for typographical errors or variations in name spelling or form. More advanced methods for comparing the demographics would have likely found much higher rates of duplicates or potential duplicates. Finally, we did not evaluate how the approaches implemented for managing duplicate records at each site were implemented, possibly explaining our failure to detect an effect of the approaches. For instance, some methods of training registrars may be more effective than other methods; thus, the number of methods employed by a given institution may not be a good measure of effectiveness at preventing or eliminating duplicate records.

9.1.6 CONCLUSION

Despite a number of benefits that EHRs present for healthcare institutions, adoption of various methods for preventing, detecting and removing duplicate records, in addition to methods for mitigating errors in our sample was low; duplicate or potentially duplicate records persist, as indicated by high rates of matching patient identifiers in the evaluated institutions. Further efforts are necessary to improve management of duplicate or potentially duplicate records and minimise the risk for patient harm.

REFERENCES

1. Singh H, Naik AD, Rao R, et al. Reducing diagnostic errors through effective communication: harnessing the power of information technology. J Gen Intern Med 2008;23:489–94.
2. Sittig DF, Joe JC. Toward a statewide health information technology center (abbreviated version). South Med J 2010;103:1111–14.
3. Sittig DF, Singh H. Legal, ethical, and financial dilemmas in electronic health record adoption and use. Pediatrics 2011; 127(4):e10427.
4. Smith PC, Araya-Guerra R, Bublitz C, et al. Missing clinical information during primary care visits. JAMA 2005;293:565–71.
5. Stiell A, Forster AJ, Stiell IG, et al. Prevalence of information gaps in the emergency department and the effect on patient outcomes. CMAJ 2003;169:1023–8.
6. Joffe E, Bearden CF, Byrne MJ, Bernstam EV. Duplicate patient records—implication for missed laboratory results. AMIA Annu Symp Proc 2012;2012:126975.
7. Achimugu P, Soriyan A, Oluwagbemi O, et al. Record Linkage system in a complex relational database—MINPHIS example. Stud Health Technol Inform 2010;160(Pt 2):1127–30.
8. Arellano MG, Weber GI. Issues in identification and linkage of patient records across an integrated delivery system. J Healthc Inf Manag 1998;12:43–52.
9. Duvall SL, Fraser AM, Kerber RA, et al. The impact of a growing minority population on identification of duplicate records in an enterprise data warehouse. Stud Health Technol Inform 2010;160(Pt 2):1122–6.
10. McClellan MA. Duplicate medical records: a survey of twin cities healthcare organizations. AMIA Annu Symp Proc 2009;2009:421–5.
11. Miller PL, Frawley SJ, Sayward FG. Exploring the utility of demographic data and vaccination history data in the deduplication of immunization registry patient records. J Biomed Inform 2001;34:37–50.
12. Sauleau EA, Paumier J-P, Buemi A. Medical record linkage in health information systems by approximate string matching and clustering. BMC Med Inform Decis Mak 2005;5:32.
13. Waien SA. Linking large administrative databases: a method for conducting emergency medical services cohort studies using existing data. Acad Emerg Med 1997;4:1087–95.
14. Thornton SN, Hood SK. Reducing duplicate patient creation using a probabilistic matching algorithm in an open-access community data sharing environment. AMIA Annu Symp Proc 2005;2005:1135.
15. Duvall SL, Fraser AM, Rowe K, Thomas A, Mineau GP. Evaluation of record linkage between a large healthcare provider and the Utah Population Database. J Am Med Inform Assoc 2012;19(1e):e549.
16. Grannis SJ, Overhage JM, Hui S, et al. Analysis of a probabilistic record linkage technique without human review. AMIA Annu Symp Proc 2003;2003:259–63.
17. Grannis SJ, Overhage JM, McDonald C. Real world performance of approximate string comparators for use in patient matching. Stud Health Technol Inform 2004; 107(Pt 1):43–7.

18. Jurczyk P, Lu JJ, Xiong L, et al. FRIL: a tool for comparative record linkage. AMIA Annu Symp Proc 2008;2008:440–4.
19. Márquez Cid M, Chirlaque MD, Navarro C. DataLink record linkage software applied to the cancer registry of Murcia, Spain. Methods Inf Med 2008;47:448–53.
20. Bittle MJ, Charache P, Wassilchalk DM. Registration-associated patient misidentification in an academic medical center: causes and corrections. Jt Comm J Qual Patient Saf 2007; 33:25–33.
21. DuVall SL, Kerber RA, Thomas A. Extending the Fellegi-Sunter probabilistic record linkage method for approximate field comparators. J Biomed Inform 2010;43:24–30.
22. Henneman PL, Fisher DL, Henneman EA, et al. Patient identification errors are common in a simulated setting. Ann Emerg Med 2010;55:503–9.
23. Lee ACW, Leung M, So KT. Managing patients with identical names in the same ward. Int J Health Care Qual Assur Inc Leadersh Health Serv 2005;18:15–23.
24. O'Neill KA, Shinn D, Starr KT, et al. Patient misidentification in a pediatric emergency department: patient safety and legal perspectives. Pediatr Emerg Care 2004;20:487–92.
25. Ranger CA, Bothwell S. Making sure the right patient gets the right care. Qual Saf Health Care 2004;13:329.
26. Sideli RV, Friedman C. Validating patient names in an integrated clinical information system. Proc Annu Symp Comput Appl Med Care 1991;1991:588–92.
27. Yancey WE. Expected Number of Random Duplications Within or Between Lists. JSM 2010;2010:2938–46.
28. Gray JE, Suresh G, Ursprung R, et al. Patient misidentification in the neonatal intensive care unit: quantification of risk. Pediatrics 2006;117:e43–47.
29. Henneman PL, Fisher DL, Henneman EA, et al. Providers do not verify patient identity during computer order entry. Acad Emerg Med 2008;15:641–8.
30. Schulmeister L. Patient misidentification in oncology care. Clin J Oncol Nurs 2008;12:495–8.
31. Magrabi F, Ong M-S, Runciman W, Coiera E. Using FDA reports to inform a classification

SAFER SELF-ASSESSMENT: PATIENT IDENTIFICATION

SAFER Guides

OVERVIEW

Processes related to patient identification are complex and vulnerable to breakdown. In the EHR-enabled healthcare environment, we rely upon technology to help support and manage these complex identification processes and thus EHRs should optimize how information related to patient identification is displayed and communicated. Technology configurations alone cannot ensure accurate patient identification. Staff must also be supported with adequate training and procedures. This self-assessment is intended to increase awareness of EHR system characteristics related to design, configuration, and implementation decisions related to patient identification. This assessment can help identify and evaluate where breakdowns related to patient identification may occur in your healthcare delivery system. It focuses on the processes related to creation of new patients in the EHR, patient registration, retrieval of information on previously registered patients and other types of patient identification processes in the EHR with the goal being to mitigate problems that arise from duplicative records and patient mix-ups. Thoughtful use of this assessment by EHR users is intended to stimulate implementation of the recommended practices, as well as sustain those that are already present. When assessing EHRs at repeated intervals, (such as initially, annually and when changes are made), the assessment can be used to establish a baseline for measuring the effect of interventions designed to improve the safety of patient identification. The assessment works for ambulatory physician practices and other outpatient settings as well as for hospitals.

EXPECTATIONS

Healthcare professionals should use this assessment to aid in identifying and prioritizing patient safety issues related to EHR enabled patient identification. For example, you should consider both the frequency and severity of a safety event that might result in absence of these practices. We anticipate this to be a useful tool in ongoing safety and risk management programs, allowing you to address new risks that arise in EHR-enabled healthcare settings and helping you take advantage of the safety benefits of EHR-enabled healthcare settings. Please refer to the Guide for additional information, including the specific risks and rationales addressed by the recommended practices, and example strategies implemented in other clinical settings to support the recommended practices.

Legend	Key Facilitators of Practice Implementation
C	Clinicians, support staff, and/or clinical administration (e.g., Health Information Management and Risk Managers)
Dx	Diagnostic services, such as laboratory or radiology—could be local or remote
Ev	EHR vendor
IT	IT support staff, could be local or contracted. Responsible for maintaining the EHR and infrastructure
Rx	Pharmacy – could be local or remote

PATIENT IDENTIFICATION

Wrong patient errors are likely one of the most common types of errors in the modern EHR-enabled healthcare system. Accurate identification of the patient during registration coupled with on-going selection and use of the correct electronic record for each patient is the cornerstone upon which the entire electronic healthcare record (EHR) is based. Failure to identify that an existing patient record exists for a patient during registration can result in the creation of a duplicate record. A far more dangerous problem occurs when the wrong patient's record is used during the data review process or when recording new data. These so-called, co-mingled (or overlay) records are much more difficult to detect and once detected are very difficult to fix. A number of approaches to managing duplicate records have been described. The first approach is prevention, where institutions try to keep duplicate records from occurring. Prevention approaches include effective training, centralized registration, and notifications to users when creating a new record that is similar to an existing record, and use of master patient index (MPI) technology. When prevention is not implemented or is insufficient, institutions must have methods in place to detect and resolve duplicate records. Detection methods include automated (e.g., deterministic or probabilistic) and manual reviews for similarity. After potential duplicate records have been detected and confirmed as actual duplicates, they must be merged. Finally, institutions should adopt error mitigation approaches, to prevent or reduce patient harm when duplicate or potentially duplicate patient records persist within systems. Methods for mitigating errors include notifying users who access a record that is similar to another record,

Assessment of Patient Identification Related Practices

alternating row colors in patient lists, requiring photo or biometric (e.g., vein pattern matching) identification during patient registration, and including an up-to-date picture of the patient in the record.

Recommended Practices	Rationale for Practice or Risk Addressed	Examples of Potentially Useful Practices/Scenarios
Phase 1 – Make Health IT Safer		
Principle: Data Availability (EHRs and the data contained within them are available to authorized individuals where and when required to support healthcare delivery and business operations.)		
1. An enterprise-wide master patient index that includes patient's demographic information and medical record number(s) from different parts of the same organization (and, if available, from external organizations) is used to identify patients before importing data. IT	Duplicate patient records are a common problem and can cause harm when clinicians lack complete records. Likewise, when two patients' records are commingled harm can result. An enterprise-wide master patient index reduces the occurrence of duplicate patient records by increasing the likelihood that patients with previous encounters are identified.	• The master patient index employs a probabilistic matching algorithm that uses patient's first and last names, date of birth, gender, and zip code or telephone number or social security number. • Organizations have policies and procedures to identify and prevent duplicate patient records and to integrate unintentional duplicate records into one complete record. • Organizational policies address how to ensure correct patient identification of information from external sources, and monitor compliance with those policies. • Organizations update policies on patient identification related to the master patient index as best practices change.
2. Clinicians can select patient records from electronically generated lists based on specific criteria (e.g., user, location, time, service). Ev, IT	Selecting a patient from a short list of relevant patients reduces the risk of selecting the wrong patient.	• Patient lists can be automatically generated in several formats: Person-specific (e.g., all patients a clinician is responsible for), location-specific (e.g., all patients on a particular nursing unit or clinic), time-specific (e.g., all patients on today's schedule), and service-specific (e.g., all patients being cared for by a particular specialty or service). • Clinicians can view (read), edit (write: create, modify, delete), and use (execute: select a patient) patient lists.

Recommended Practices	Rationale for Practice or Risk Addressed	Examples of Potentially Useful Practices/Scenarios
		• Patient lists should by sorted in a clinically relevant order by default (e.g., by room number or appointment time), rather than alphabetically, to reduce the chance of look-alike or sound-alike names appearing close together.
		• There are 2 or more patient identifiers included with each patient on the list (e.g., name & date of birth, Medical record number, gender).

Phase 1 – Make Health IT Safer

Principle: Data Integrity (Data are accurate, consistent and not lost, altered or created inappropriately.)

3. Information required to accurately identify the patient is clearly displayed on all computer screens, wristbands, and printouts. Ev, IT	Providing medical services to the wrong patient is one of the most common preventable sources of patient harm. Steps should be taken to ensure that the person using an EHR to care for a patient is addressing the intended patient. Doing so reduces the risk of wrong patient errors.	• Organizational policies and all computer-generated displays incorporate the following information to facilitate patient identification: o LAST name o First name o Date of birth (with calculated age) o Gender o Medical record number o in-patient location (or home address) o Recent photograph (recommended) o Responsible physician (optional) • Organizational policies and workflows incorporate use of the EHR into ensuring correct patient identification
4. Patient names on adjacent lines in the EHR display are visually distinct. Ev, IT	Keeping patient names visually distinct in the EHR reduces the likelihood of unintentionally selecting the wrong patient. This is a basic good usability practice.	• On all patient lists containing two or more patients with the same last name, the names in common are displayed in a visually distinct manner (e.g., bold, italics, different color).

Assessment of Patient Identification Related Practices 261

Recommended Practices	Rationale for Practice or Risk Addressed	Examples of Potentially Useful Practices/Scenarios
		• Use alternate line colors for adjacent patients
5. Medical record numbers incorporate a "check digit" to help prevent data entry transposition errors. Ev, IT	A check digit greatly reduces data entry transposition errors.	• Organizational policies optimize automated processes in the EHR to prevent common errors, including transposition errors, which can result in poor patient identification
		• The "Verhoeff algorithm" works with strings of decimal digits of any length and detects all single-digit errors and all transposition errors involving two adjacent digits.
6. Users are warned when they attempt to create a new record for a patient (or look-up a patient) whose first and last name is the same as another patient. Ev, IT	Using automated EHR processes to prevent duplicate records can prevent unintentional human errors that could lead to patient harm. Creating a duplicate (split) record or commingling two different patient records results in a serious patient safety risk.	• System generates a pop-up alert when a user attempts to create a record for a new patient or looks up an existing patient with the same first and last name as an existing patient.
		• System generates an alert when a user attempts to create a record for a new patient or looks up an existing patient with a similar sounding first and last name as an existing patient, using a phonetic algorithm such as Soundex.
		• System monitors for similar names (nicknames), or changed last names (e.g., marriage, divorce, adoption), when other demographics match.
		• Alert provides additional demographic information context for the existing patient to help the user confirm or rule out that it is the same patient.

Phase 2 – Safer Application and Use of IT

Principle: Complete/Correct EHR Use (Correct system usage [i.e., features and functions used as designed, implemented, and tested] is required for mission-critical clinical and administrative processes throughout the organization.)

Recommended Practices	Rationale for Practice or Risk Addressed	Examples of Potentially Useful Practices/Scenarios
7. Patients are registered using a centralized, common database using standardized procedures. C, Ev, IT	Nonstandard registration practices and lack of access to a common database are common causes of duplicate medical records on the same patient.	• The organization requires a picture ID or uses biometric authentication (e.g., iris or vein scan) when authenticating new patients.
		• Organizational policy establishes standardized registration procedures involving the EHR and a common database to serve as the "source of truth" on whether a record already exists on a person who presents for services.
		• The organization requires a picture ID19 or uses biometric authentication (e.g., iris or vein scan) when authenticating new patients.
		• Registration clerks are trained to look up patients using the enterprise master patient index before creating a new record.
		• When new patient records are being created during the registration process, the registrar is prompted to consider other potential matches in the existing database.
8. The user interfaces of the training, test, and read-only backup versions of the EHR are clearly different from the production ("live") version to prevent incorrect entry or review of patient information in the wrong system. Ev, IT	If a clinician logs into and begins using the training, test, or read-only backup versions of the EHR by mistake, any information they attempt to enter will be lost.	• The screen background color on the production ("live") EHR is different from all other EHR environments.
		• EHR users are trained to understand the meaning of the visual differences between the different environments

Assessment of Patient Identification Related Practices

Recommended Practices	Rationale for Practice or Risk Addressed	Examples of Potentially Useful Practices/Scenarios
9. The organization has a process to assign a "temporary" unique patient ID (which is later merged into a permanent ID) in the event that either the patient registration system is unavailable or the patient is not able to provide the required information. Ev, IT	Inevitably, in certain cases, care must be delivered to patients who are not yet registered. Processes must be in place to ensure that they soon have a permanent ID and to merge records to avoid duplicate or incomplete records.	• A process (automated or manual, such as naming conventions) is in place to assign temporary IDs to newborns and patients arriving at the Emergency Department unable to provide their demographic information. • Staff members are trained in areas where temporary IDs may be required to ensure that temporary records are integrated into permanent ones. • Any downstream use of a temporary ID, such as in billing or in transfers between facilities, is tracked and corrected in all electronic systems, including at transfer facilities. • Organizations monitor resolution of temporary IDs.
10. Patient identity is verified at key points or transitions in the care process (e.g., rooming patient, vital sign recording, order entry, medication administration, and check-out). C	To avoid wrong patient errors, care must be taken to check the patient's identification at all critical points in the healthcare process and to ensure that EHR use is integrated into workflows that support correct patient identification.	• Before opening a specific patient record or signing an order, the user is shown a picture, or the name, gender and age of the patient. • Clinicians are asked to "re-enter" the patient's initials before signing an order. • Workflow related to verification of patient identity is evaluated to optimize use of the EHR to prevent wrong patient errors.

Phase 2 – Safer Application and Use of IT

Principle: System Usability (All EHR features and functions required to manage the treatment, payment, and operations of the healthcare system are designed, developed, and implemented in such a way to minimize the potential for errors. In addition all information in the system must be clearly visible, understandable, and actionable to authorized users.)

Recommended Practices	Rationale for Practice or Risk Addressed	Examples of Potentially Useful Practices/Scenarios
11. The EHR limits the number of patient records that can be displayed on the same computer at the same time to one, unless all subsequent patient records are opened as "Read Only" and clearly differentiated to the user. Ev, IT	Distractions while documenting or reviewing information in the EHR are common. EHRs should be designed to reduce the likelihood of working with the wrong patient's record as the result of distractions. When working on multiple patients, potential gains in efficiency are outweighed by the risks associated with entering or reviewing data on the wrong patient.	• Clinicians are engaged in developing EHR configuration and policies to prevent errors due to distractions and the resulting danger of working on the wrong patient chart when more than one is open. • Workflow is evaluated to ensure that clinicians are able to respond to urgent situations in which they may need to look at a new record without completing review of a first patient. The practice environment should be designed to minimize the need to open and actively use more than one patient's records on the same computer. • Before allowing the user to change the current patient, the system checks that all entered data has been saved (i.e., signed) before allowing the system to display a different patient's data.
12. Patients who are deceased are clearly identified as such. Ev, IT	In many instances, selection of a deceased patient represents a "wrong patient" error. Clinicians should be reminded that the patient they have selected is dead.	• The system displays either a pop-up alert when opening the record or a different background color for the deceased patient header in the EHR.
13. The use of test patients in the production (i.e., "live") environment is carefully monitored. When they do exist, they have unambiguously assigned "test" names (e.g., including numbers or multiple ZZ's) and are clearly identifiable as test patients (e.g., different background color for patient header). IT	Test patients in the production system are necessary to facilitate end-to-end testing, but care must be taken to ensure that they are not mistaken for "real" patients.	• Test patients should have names that clearly identify them as such: BWH17, ZZZOrders or MGH23zz, ZResults (examples are Last, First). • "Cute" names, e.g., "Marcus Welby" or "Jim Test" should not be used since there are real patients with those names.

Recommended Practices	Rationale for Practice or Risk Addressed	Examples of Potentially Useful Practices/Scenarios
Phase 3 – Leverage IT to Facilitate Oversight and Improvement of Patient Safety		
Principle: Safety Surveillance and Optimization (Monitor, detect and report on safety-critical clinical and administrative aspects of EHRs and healthcare processes and make iterative refinements to optimize safety.)		
14. The organization regularly monitors their patient database for erroneous patient identification errors. ,10 Ev, IT	Organizations must be prepared to monitor their system for potential patient ID errors and to investigate their causes.	• The order – retract – reorder algorithm can be used to estimate erroneous orders due to patient ID errors. • The "inconsistent gender algorithm" can be used to estimate the number of erroneous freetext notes due to patient ID errors. • Duplicate records are detected and merged. • Industry standards for duplicate error rates are available. The organization consistently monitors its own duplicate error rate, and ensures that it remains at or below industry standards

REFERENCES

1. Brigham and Women's Hospital video uses slapstick to promote patient safety. 03/07/2013 Available at: http://www.boston.com/whitecoatnotes/2013/03/07/brighamvideo/niv40v37wIxbqvim5eiKEJ/story.html?comments=all#add-comment
2. Bittle MJ, Charache P, Wassilchalk DM. Registration-associated patient misidentification in an academic medical center: causes and corrections. Jt Comm J Qual Patient Saf. 2007 Jan;33(1):25–33.
3. DuVall SL, Kerber RA, Thomas A. Extending the Fellegi-Sunter probabilistic record linkage method for approximate field comparators. J Biomed Inform. 2010 Feb;43(1):24–30.
4. Miller PL, Frawley SJ, Sayward FG. Exploring the utility of demographic data and vaccination history data in the deduplication of immunization registry patient records. J Biomed Inform. 2001 Feb;34(1):37–50.
5. Lee ACW, Leung M, So KT. Managing patients with identical names in the same ward. Int J Health Care Qual Assur Inc Leadersh Health Serv. 2005;18(1):15–23
6. Arellano MG, Weber GI. Issues in identification and linkage of patient records across an integrated delivery system. J Healthc Inf Manag. 1998;12(3):43–52.
7. Henneman PL, Fisher DL, Henneman EA, Pham TA, Campbell MM, Nathanson BH. Patient identification errors are common in a simulated setting. Ann Emerg Med. 2010 Jun;55(6):503–9

8. O'Neill KA, Shinn D, Starr KT, Kelley J. Patient misidentification in a pediatric emergency department: patient safety and legal perspectives. Pediatr Emerg Care. 2004 Jul;20(7):487–92.
9. Ranger CA, Bothwell S. Making sure the right patient gets the right care. Qual Saf Health Care. 2004 Oct;13(5):329.
10. Sideli RV, Friedman C. Validating patient names in an integrated clinical information system. Proc Annu Symp Comput Appl Med Care. 1991;588–92.
11. McCoy AB, Wright A, Kahn MG, Shapiro JS, Bernstam EV, Sittig DF. Matching identifiers in electronic health records: implications for duplicate records and patient safety. BMJ Qual Saf. 2013 Jan 29.
12. AHIMA. "Reconciling and Managing EMPIs (Updated)." Journal of AHIMA 81, no.4 (April 2010): 52-57. Available at: http://library.ahima.org/xpedio/groups/public/documents/ahima/bok1_046942.hcsp?dDocName=bok1_046942
13. Smith JA. AHIMA. "Fundamentals for Building a Master Patient Index/Enterprise Master Patient Index (Updated)." Journal of AHIMA (Updated September 2010).
14. Understanding file permissions on Unix: a brief tutorial. Available at: http://www.dartmouth.edu/~rc/help/faq/permissions.html
15. Valenstein PN, Sirota RL.Identification errors in pathology and laboratory medicine. Clin Lab Med. 2004 Dec;24(4):979-96, vii.
16. NHS CUI Programme Team, National Health Service Common User Interface (CUI) Design Guide Workstream – Design Guide Entry – Patient Banner v4.0.0.0 Baseline. Last modified on 25 June 2009 Available at: http://www.cuisecure.nhs.uk/CAPS/Patient%20Identification1/Patient%20Banner.pdf
17. Kirtland J. Identification Numbers and Check Digit Schemes. The Mathematical Association of America; 1st edition (January 15, 2001) pp 174.
18. Salomon, David (2005). Coding for Data and Computer Communications. Springer. p. 56. ISBN 0-387-21245-0.
19. Hyman D, Laire M, Redmond D, Kaplan DW. The use of patient pictures and verification screens to reduce computerized provider order entry errors. Pediatrics. 2012 Jul;130(1):e211-9. doi: 10.1542/peds.2011-2984. Epub 2012 Jun 4.
20. Sittig DF, Ash JS, Zhang J, Osheroff JA, Shabot MM. Lessons from "Unexpected increased mortality after implementation of a commercially sold computerized physician order entry system". Pediatrics. 2006 Aug;118(2):797-801.
21. Adelman JS, Kalkut GE, Schechter CB, Weiss JM, Berger MA, Reissman SH, Cohen HW, Lorenzen SJ, Burack DA, Southern WN.
22. Understanding and preventing wrong-patient electronic orders: a randomized controlled trial. J Am Med Inform Assoc. 2012 Jun 29.
23. Paparella SF. Accurate patient identification in the emergency department: meeting the safety challenges. J Emerg Nurs. 2012 Jul;38(4):364-7. doi: 10.1016/j.jen.2012.03.009.
24. Sittig DF, Teich JM, Yungton JA, Chueh HC. Preserving context in a multi-tasking clinical environment: a pilot implementation. Proc AMIA Annu Fall Symp. 1997:784-8.
25. Wilcox AB, Chen YH, Hripcsak G.Minimizing electronic health record patient-note mismatches. J Am Med Inform Assoc. 2011 Jul-Aug;18(4):511-4. Epub 2011 Apr 12.

CHAPTER 10

ASSESSMENT OF COMPUTER-BASED PROVIDER ORDER ENTRY WITH CLINICAL DECISION SUPPORT

DEVELOPMENT AND FIELD TESTING OF A SELF-ASSESSMENT GUIDE FOR COMPUTER-BASED PROVIDER ORDER ENTRY

Carl V. Vartian, Hardeep Singh, Elise Russo, and Dean F. Sittig

10.1.1 BACKGROUND AND SIGNIFICANCE

The use of electronic health records (EHRs) and computerized provider order entry (CPOE) with advanced clinical decision support (CDS) has the potential to increase efficiency, reduce harm, and improve patient safety (Amarasingham, Plantinga, ener-West, Gaskin, & Powe, 2009; Ammenwerth, Schnell-Inderst, Machan, & Siebert, 2008; Bates et al., 1999; Bates et al., 2001; Bobb et al., 2004; Franklin, O'Grady, Donyai, Jacklin, & Barber, 2007; Mekhjian et al., 2002; Sittig & Stead, 1994; Sittig & Singh,

---

Vartian, C., H. Singh, M.E. Debakey, E. Russo, and D. F. Sittig. 2014. "Development and Field Testing of a Self-Assessment Guide for Computer-Based Provider Order Entry." Journal of Healthcare Management *59 (5): 338–352, Chicago: Health Administration Press.*

2012; Wolfstadt et al., 2008). Despite these potential benefits, several barriers limit the adoption of these technologies, including high costs, major time commitments for training personnel, interruptions in workflow, and medical practitioners' reluctance to move away from traditional paper records (Kuperman & Gibson, 2003; Ash, Gorman, & Hersh, 1998; Wang & Huang, 2012).

To encourage widespread implementation of fully functional EHRs (i.e., EHRs that contain integrated CPOE and CDS applications), Congress passed the Health Information Technology for Economic and Clinical Health (HITECH) Act as part of the 2009 American Recovery and Reinvestment Act (Electronic Health Record Incentive Program, 2010). In 2010, HITECH allocated financial incentives totaling almost $27 billion to hospitals and eligible professionals (e.g., doctors of medicine or osteopathy, doctors of dental surgery or dental medicine, doctors of podiatric medicine, doctors of optometry, and chiropractors) who are able to demonstrate the "meaningful use" of EHRs, a critical component of which is CPOE with CDS (Wright, Feblowitz, Samal, McCoy, & Sittig, 2014). In view of the time limits on these incentives and the eventual financial penalties for non-adopters, since May 2012, 62,226 of the estimated 509,328 eligible physicians have attested to meaningful use of EHRs under the Medicare program (Vest, Yoon, & Bossak, 2013; Wright, Henkin, et al., 2013). Similarly, whereas only 27% of hospitals used CPOE in 2008, by 2012 the figure was 72% (Charles, King, Furukawa, & Patel, 2013).

CPOE has been described primarily as an innovation to improve patient safety, yet the implementation of CPOE within EHRs also has potential to introduce novel and unexpected risks. For example, Horsky, Kuperman, and Patel (2005) identified a series of errors in potassium chloride ordering resulting from a confluence of factors, including (1) misunderstandings about the patient's current potassium level due to ineffective display of recent laboratory results, (2) confusing displays of current orders and recent medication administrations, (3) misunderstandings surrounding the different meanings of "total volume" when ordering time-limited, continuous intravenous drips and amount-limited, intravenous bolus injections, and (4) use of free-text versus coded (i.e., computer-understandable) data entry. The effects of introducing fully functional EHRs into highly complex healthcare organizations are difficult to fully anticipate. It is not surprising that the accelerated pace of EHR implementation has given way to increasing reports of unintended negative consequences resulting from the use of these technologies (Allen & Sequist, 2012; Ash, Sittig, Poon et al., 2007; Ash, Sittig, Dykstra, et al., 2007; Ash, Sittig, Dykstra, Campbell, & Guappone, 2009; Berger & Kichak, 2004; Caldwell & Power, 2012; Campbell, Sittig, Ash, Guappone, & Dykstra, 2006; Campbell, Sittig, Guappone, Dykstra, & Ash, 2007; FitzHenry et al., 2007; Gandhi et al., 2005; Han et al., 2005; Horsky et al., 2005; Koppel et al., 2005; Koppel et al., 2008; Koppel, Wetterneck, Telles, & Karsh, 2008; Metzger, Welebob, Bates, Lipsitz, & Classen, 2010; Nanji et al., 2011; Nebeker, Hoffman, Weir, Bennett, & Hurdle, 2005; Singh et al., 2009; Zhan, Hicks, Blanchette, Keyes, & Cousins, 2006). Unfortunately, unexpected consequences often come to the attention of healthcare personnel only after errors and patient safety hazards have emerged. These EHR-related safety concerns must be addressed by taking into account the complex sociotechnical context in which EHRs are deployed (Sittig & Singh, 2010).

In order to address the safety concerns raised by the rapid adoption of EHRs by hospitals, the Office of the National Coordinator for Health Information Technology (ONC) convened an expert panel to investigate the potential for unintended negative consequences and sponsored the recent Institute of Medicine report Health IT and Patient Safety: Building Safer Systems for Better Care (2011). These efforts also spurred the Safety Assurance Factors for EHR Resilience (SAFER) project. The goal of the SAFER project was to develop self-assessment guides that healthcare organizations can use to proactively assess potential EHR-related safety problems in nine high-risk areas, one of which is CPOE, whether used in the inpatient or outpatient setting. The general plan for development, refinement, and beta-testing of the SAFER assessment guides has been described elsewhere (Singh, Ash, & Sittig, 2013). The SAFER guides are freely available on the ONC's website, http://www.healthit.gov/policy-researchers-implementers/safer.

In this paper, we describe the development and field testing of recommendations for CPOE-specific risk assessment in the SAFER project.

10.1.2 MATERIALS AND METHODS

10.1.2.1 CPOE GUIDE DEVELOPMENT

The rationale for the development of the SAFER guides was to enable proactive self-assessment to build system resilience around EHR safety. The development process for the CPOE-specific and other guides followed a series of common steps. We first conducted a literature search on CPOE and its effects on patient safety using medical subject headings (MeSH) search terms such as medical order entry systems and patient safety, along with the following text words: (computerized or computer-based) (physician or provider) order entry and (patient safety or errors or adverse events). All articles were reviewed for potential relevance based on the title and abstract. All relevant articles were subsequently reviewed in full. We also consulted with subject matter experts in informatics, pharmacy, patient safety, human factors engineering, and usability for additional literature and potential topics to be included in the guide. Based on this initial assessment, we drafted a set of approximately 250 items, each representing practices relevant to the safe use of CPOE. Three investigators (CVV, HS, DFS) with backgrounds in clinical medicine, patient safety, and informatics worked together as a team to review and condense these items, removing redundant items as well as those related to other types of healthcare outcomes (e.g., efficiency). We then conducted preliminary validation of the items through five site visits lasting 1 to 3 days at small and large ambulatory practices and hospitals. Site visits included one-on-one interviews with a variety of personnel involved in the use of CPOE, such as clinicians, information technology professionals, pharmacists, informaticians, and quality improvement leaders. We made further revisions to the items on the basis of feedback from these informants, with a goal of maximizing the usefulness and interpretability of the items by involving individuals with differing types and degrees of expertise. The resultant draft consisted of 22 checklist-type items that represented CPOE-related safety practices, with additional detailed descriptions of these practices and examples of how to operationalize them included in the Guide (page 285). For each item, respondents could indicate the degree of implementation of each practice at their respective sites. Possible responses included "not implemented," "partially implemented in some areas," "fully implemented in some areas," and "fully implemented in all areas."

10.1.2.2 FIELD TESTING

We field tested the CPOE guide with chief medical informatics officers (CMIOs) at various institutions across the country (Leviss, Kremsdorf, & Mohaideen, 2006). This group represented a highly knowledgeable population of informants, given their heavy involvement in CPOE implementation (Leviss et al., 2006). Following Institutional Review Board approval, we purposefully recruited informants through the listserv of the Association of Medical Directors of Information Systems (AMDIS). One of the authors (CVV), as an AMDIS member, regularly monitored postings by the CMIO community and identified those who expressed some interest in CPOE or were involved in CPOE-related activities. Potential participants were contacted initially via e-mail with an explanation of the SAFER project and an invitation to evaluate the CPOE guide. We recruited nine individuals; no compensation was provided. Participants were asked to complete the assessment within the CPOE guide and then complete a structured interview by phone. One of the authors (CVV) conducted all the interviews, each lasting approximately 30 minutes. In addition to inquiring about the characteristics of the respondent's facility (e.g., bed size, EHR platform, teaching status, etc.), the interviewer administered nine structured interview questions (see Table 10.1.1). Responses were collated and presented as frequencies within each of the response categories. Free-text responses were reviewed for common themes and used to refine the content, organization, implementation, and use of the guides.

10.1.3 RESULTS

10.1.3.1 INFORMANT AND FACILITY CHARACTERISTICS

The nine respondents represented pediatric facilities (n = 2) and tertiary care adult, acute care hospitals (n = 7) that were geographically distributed across eight states in several regions of the United States. Institutions ranged in size from approximately 85 beds at one pediatric facility to 1,100 beds at an acute care hospital. Seven were teaching hospitals. The EHR platforms used at these facilities were Allscripts (n = 3), Epic (n = 3), Meditech (n = 2), and Cerner (n = 1). All systems were in clinical use, and all clinicians were trained in the CPOE features and functions.

EHR and CPOE system configuration and implementation processes across the nine facilities were widely heterogeneous. Most involved ancillary systems (i.e., pharmacy, laboratory, or radiology) from multiple vendors that were interfaced to create a best-of-breed, end-to-end solution. In addition, drug-patient age checking and dose-range checking were simply unavailable in some institutions, given the current state of their EHR systems.

10.1.3.2 FEASIBILITY OF CPOE GUIDE COMPLETION

The modal completion time was 10 to 15 minutes, with a range of 5 to 30 minutes. Informants largely indicated that they felt comfortable answering the questions themselves. Two individuals sought additional input from pharmacy personnel. None of the informants indicated that any of the items were redundant with one another. About half the informants referred to the back of the guide for additional information and examples while completing the items.

10.1.3.3 ENDORSEMENT OF CPOE-RELATED SAFETY PRACTICES

Table 10.1.2 displays CMIOs' responses to each of the 22 recommended practices; responses varied across the entire spectrum from non- to full implementation. Only five recommended practices, mostly involving basic CPOE functionality (items 4 and 5) and allergy checking (items 1 and 8), were fully implemented at all sites. Respondents suggested several additions to the recommended practices; these included items pertaining to (1) training requirements, such as "providers should be trained before they can enter orders"; (2) alert fatigue; that is, "interruptive alerts are used with

Computer-Based Provider Order Entry with Clinical Decision Support 273

discretion and only for certain high-risk, high-priority conditions"; (3) clinical decision support, as in "CDS interventions should be included in the CPOE guide"; (4) downtime, such as "downtime procedures should be included in the CPOE guide"; (5) unintended consequences; for example, "potential unintended consequences of CPOE should be monitored"; (6) monitoring, as in "the ability to track lifetime exposure to radiation should be included"; (7) alerts, such as "non-medication duplicate order alerts should be included"; (8) mobility, or the ability to carry their EHR around with them (e.g., by iPhone or iPad), should be addressed; (9) integration across systems, such as "platform integration across the continuum of care (e.g., outpatient, emergency room, inpatient) should be emphasized"; and (10) orderable/test name synonyms or aliases, such as "CBC for complete blood count," should be mentioned.

Respondents raised specific concerns for drug–condition checking and drug-interaction-related alerts. They cited excessive firing of alerts as a common phenomenon and also reported that clinical decision support in their EHRs was overly simplistic. This led to such excessive levels of "noise" that many of these alerts were deactivated by the organization's clinical leadership.

Consensus around best practices was surprisingly low. For example, less than half of all respondents thought that requiring a physician to re-enter his or her password to "sign" an order was beneficial. Furthermore, respondents commented that the question on corollary orders was vague, because it used the term "certain medications" without further clarification. They believed it was unclear how many corollary orders they would need to have in place before answering "4: fully implemented in all areas for all patients, processes, and staff." Because not all medications require corollary orders, it is not feasible to recommend that this occur for all types of medications.

10.1.3.3.1 PERCEIVED USEFULNESS AND APPLICABILITY OF THE GUIDE

When asked whether the guide was useful, all informants responded in the affirmative. One informant remarked specifically about the potential difficulty in implementing the recommended practices in community hospitals lacking dedicated informatics personnel. Other specific comments about the usefulness of the items included:

- "Better confidence that we're doing the right thing"
- "It represents best practices"
- "It made me think a lot about what we're doing and does it make things better"
- "If the community agrees on a standard, it gives the hospital ways to compare themselves"

Notably, however, many informants questioned whether all the items represented critical patient safety practices, such as, taking the Leapfrog test of clinical decision support. Some indicated that there were no immediate plans to implement certain items at their institutions. Additionally, one CMIO observed, "If it is a certified EHR, much of it [the recommended practices] is already available."

When asked about the perceived purpose of the guide, informants largely focused on its use for checking on the implementation of important or accepted practices. One informant viewed the guide as a means to "identify glaring holes and opportunities for improvement." When asked to comment on the appropriate frequency of re-assessment, six CMIOs indicated that retaking the survey annually was reasonable, though one of these thought it might be less frequent if "at steady state" (i.e., CPOE has been fully implemented in all locations and physician utilization for all order types is stable at greater than 90%). One CMIO suggested a 2-year re-assessment timeline, whereas another suggested variable re-assessment intervals "depending on new regulations or safety concerns." Finally, one informant stated, "We wouldn't do it again, but it would be useful to submit data to a national processing center to get comparisons with how our hospital is doing compared to other institutions around the country."

Finally, when asked to identify additional opportunities for improving the guide, seven of the nine informants offered no further comments. One CMIO suggested that "the introduction might ask pharmacy to fill out some of the answers." Another CMIO suggested an additional question: "Have you had any patient safety events related to CPOE?"

The refined version of the combined CPOE and CDS SAFER Guide is included as the CPOE SAFER guide (pg 285).

10.1.4 DISCUSSION

We developed and field tested a guide for proactive self-assessment of CPOE-related safety practices. On testing the guide with nine CMIOs at hospitals of varying sizes and geographical locations, we found that the completion of the self-assessment was feasible, requiring less than 30 minutes. Although the assessment was designed to be completed by a multidisciplinary team, all but two respondents were able to answer the questions without additional input. Informants generally found the CPOE guide to be useful in outlining best practices, and most believed that an annual re-assessment was reasonable.

Our field testing indicated that CMIOs varied considerably in their opinions as to what constitutes safe and effective CPOE use. Although none of the recommended practices were considered redundant, informants had divergent opinions about their importance in many cases. For example, the requirement to enter a unique personal identification number (PIN) or login password to authenticate orders was nearly evenly split between full and non-implementation. The five CMIOs whose institutions did not require this practice believed that re-entering a PIN or password was redundant to logging into the EHR and was "a waste of time" for the provider, whereas others believed that this redundancy provided an important final check of order accuracy and the provider's ordering authority. Respondents were also divided in their perceptions of the utility of the Leapfrog Test, which "evaluates the ability of implemented CPOE systems to prevent the occurrence of medication errors that have a high likelihood of leading to adverse drug events" (Kilbridge, Welebob, & Classen, 2006; Metzger et al., 2010). The availability of corollary orders (item 15) was also problematic, with the majority of institutions not having fully implemented this practice. Unfortunately, to our knowledge there is no "official" list of medications that warrant corollary orders, although there is a list that has been successfully implemented (Overhage, Tierney, Zhou,

& McDonald, 1997). Our findings underscore the importance of continual updates of EHR-related best practices as the use of and experience with EHRs grows.

Our field testing revealed useful feedback to improve the future implementation and use of the guide. We envisioned the guide might require input from multiple stakeholders. However, we learned that when multiple stakeholders are unable to participate in the self-assessment process, one individual (perhaps with knowledge and experience levels similar to CMIOs) may still be able to generate useful information. Furthermore, the original design of the SAFER project was to develop nine self-assessment guides, including one for CPOE and one addressing CDS. Based on the feedback from the CMIOs and the inherently close relationship between CPOE and embedded CDS, we combined the CDS and CPOE self-assessment guides, resulting in the addition of seven recommended, CDS-focused practices to the original CPOE guide. Because most of the clinical and patient safety benefits of CPOE are attributable to the effectiveness of CDS, it would be best to assess their safety and effectiveness together.

We also obtained useful information on improving the content of the guide items. In addition to holding varied opinions about the merits of recommended practices, many respondents considered our list incomplete and suggested additional topics not included in the original guide. Several additional topics that were mentioned (e.g., training, alert fatigue, and unintended consequences such as downtimes) were addressed in the examples that were included in the appendix of the original guide, so it is possible that they were overlooked by those who, by their own admission, may have only skimmed the contents of that section. Finally, the issue of downtime procedures is covered in a separate SAFER guide entitled "Contingency Planning for Electronic Health Record-Based Care Continuity" (Sittig, Gonzalez, & Singh, 2014). Due to the complex interactions of the many dimensions of EHR-enabled healthcare, it is likely that institutions or practices will need to use one or more of the additional eight guides to comprehensively assess their EHR-based system.

Issues related to wording came up often, and we used this feedback to improve some of the item content. However, we needed to weigh some of the feedback on item wording, and we did not always make changes to specific items. For example, several respondents noted that certain recommended practices were worded as compound questions (e.g., "Evidence-based order sets are available for common tasks and are updated on a regular basis, and usage is monitored"). They wondered how this question could be answered if the order sets are in place but are not updated or usage is not monitored. Our team found this aspect of self-assessment to be challenging but believed that some compound questions would be necessary. This is because best practices often come as a "bundle," wherein each bundled item needs to be satisfied independently in order to make the bundle effective. We further reviewed the use of compound questions, including "Key metrics related to order entry are defined, measured, reported, and acted upon." This was again determined to be a best practice bundle where each of the four facets is vital. We thus learned that future respondents need to be instructed that they must be performing all aspects of compound or bundled recommendations to receive "credit" for that practice.

Finally, there were several respondents who noted that while they agreed with the importance of several recommendations (e.g., dose–range and drug–patient age checking), their EHRs as currently designed and developed were simply not able to perform those functions. In fact, this type of scenario was not uncommon during our preliminary validations of the various SAFER guides. Therefore, we note that to fully comply with the SAFER guide, EHR developers will need to identify and address specific functionality issues within their EHR offerings. This type of effort could gain momentum if a national body, such as the U.S. Department of Health and Human Services in conjunction with the ONC, could include specific CDS functionality in future EHR certification efforts by the ONC's Authorized Testing and Certification Bodies.

Several limitations of this study are worthy of mention. Although many medical informaticists are members of AMDIS, there are undoubtedly others who do not belong to this organization. Limiting our informants to those who have posted on the AMDIS listserv likely introduced a selection bias. Similarly, by only querying hospital CMIOs, we excluded feedback from institutions without a dedicated CMIO. While having a dedicated CMIO is a highly recommended practice, we acknowledge that, presently, most hospitals do not have such an individual (Wright, Ash, et al., 2013). CMIOs are more likely to exist in organizations with a longer history of

EHR deployment, yet it was the more recent and rapid adoption of EHRs that prompted the ONC to convene their expert panel and ultimately produce the SAFER guides. Seven of the nine institutions queried in this survey are teaching hospitals, and it is possible they are not representative of community facilities across the country. As these guides are being developed for use by a wide range of hospitals, including those without a CMIO and possibly with a recently installed EHR, we might have excluded potentially valuable comments from audiences that would be likely future users of this guide. However, we anticipate that the guides we created will be useful sources of knowledge of best practices regardless of their ultimate users.

TABLE 10.1.1: CPOE Structured Interview Guide

---

1. How long did it take you to complete the CPOE guide?
2. Do you feel like the team that completed the guide was the correct team to do so? If no, how would you recommend identifying the correct team?
3. Do you feel that any important practices were left out? If yes, what was left out?
4. Were any of the practices redundant? If yes, provide more detail.
5. Do you feel that the guide was useful? In what ways do you find the guide useful?
6. What do you feel is the purpose of the guide?
7. Did you refer to the appendix for additional information/examples when completing the guide? If yes, did you find the appendix useful?
8. How frequently would your team be willing to complete the guide?
9. Do you have any other suggestions for how the guide could be improved?

In conclusion, we have developed and field tested the SAFER CPOE self-assessment guide and found that the completion of the self-assessment is feasible and usually does not require much more than the CMIO's own knowledge of the organization. Although the 22 recommended safety practices were previously identified through a rigorous content review and expert opinion process, few were consistently implemented across facilities, and it was apparent that our informants held divergent opinions as to what constitutes "best practices" for CPOE-related safety. As CMIOs are likely to be the individuals advising their organizations, it would appear that full adoption of many recommended practices is unlikely until there is greater consensus about the merits of these and other safety practices. Additionally, EHR vendors will need to prioritize safe functionality of CPOE and CDS within the design of these systems so that best practices can have maximal reach and improve patient care. A CPOE-specific guide such as ours may provide a practical and actionable tool to stimulate further discussion and the development of best practices in this area.

TABLE 10.1.2: Frequencies of Endorsement of Practices to Improve Safety of CPOE. The numbers in each column represent the total number of sites with that response.

	Recommended Practice	Not implemented	Partially implemented in some areas	Fully implemented in some areas	Fully implemented in all areas
1.	Coded allergen and reaction information (or No Known Allergies [NKA]) are entered and updated in the EHR prior to order entry.	-	-	-	9
2.	Evidence-based order sets are available for common tasks/conditions and are updated on a regular basis, and usage is monitored.	-	1	2	6
3.	User-entered orderable items are matched to (or can be looked up from) a list of standard terms.	-	1	-	8
4.	EHR can cancel and acknowledge receipt of an order with lab, radiology, and pharmacy.	-	-	-	9
5.	EHR is used for ordering medications, diagnostic tests, and procedures.	-	-	-	9
6.	There is minimal use of free-text order-entry (i.e., data are entered and stored in coded form).	-	-	-	9

TABLE 10.1.2: *Cont.*

	Recommended Practice	Not implemented	Partially implemented in some areas	Fully implemented in some areas	Fully implemented in all areas
7.	Order entry information is electronically communicated (i.e., via the computer/mobile messaging) to the appropriate people responsible for carrying out the order.	-	-	1	8
8.	Drug–allergy interaction checking occurs at entry of new medication orders or new allergies.	-	-	-	9
9.	Duplicate checking occurs for certain orders (excluding PRN medications).	-	-	1	8
10.	Drug–condition checking occurs for important interactions between drugs and selected conditions.	2	1	2	4
11.	Drug–patient age checking occurs for important age-related interactions.	2	2	2	3
12.	Dose-range checking occurs before medication orders are submitted for dispensing (e.g., maximum dose amoxicillin 2-g oral tablets).	1	2	1	5
13.	Only the most significant and actionable drug–drug interaction-related alerts, as determined by the facility, are presented to providers.	1	-	1	7
14.	Clinicians are required to re-enter their password, or a unique PIN, to "sign" or authenticate an order.	5	-	-	4
15.	Corollary (or consequent) orders are automatically suggested by certain medication entries and are linked to and carried forward with the original order.	-	5	3	1

TABLE 10.1.2: *Cont.*

	Recommended Practice	Not implemented	Partially implemented in some areas	Fully implemented in some areas	Fully implemented in all areas
16.	Users can access clinical reference materials, including institution-specific knowledge links, directly from the EHR.	-	1	3	5
17.	The Leap Frog Test is taken to ensure safety of CDS.	5	-	-	4
18.	Critical patient information is visible during the order-entry process.	-	1	-	8
19.	The clinician is notified (e.g., by icon to signify non-formulary medication or send-out test) when additional steps (electronic or manual) are needed to complete the order being requested.	1	1	-	7
20.	There is minimal use of abbreviations and acronyms, and when they are used, they are clearly spelled out in all on-screen or printed information displays.	-	2	-	7
21.	Additional safeguards prevent errors related to prescribing of high-risk medications in the EHR.	-	1	1	7
22.	Key metrics related to order-entry use are defined, measured, reported, and acted upon.	-	3	-	6

REFERENCES

1. Allen, A. S., & Sequist, T. D. (2012). Pharmacy dispensing of electronically discontinued medications. Annals of Internal Medicine, 157(10), 700–705.

2. Amarasingham, R., Plantinga, L., ener-West, M., Gaskin, D. J., & Powe, N. R. (2009). Clinical information technologies and inpatient outcomes: A multiple hospital study. Archives of Internal Medicine, 169(2), 108–114.
3. Ammenwerth, E., Schnell-Inderst, P., Machan, C., & Siebert, U. (2008). The effect of electronic prescribing on medication errors and adverse drug events: A systematic review. Journal of the American Medical Informatics Association, 15(5), 585–600.
4. Ash, J. S., Gorman, P. N., & Hersh, W. R. (1998). Physician order entry in U.S. hospitals. American Medical Informatics Association Annual Symposium Proccedings, 235–239.
5. Ash, J. S., Sittig, D. F., Dykstra, R., Campbell, E., & Guappone, K. (2009). The unintended consequences of computerized provider order entry: Findings from a mixed methods exploration. International Journal of Medical Informatics, 78(Suppl. 1), S69–S76.
6. Ash, J. S., Sittig, D. F., Dykstra, R. H., Guappone, K., Carpenter, J. D., & Seshadri, V. (2007). Categorizing the unintended sociotechnical consequences of computerized provider order entry. International Journal of Medical Informatics, 76(Suppl. 1), S21–S27.
7. Ash, J. S., Sittig, D. F., Poon, E. G., Guappone, K., Campbell, E., & Dykstra, R. H. (2007). The extent and importance of unintended consequences related to computerized provider order entry. Journal of the American Medical Informatics Association, 14(4), 415–423.
8. Bates, D. W., Cohen, M., Leape, L. L., Overhage, J. M., Shabot, M. M., & Sheridan, T. (2001). Reducing the frequency of errors in medicine using information technology. Journal of the American Medical Informatics Association, 8(4), 299–308.
9. Bates, D. W., Teich, J. M., Lee, J., Seger, D., Kuperman, G. J., Ma'Luf, N, Boyle D, Leape L. (1999). The impact of computerized physician order entry on medication error prevention. Journal of the American Medical Informatics Association, 6(4), 313–321.
10. Berger, R. G., & Kichak, J. P. (2004). Computerized physician order entry: Helpful or harmful? Journal of the American Medical Informatics Association, 11(2), 100–103.
11. Bobb, A., Gleason, K., Husch, M., Feinglass, J., Yarnold, P. R., & Noskin, G. A. (2004). The epidemiology of prescribing errors: The potential impact of computerized prescriber order entry. Archives of Internal Medicine, 164(7), 785–792.
12. Caldwell, N. A., & Power, B. (2012). The pros and cons of electronic prescribing for children. Archives of Disease in Childhood, 97(2), 124–128.
13. Campbell, E. M., Sittig, D. F., Ash, J. S., Guappone, K. P., & Dykstra, R. H. (2006). Types of unintended consequences related to computerized provider order entry. Journal of the American Medical Informatics Association, 13(5), 547–556.
14. Campbell, E. M., Sittig, D. F., Guappone, K. P., Dykstra, R. H., & Ash, J. S. (2007). Overdependence on technology: An unintended adverse consequence of computerized provider order entry. American Medical Informatics Association Annual Symposium Proccedings, 94–98.
15. Charles, D., King, J., Furukawa, M. F., & Patel, V. (2013). Hospital adoption of electronic health record technology to meet meaningful use objectives: 2008–2012 (ONC Data Brief No. 10). Washington, D.C.: Office of the National Coordinator for Health Information Technology.

16. FitzHenry, F., Peterson, J. F., Arrieta, M., Waitman, L. R., Schildcrout, J. S., & Miller, R. A. (2007). Medication administration discrepancies persist despite electronic ordering. Journal of the American Medical Informatics Association, 14(6), 756–764.
17. Franklin, B. D., O'Grady, K., Donyai, P., Jacklin, A., & Barber, N. (2007). The impact of a closed-loop electronic prescribing and administration system on prescribing errors, administration errors and staff time: A before-and-after study. BMJ Quality & Safety, 16(4), 279–284.
18. Gandhi, T. K., Weingart, S. N., Seger, A. C., Borus, J., Burdick, E., Poon, E. G., Leape L.L., Bates D.W. (2005). Outpatient prescribing errors and the impact of computerized prescribing. Journal of General Internal Medicine, 20(9), 837–841.
19. Han, Y. Y., Carcillo, J. A., Venkataraman, S. T., Clark, R. S., Watson, R. S., Nguyen, T. C., Bayir H, Orr R.A. (2005). Unexpected increased mortality after implementation of a commercially sold computerized physician order entry system. Pediatrics, 116(6), 1506–1512.
20. Horsky, J., Kuperman, G. J., & Patel, V. L. (2005). Comprehensive analysis of a medication dosing error related to CPOE. Journal of the American Medical Informatics Association, 12(4), 377–382.
21. Institute of Medicine. Health IT and patient safety: Building safer systems for better care. (2011, November). Retrieved from http://www.iom.edu/Reports/2011/Health-IT-and-Patient-Safety-Building-Safer-Systems-for-Better-Care.aspx
21a. Institute of Medicine. Health IT and Patient Safety: Building Safer Systems for Better Care. The National Academies Press, Washington DC. (2012).
22. Kilbridge, P. M., Welebob, E. M., & Classen, D. C. (2006). Development of the Leapfrog methodology for evaluating hospital implemented inpatient computerized physician order entry systems. Quality and Safety in Health Care, 15(2), 81–84.
23. Koppel, R., Leonard, C. E., Localio, A. R., Cohen, A., Auten, R., & Strom, B. L. (2008). Identifying and quantifying medication errors: Evaluation of rapidly discontinued medication orders submitted to a computerized physician order entry system. Journal of the American Medical Informatics Association, 15(4), 461–465.
24. Koppel, R., Metlay, J. P., Cohen, A., Abaluck, B., Localio, A. R., Kimmel, S. E., Strom BL. (2005). Role of computerized physician order entry systems in facilitating medication errors. Journal of the American Medical Association, 293(10), 1197–1203.
25. Koppel, R., Wetterneck, T., Telles, J. L., & Karsh, B. T. (2008). Workarounds to barcode medication administration systems: Their occurrences, causes, and threats to patient safety. Journal of the American Medical Informatics Association, 15(4), 408–423.
26. Kuperman GJ, Gibson RF. (2003) Computer physician order entry: benefits, costs, and issues. Annals of Internal Medicine. 139(1), 31-9.
27. Leviss, J., Kremsdorf, R., & Mohaideen, M. F. (2006). The CMIO—A new leader for health systems. Journal of the American Medical Informatics Association, 13(5), 573–578.
28. Medicare and Medicaid programs; Electronic health record incentive program, 75. Fed. Reg. 44313 (July 28, 2010) (42 C.F.R. pts. 412, 413, 422, & 495). Available at: http://www.gpo.gov/fdsys/pkg/FR-2010-07-28/pdf/2010-17207.pdf

29. Mekhjian, H. S., Kumar, R. R., Kuehn, L., Bentley, T. D., Teater, P., Thomas, A., Payne B, Ahmad A. (2002). Immediate benefits realized following implementation of physician order entry at an academic medical center. Journal of the American Medical Informatics Association, 9(5), 529–539.
30. Metzger, J., Welebob, E., Bates, D. W., Lipsitz, S., & Classen, D. C. (2010). Mixed results in the safety performance of computerized physician order entry. Health Affairs (Millwood), 29(4), 655–663.
31. Nanji, K. C., Rothschild, J. M., Salzberg, C., Keohane, C. A., Zigmont, K., Devita, ., Bates D.W., Poon E.G. (2011). Errors associated with outpatient computerized prescribing systems. Journal of the American Medical Informatics Association, 18(6), 767–773.
32. Nebeker, J. R., Hoffman, J. M., Weir, C. R., Bennett, C. L., & Hurdle, J. F. (2005). High rates of adverse drug events in a highly computerized hospital. Archives of Internal Medicine, 165(10), 1111–1116.
33. Overhage, J. M., Tierney, W. M., Zhou, X. H., & McDonald, C. J. (1997). A randomized trial of "corollary orders" to prevent errors of omission. Journal of the American Medical Informatics Association, 4(5), 364–375.
34. Singh, H., Ash, J. S., & Sittig, D. F. (2013). Safety Assurance Factors for Electronic Health Record Resilience (SAFER): study protocol. BMC Medical Informatics and Decision Making, 13, 46.
35. Singh, H., Mani, S., Espadas, D., Petersen, N., Franklin, V., & Petersen, L. A. (2009). Prescription errors and outcomes related to inconsistent information transmitted through computerized order entry: A prospective study. Archives of Internal Medicine, 169(10), 982–989.
36. Sittig, D. F., Gonzalez, D., & Singh, H. (2014). Contingency planning for electronic health record-based care continuity: A survey of recommended practices. Manuscript under review.
37. Sittig, D. F., & Singh, H. (2010). A new sociotechnical model for studying health information technology in complex adaptive healthcare systems. Quality and Safety in Health Care, 19(Suppl. 3), i68–i74.
38. Sittig, D. F., & Singh, H. (2012). Electronic health records and national patient-safety goals. New England Journal of Medicine, 367(19), 1854–1860.
39. Sittig, D. F., & Stead, W. W. (1994). Computer-based physician order entry: The state of the art. Journal of the American Medical Informatics Association, 1(2), 108–123.
40. Vest, J. R., Yoon, J., & Bossak, B. H. (2013). Changes to the electronic health records market in light of health information technology certification and meaningful use. Journal of the American Medical Informatics Association, 20(2), 227–232.
41. Wang, C. J., & Huang, A. T. (2012). Integrating technology into health care: What will it take? Journal of the American Medical Association, 307(6), 569–570.
42. Wolfstadt, J. I., Gurwitz, J. H., Field, T. S., Lee, M., Kalkar, S., Wu, W., Rochon P.A. (2008). The effect of computerized physician order entry with clinical decision support on the rates of adverse drug events: A systematic review. Journal of General Internal Medicine, 23(4), 451–458.
43. Wright, A., Ash, J., Erickson, J. L., Wasserman, J., Bunce, A., Stanescu, A...., Sittig, D. F. (2014). A qualitative study of the activities performed by people involved in

clinical decision support: Recommended practices for success. Journal of the American Medical Informatics Association, 21(3), 464–472.
44. Wright, A., Feblowitz, J., Samal, L., McCoy, A. B., & Sittig, D. F. (2014). The Medicare electronic health record incentive program: Provider performance on core and menu measures. Health Services Research, 49(1 Pt. 2), 325–346.
45. Wright, A., Henkin, S., Feblowitz, J., McCoy, A. B., Bates, D. W., & Sittig, D. F. (2013). Early results of the meaningful use program for electronic health records. New England Journal of Medicine, 368(8), 779–780.
46. Zhan, C., Hicks, R. W., Blanchette, C. M., Keyes, M. A., & Cousins, D. D. (2006). Potential benefits and problems with computerized prescriber order entry: Analysis of a voluntary medication error-reporting database. American Journal of Health-System Pharmacy, 63(4), 353–358.

COMPUTERIZED PROVIDER ORDER ENTRY WITH CLINICAL DECISION SUPPORT

SAFER Guides

Recommended Practices	Rationale for Practice or Risk Addressed	Examples of Potentially Useful Practices/Scenarios
Phase 1 – Make Health IT Safer		
Principle: Data Availability (EHRs and the data contained within them are available to authorized individuals where and when required to support healthcare delivery and business operations.)		
1. Coded allergen and reaction information (or No Known Allergies [NKA]) are entered and updated in the EHR prior to any order entry.39 C, Ev	One of the main purposes of CDS is automated drug/allergy checking, which requires coded entry of allergies in the EHR.	• Users are reminded to enter patients' allergies or "no known allergies" before entering any medication orders. • A standard, controlled vocabulary of allergens and reactions (e.g., SNOMED-CT) is available and used. • There is a defined hierarchy of authority to edit or remove allergy-related information from a patient's EHR. • The EHR system permits entry of medication intolerances, separate from true allergies.

Recommended Practices	Rationale for Practice or Risk Addressed	Examples of Potentially Useful Practices/Scenarios
2. Evidence-based order sets are available in the EHR for common tasks/ conditions and are updated regularly.38 C, Dx, Ev, Rx	Order sets minimize errors of omission through standardization. Requiring clinicians to enter each of the individual orders for routine clinical practices increases risk of overlooking one or more items.	• Order sets for medications, diagnostic tests, and procedures are developed on the basis of Institute For Safe Medical Practices guidelines. [40] • Order sets exist for top the 10 most common clinical conditions (e.g., management of chest pain), procedures (e.g., insulin administration and monitoring), and clinical services (e.g., admission to labor & delivery). [41] • Clinical content is developed or modified based on evidence from authoritative sources, such as those in the ARHQ CDS initiative or by specialists within the organization. • EHR developer-provided clinical content is based on authoritative sources and is updated whenever those sources are updated. • Order sets for medications include complete pre-written medication orders (aka, order sentences) that include dose, dose form when necessary, route of administration, frequency, and a PRN flag and indication, if appropriate. [39] • Pre-written medication orders use doses that are weight- based, when appropriate. • Personalized order sets are not used. If an institution permits them, there is an annual review process, (e.g., clinical quality committee or medical director approval). • Medications requiring complex dosing guidelines e.g., insulin sliding scale, are standardized and available electronically. • CPOE list of orderable items (i.e., medication dictionary or orderable catalog) includes all formulary medications.

Recommended Practices	Rationale for Practice or Risk Addressed	Examples of Potentially Useful Practices/Scenarios
		• CPOE list of orderable items includes acceptable, non-formulary medications, which are clearly marked, that users can order for out of formulary fulfillment.
		• Prescribing systems for children use weight-based dosing recommendations, age-appropriate dosing calculators and dose-range checking, and pediatric-specific drug-drug interaction alerts.
3. User entered orderable items are matched to (or can be looked up from) a list of standard terms. 42 C, Dx, Ev, Rx	CDS is important to patient safety. CDS can be supported by orders of standardized items, but not on free text orders	• Users can look-up all orderable items (e.g., medications, laboratory and radiology tests) and pick terms from lists instead of entering free-text. This should support various word orders (e.g., "abdominal ultrasound" or "ultrasound, abdominal"), various names (e.g., generic or brand, synonym), and should be able to be browsed alphabetically. [43]

Phase 1 – Make Health IT Safer

Principle: Data Integrity (Data are accurate, consistent and not lost, altered or created inappropriately.)

4. The EHR can facilitate both cancellation and acknowledgement of receipt of an orders for laboratory, radiology, and pharmacy.38 Dx, Ev, Rx, IT	Communication errors, especially related to medication orders and diagnostic services, are frequent occurrences. Order tracking can reduce these errors.	• The user can look up whether the lab has received the specimen for testing or not • When medication orders are canceled, information is received and acted upon appropriately by the responsible pharmacy. • The 2-way interfaces that facilitate order tracking are tested pre- and post- go-live.
5. CDS alerts are displayed in the relevant clinical context.44-49 Ev, C, IT	CDS to improve diagnostic or therapeutic decision-making should be accessible in real time at the point of care, otherwise, the advice generated may be useless or under-utilized.50 Risks include information overload and clinician dissatisfaction.3,31,32	• A process is in place to identify and remove alerts that do not make sense in the particular clinical context. In some cases the process may require communication with the EHR developer.

Recommended Practices	Rationale for Practice or Risk Addressed	Examples of Potentially Useful Practices/Scenarios
		• Ambulatory alerts for cancer screening protocols should not be presented in the inpatient setting. [51,52]
		• Alerts for diabetic foot screening should not be presented on patients with bi-lateral below the knee amputations.
6. CDS incorporates current "best practices" and guidelines from authoritative sources, such as national organizatons and medical specialty professional associations.53 C, Ev, IT	Out of date or incorrect knowledge provided by the CDS system may be harmful.3,31,32	• For organizations that rely on EHR developer-provided CDS, a process is in place to ensure that CDS is based on authoritative sources and is regularly updated.
		• The expertise supporting CDS is demonstrated to EHR users before adoption.
		• Examples of authoritative sources include AHRQ's CDS Initiative and professional associations.
		• Colon cancer screening reminder follows U.S. Preventive Services Task Force guidelines [54]
		• Vaccination reminders use the latest recommendations from the Advisory Committee on Immunization Practices [55]

Phase 2 – Safer Application and Use of IT

Principle: Complete/Correct EHR Use (Correct system usage [i.e., features and functions used as designed, implemented, and tested] is required for mission-critical clinical and administrative processes throughout the organization.)

7. Clinicians are trained and tested on CPOE operations before being issued login credentials. C, Dx, Ev, IT, Rx	• CPOE is a complex tool. In order to maximize its safe and effective use, clinicians must trained rigorously and should not be expected to "learn the basics on the job."	• Incentives such as continuing education (CME or CEU) credits are awarded for clinicians getting trained on CPOE.
		• Clinicians are required to demonstrate basic CPOE skills before getting their login credentials. [56]
		• Organizations evaluate whether specialized CPOE training should be required in high risk areas.
		• Training is reinforced periodically especially with changes/upgrades.

Computer-Based Provider Order Entry with Clinical Decision Support

Recommended Practices	Rationale for Practice or Risk Addressed	Examples of Potentially Useful Practices/Scenarios
8. Clinicians are engaged in implementing, reviewing and updating CDS related interventions.53,57-61 C, Rx, Dx	• Failure to include clinicians in decisions that affect their clinical work environment, their decision-making capabilities, or how their decisions are communicated and recorded significantly increases the risk of hazardous events. CDS systems can be optimized through monitoring of use, overrides, and clinical satisfaction.	• Clinicians are involved in making the content consistent with updated guidelines and algorithms. There is an internal regulatory process (that involves clinicians) to evaluate and prioritize CDS for priority clinical conditions. [53,60-63] • Clinicians are involved in making (and keeping) the CDS content consistent with updated guidelines and algorithms. There is a process (that involves clinicians) to manage, • evaluate, and prioritize CDS updates. [53,60-63] • Clinician-provided feedback is reviewed and used for refinement and maintenance of CDS and the relevant clinical content. [53,59-61,63] • Clinician overrides (i.e., decisions not to follow a computer-generated suggestion) for high-priority CDS elements are logged and available for review and reporting. [64-66] • For EHR developer provided or controlled CDS, a process is in place to communicate about the need for CDS improvements with the developer.
9. EHR is used for ordering medications, diagnostic tests, and procedures for which CPOE is available. [38] (MU) C, Dx, Ev, IT, Rx	• While full use of CPOE with advanced clinical decision support has been shown to reduce errors, [50] partial use of CPOE can introduce errors.	• Except in unusual situations providers are required to enter their orders into the CPOE system. • Exceptions (e.g., emergency orders in resuscitation situations) are clearly defined, and processes are in place (and followed) for their proper documentation in the EHR.
10. There is minimal use of free-text order-entry. Orders are entered and stored in standardized, coded form. [38,67] C, Ev, IT	Free-text data can introduce errors if it is inconsistent with structured data or is not used or communicated properly. Free-text orders cannot be effectively supported with CDS.	• Organizational policy addresses safety precautions to be undertaken when free text ordering is allowed.

Recommended Practices	Rationale for Practice or Risk Addressed	Examples of Potentially Useful Practices/Scenarios
		• When medications are entered using standardized, coded terms, corresponding narrative text is minimized. Processes are in place to ensure timely use and review of any narrative text.
		• When medications must be ordered using free text, as constrained by organizational policy, a pharmacist reviews the order to identify and address any drug-drug or drug-allergy interactions.
11. Order entry information is electronically communicated, such as through the computer/mobile messaging, to the people responsible for carrying out the order. [68] (MU) C, Ev, IT	To have effective CPOE, orders must be electronically communicated. An automated process minimizes lapses in communication.	• Nurses are notified via the EHR when new results or orders are entered into the system for one of their patients (e.g., when they log-in to the system an alert tells them that new orders are available, or they are sent an informative page or text message). [69]
		• Orders that are not acknowledged by the individual responsible for carrying out the orders within 4 hours are automatically sent to a appropriate supervisor. [70]
		• Workflow is evaluated to ensure that all electronic orders go to the intended recipient and that person documents their actions in the EHR.
12. Interruptive alerts, such as pop-ups at the time of ordering, are used with discretion and only for high-risk, high-priority conditions. [44-49,60] EV, IT	Excessive use of interruptive alerts creates clinician dissatisfaction and reduces their effectiveness, causing clinicians to miss important alerts.29	• For low priority conditions, passive alerts that do not force an interruption of the workflow are available. [47]
		• High risk, high priority conditions that justify interruptive alerts are identified by clinicians and are subject to review.
		• Interruptive alerts at the point-of-care are used only after considering other available options. [71]

Recommended Practices	Rationale for Practice or Risk Addressed	Examples of Potentially Useful Practices/Scenarios
13. Drug-allergy interaction checking occurs during the entry of new medication orders and new allergies. [50,67] (MU) C, Ev	Interaction checking minimizes the risk of adverse drug events related to allergies.	• Checking occurs when an ACE inhibitor is prescribed to ensure that a patient with a history of ACE inhibitor-induced angioedema is protected. • Allergy checking also occurs whenever a new allergy is entered into the system.
14. Duplicate checking occurs for high-risk medication, diagnostic test, or procedure orders (excluding as needed "PRN" medications). [50,67] C, Ev	Duplicate order checking reduces the risk of inadvertent drug overdoses and unnecessary tests and procedures. [50,67]	• Therapeutic duplication checking occurs before new high-risk medication orders are submitted (e.g., two orders for the same or different beta-blockers are ordered). • Duplicate checking occurs before high-risk diagnostic tests or procedures are ordered. [72] • Duplicate checking does not include PRN (i.e., As needed) medication orders. • PRN orders should not include "overlapping" criteria (e.g., for pain 1-3 – give aspirin AND for pain 2-4 give vicodin).
15. Drug-condition checking occurs for important interactions between drugs and selected conditions. [50] C, Ev	Electronic drug-condition checking reduces the risk of preventable adverse drug events related to specific conditions.	• Drug-condition interaction checking occurs when new medications are ordered or new conditions are identified (e.g., Accutane or tetracycline prescribed for a pregnant woman).
16. Drug-patient age checking occurs for important age-related interactions. [13] C, Ev	Drug-patient age checking reduces the risk of preventable age-related adverse drug events.	• Drug-patient age interaction checking occurs when new medication orders are submitted for dispensing (e.g., medications contraindicated in the elderly). • Changes in frequency, dose, or substitutions are suggested for more age-appropriate strategies.
17. Dose range checking (such as maximum single dose or daily dose) occurs before medication orders are submitted for dispensing. [50,73] C, Ev	Dose range checking reduces the risk of medication overdose.	• Renal dose adjustment suggestions along with information on the patient's renal status are clearly displayed prospectively for relevant medications.

Recommended Practices	Rationale for Practice or Risk Addressed	Examples of Potentially Useful Practices/Scenarios
		• Patient context (age, renal function) dynamically changes the defaults prospectively.
		• Maximum single dose and maximum daily dose are independently checked.
		• Dose limits are age and body size appropriate.
18. A process is in place to review interactions so that only the most significant interaction-related alerts, as determined by the organization, are presented to clinicians. [46,47] (MU) C, Ev	Tiered alerting by severity (significance) is associated with higher compliance rates of Drug-drug interaction alerts.	• Less significant alerts are presented as information only, rather than interruptive alerts. [46] • Alerts are modified in a dynamic fashion based on feedback from the users and monitoring of user behavior.
19. Clinicians are required to re-enter their password, or a unique PIN, to "sign" or authenticate an order. C, Ev	Explicit order authentication reduces the risk of inadvertently entering orders under the wrong identity when someone else is logged in. It gives users an additional opportunity to confirm that the orders they entered are correct, and prevents them from inadvertently signing orders they did not intend to sign.	• An explicit authentication process occurs in addition to their original login for access to the EHR
20. Corollary (or consequent) orders are automatically suggested when appropriate and are linked together, so that changes are reflected when the original order is rescheduled, renewed, or discontinued. [74] Ev	Automatically suggested linked orders reduce order inconsistencies by managing closely associated orders in tandem.	• Examples include: Prothrombin time monitoring when Warfarin is prescribed, or drug level measurement with Vancomycin or aminoglycoside orders. [74] • Corollary orders are deleted whenever the main order is deleted (e.g., if colonoscopy is cancelled, bowel prep is also cancelled).
21. Users can access clinical reference materials, directly from the EHR, including organization-specific information when available. [42,53,59,60,62,75] Ev, IT	Ready access to information can reduce the risk of errors. CDS to improve diagnostic or therapeutic decision-making should be accessible in real time at the point of care; otherwise, the advice generated may be useless or underutilized. [50]	• Medication monographs (such as Micromedex), dosing calculators, diagnostic guides, laboratory reference materials, image atlases, anatomical diagrams, patient education materials, and disease-specific treatment guidelines are directly accessible from the order entry screen or module. [76]

Recommended Practices	Rationale for Practice or Risk Addressed	Examples of Potentially Useful Practices/Scenarios
22. CPOE and CDS functionality are tested to ensure proper operation before go-live and with test patients in the production system before clinical use. C, Ev, IT	Appropriate testing reduces the risk of errors associated with inappropriate CDS or CPOE system behavior.	• Leap Frog Test is taken to ensure safety of CDS. [77-79] • CDS interventions are evaluated to ensure correct firing of alerts and reminders. [80]

Phase 2 – Safer Application and Use of IT

Principle: System Usability (All EHR features and functions required to manage the treatment, payment, and operations of the healthcare system are designed, developed, and implemented in such a way to minimize the potential for errors. In addition, pertinent information should be easily accessible by authorized users, visible, understandable, prioritized and organized by relevance to the specific user.)

23. Questions presented to the user by CPOE and CDS are unambiguous. [50,81] Ev, IT	Misunderstanding queries posed by the system can lead to risks of errors and adverse events. [82]	• There are policies and procedures to evaluate the clarity of questions posed to users. • Questions should be kept simple and focused. For example, "Is IV contrast contraindicated?" may be confusing. It might be better to ask: Is IV contrast safe to administer? Yes, safe. No, not safe. • Avoid negatively and poorly worded questions such as "Do you want to cancel this alert? Yes, No, Cancel."
24. CPOE and CDS implementation and use are supported by usability testing based on best practices from human factors engineering. [83] C, Dx, Ev, IT, Rx	Risks of untested usability include decreased clinician efficiency and clinician dissatisfaction, as well as errors and adverse events due to unintended consequences of CDS use.	• Major CDS and CPOE changes/ interventions are tested with representative end users.83 • Clinician-reported hazards associated with CPOE and CDS due to poor usability are regularly communicated to someone in a position to make improvements. Follow-up is monitored
25. Critical patient information is visible during the order entry process. [84] Ev	Ensuring that critical data is visible in the EHR minimizes errors related to misidentification or failing to account for common clinical issues	• Pertinent clinical information (age, weight, allergies, pregnancy status, creatinine clearance/GFR) as well as identifying patient information is displayed on or behind the ordering screen with no scrolling required to view all the pertinent clinical data. [84]

Recommended Practices	Rationale for Practice or Risk Addressed	Examples of Potentially Useful Practices/Scenarios
26. The clinician is informed during the ordering process when additional steps are needed to complete the order being requested. Dx, Ev, Rx	Clinicians may not be aware that an order will not be completed without additional steps, leading to delays in performing the order.	
27. Use of abbreviations and acronyms is minimized and standardized. [85-87] C, Ev	Acronyms and abbreviations are a source of errors in both paper and electronic records. Minimizing and standardizing use of acronyms and abbreviations reduces the risk of errors related to misunderstanding.	• Organizational policies on the use of abbreviations and acronyms incorporate and are consistent with their use in EHRS. • Use of abbreviations and acronyms is consistent with industry best practices. • Abbreviations such as qd or qid are avoided
28. Additional safeguards, such as double check by a second specialist, are implemented in the EHR before high-risk medications are prescribed. Ev, Rx	Medication errors are the most common type of error that reach patients and cause harm. For high-risk medications, additional safeguards are justified to reduce the likelihood of harm	• A clinician- or specialist-driven process is in place to identify high risk medications that justify additional safeguards and to integrate those safeguards into the EHR. • Chemotherapy agents require special authorization and are displayed in a visually distinct way (e.g., different color, italics, etc.). • TALLman lettering is used to reduce CPOE errors from orthographically similar medication names (i.e., look-alike or sound-alike medication names; acetaZOLAMIDE and acetoHEXAMIDE). [88-90]

Phase 3 – Leverage IT to Facilitate Oversight and Improvement of Patient Safety

Principle: Safety Surveillance and Optimization (Monitor, detect and report on safety-critical clinical and administrative aspects of EHRs and healthcare processes and make iterative refinements to optimize safety.)

29. Key metrics related to order entry and clinical decision support (e.g., override rates) are defined, measured, reported and acted upon. [38,91] C, Ev, IT	Well-designed and correctly used CPOE and CDS can reduce the most common errors that harm patients. Monitoring and oversight of the performance and clinician use of CPOE and CDS functionality allows optimization of a powerful driver of improved patient safety in an EHR-enabled health care system.	Key CPOE safety indicators, such as the following, are monitored and reported to leadership on a periodic basis: • Rates of preventable ADEs • CPOE use rate • Frequency (volume) of orders that generate an alert

Recommended Practices	Rationale for Practice or Risk Addressed	Examples of Potentially Useful Practices/Scenarios
		• Override rate (% of alerts that are overridden) in comparison to alert volume
		• Median turn-around time for STAT laboratory or radiology results.
		• Percent of all orders requiring modification by someone other than the ordering provider
		• Alerts with the highest percentage of overrides are evaluated on at least a quarterly basis for effectiveness and turned off if deemed unacceptable.
		• Usage of evidence-based order sets is monitored
		• Clinician satisfaction with CDS alert functionality.

REFERENCES

1. Ammenwerth E, Schnell-Inderst P, Machan C, Siebert U. The effect of electronic prescribing on medication errors and adverse drug events: a systematic review. J Am Med Inform Assoc. 2008;15:585-600.
2. Bates DW, Teich JM, Lee J et al. The impact of computerized physician order entry on medication error prevention. J Am Med Inform Assoc. 1999;6:313-321.
3. Bates DW, Cohen M, Leape LL, Overhage JM, Shabot MM, Sheridan T. Reducing the frequency of errors in medicine using information technology. J Am Med Inform Assoc. 2001;8:299-308.
4. Bobb A, Gleason K, Husch M, Feinglass J, Yarnold PR, Noskin GA. The epidemiology of prescribing errors: the potential impact of computerized prescriber order entry. Arch Intern Med. 2004;164:785-792.
5. Franklin BD, O'Grady K, Donyai P, Jacklin A, Barber N. The impact of a closed-loop electronic prescribing and administration system on prescribing errors, administration errors and staff time: a before-and-after study. Qual Saf Health Care. 2007;16:279-284.
6. Mekhjian HS, Kumar RR, Kuehn L et al. Immediate benefits realized following implementation of physician order entry at an academic medical center. J Am Med Inform Assoc. 2002;9:529-539.
7. Sittig DF, Stead WW. Computer-based physician order entry: the state of the art. J Am Med Inform Assoc. 1994;1:108-123.

8. Wolfstadt JI, Gurwitz JH, Field TS et al. The effect of computerized physician order entry with clinical decision support on the rates of adverse drug events: a systematic review. J Gen Intern Med. 2008;23:451-458.
9. Ash JS, Sittig DF, Poon EG, Guappone K, Campbell E, Dykstra RH. The extent and importance of unintended consequences related to computerized provider order entry. J Am Med Inform Assoc. 2007;14:415-423.
10. Ash JS, Sittig DF, Dykstra RH, Guappone K, Carpenter JD, Seshadri V. Categorizing the unintended sociotechnical consequences of computerized provider order entry. Int J Med Inform. 2007;76 Suppl 1:S21-S27.
11. Ash JS, Sittig DF, Dykstra R, Campbell E, Guappone K. The unintended consequences of computerized provider order entry: findings from a mixed methods exploration. Int J Med Inform. 2009;78 Suppl 1:S69-S76.
12. Berger RG, Kichak JP. Computerized physician order entry: helpful or harmful? J Am Med Inform Assoc. 2004;11:100-103.
13. Caldwell NA, Power B. The pros and cons of electronic prescribing for children. Arch Dis Child. 2012;97:124-128.
14. Campbell EM, Sittig DF, Ash JS, Guappone KP, Dykstra RH. Types of Unintended Consequences Related to Computerized Provider Order Entry. J Am Med Inform Assoc. 2006;13:547-556.
15. Campbell EM, Sittig DF, Guappone KP, Dykstra RH, Ash JS. Overdependence on technology: an unintended adverse consequence of computerized provider order entry. AMIA Annu Symp Proc. 2007;94-98.
16. FitzHenry F, Peterson JF, Arrieta M, Waitman LR, Schildcrout JS, Miller RA. Medication administration discrepancies persist despite electronic ordering. J Am Med Inform Assoc. 2007;14:756-764.
17. Gandhi TK, Weingart SN, Seger AC et al. Outpatient prescribing errors and the impact of computerized prescribing. J Gen Intern Med. 2005;20:837-841.
18. Han YY, Carcillo JA, Venkataraman ST et al. Unexpected increased mortality after implementation of a commercially sold computerized physician order entry system. Pediatrics. 2005;116:1506-1512.
19. Horsky J, Kuperman GJ, Patel VL. Comprehensive analysis of a medication dosing error related to CPOE. J Am Med Inform Assoc. 2005;12:377-382.
20. Koppel R, Metlay JP, Cohen A et al. Role of computerized physician order entry systems in facilitating medication errors. JAMA. 2005;293:1197-1203.
21. Koppel R, Leonard CE, Localio AR, Cohen A, Auten R, Strom BL. Identifying and quantifying medication errors: evaluation of rapidly discontinued medication orders submitted to a computerized physician order entry system. J Am Med Inform Assoc. 2008;15:461-465.
22. Koppel R, Wetterneck T, Telles JL, Karsh BT. Workarounds to Barcode Medication Administration Systems: Their Occurrences, Causes, and Threats to Patient Safety. J Am Med Inform Assoc. 2008;15:408-423.
23. Metzger J, Welebob E, Bates DW, Lipsitz S, Classen DC. Mixed results in the safety performance of computerized physician order entry. Health Aff (Millwood). 2010;29:655-663.
24. Nanji KC, Rothschild JM, Salzberg C et al. Errors associated with outpatient computerized prescribing systems. J Am Med Inform Assoc. 2011;18:767-773.

25. Nebeker JR, Hoffman JM, Weir CR, Bennett CL, Hurdle JF. High rates of adverse drug events in a highly computerized hospital. Arch Intern Med. 2005;165:1111-1116.
26. Singh H, Mani S, Espadas D, Petersen N, Franklin V, Petersen L. Prescription Errors and Outcomes Related to Inconsistent Information Transmitted through Computerized Order-Entry: A Prospective Study. Arch Intern Med. 2009;169:982-989.
27. Zhan C, Hicks RW, Blanchette CM, Keyes MA, Cousins DD. Potential benefits and problems with computerized prescriber order entry: analysis of a voluntary medication error-reporting database. Am J Health Syst Pharm. 2006;63:353-358.
28. Allen AS, Sequist TD. Pharmacy dispensing of electronically discontinued medications. Ann Intern Med. 2012;157:700-705.
29. Bates DW, Kuperman GJ, Wang S et al. Ten commandments for effective clinical decision support: making the practice of evidence-based medicine a reality. J Am Med Inform Assoc. 2003;10:523-530.
30. Bates DW, Pappius E, Kuperman GJ. Using informations systems to measure and improve quality. Int J Med Inform. 1999;53:226-124.
31. Garg A, Adhikari N, McDonald H et al. Effects of Computerized Clinical Decision Support Systems on Practitioner Performance and Patient Outcomes: A Systematic Review. JAMA. 2005;293:1223-1238.
32. Kawamoto K, Houlihan CA, Balas EA, Lobach DF. Improving clinical practice using clinical decision support systems: a systematic review of trials to identify features critical to success. BMJ. 2005;330:765.
33. Saxena K, Lung BR, Becker JR. Improving patient safety by modifying provider ordering behavior using alerts (CDSS) in CPOE system. AMIA Annu Symp Proc. 2011;2011:1207-1216.
34. Ash JS, Berg M, Coiera E. Some unintended consequences of information technology in health care: the nature of patient care information system-related errors. J Am Med Inform Assoc. 2004;11:104-112.
35. Bloomrosen M, Starren J, Lorenzi NM, Ash JS, Patel VL, Shortliffe EH. Anticipating and addressing the unintended consequences of health IT and policy: a report from the AMIA 2009 Health Policy Meeting. J Am Med Inform Assoc. 2011;18:82-90.
36. Harrington L, Kennerly D, Johnson C. Safety issues related to the electronic medical record (EMR): synthesis of the literature from the last decade, 2000-2009. J Healthc Manag. 2011;56:31-43.
37. Magrabi F, Ong MS, Runciman W, Coiera E. Using FDA reports to inform a classification for health information technology safety problems. J Am Med Inform Assoc. 2011.
38. Sittig DF, Singh H. Electronic health records and national patient-safety goals. N Engl J Med. 2012;367:1854-1860.
39. Kuperman GJ, Bobb A, Payne TH et al. Medication-related clinical decision support in computerized provider order entry systems: a review. J Am Med Inform Assoc. 2007;14:29-40.
40. ISMP's Guidelines for Standard Order Sets. Institute for Safe Medication Practices . 2012. Ref Type: Electronic Citation
41. Wright A, Feblowitz JC, Pang JE et al. Use of Order Sets in Inpatient Computerized Provider Order Entry Systems: A Comparative Analysis of Usage Patterns at Seven Sites. J Am Med Inform Assoc. In press.

42. Sittig DF, Singh H. Eight rights of safe electronic health record use. JAMA. 2009;302:1111-1113.
43. Rosenbloom ST, Miller RA, Johnson KB, Elkin PL, Brown SH. Interface terminologies: facilitating direct entry of clinical data into electronic health record systems. J Am Med Inform Assoc. 2006;13:277-288.
44. Bates D. Clinical decision support and the law: the big picture. St Louis University Journal of Health Law and Policy. 2012;5:319-324.
45. Hoffman S, Podgurski A. Drug-Drug interaction alerts: emphasizing the evidence. St Louis University Journal of Health Law and Policy. 2012;5.
46. Paterno MD, Maviglia SM, Gorman PN et al. Tiering drug-drug interaction alerts by severity increases compliance rates. J Am Med Inform Assoc. 2009;16:40-46.
47. Phansalkar S, van der SH, Tucker AD et al. Drug-drug interactions that should be non-interruptive in order to reduce alert fatigue in electronic health records. J Am Med Inform Assoc. 2012.
48. Ridgley M, Greenberg M. Too many alerts, too much liability: sorting through the malpractice implications of drug-drug interaction clinical decision support. St Louis University Journal of Health Law and Policy. 2012;5:257-296.
49. Strom BL, Schinnar R, Aberra F et al. Unintended effects of a computerized physician order entry nearly hard-stop alert to prevent a drug interaction: a randomized controlled trial. Arch Intern Med. 2010;170:1578-1583.
50. Sengstack P. CPOE configuration to reduce medical errors. Journal of Health Care Information Management. 2010;24:26-32.
51. Sittig DF, Singh H. Improving Test Result Follow-up through Electronic Health Records Requires More than Just an Alert. J Gen Intern Med. 2012;27:1235-1237.
52. Overview of CDS Five Rights. http://healthit.ahrq.gov/images/mar09_cds_book_chapter/CDS_MedMgmnt_ch_1_sec_2_five_rights.htm . 2009. AHRQ. 5-20-2013. Ref Type: Electronic Citation
53. Sittig DF, Wright A, Ash JS, Middleton B. A set of preliminary standards recommended for achieving a national repository of clinical decision support interventions. AMIA Annu Symp Proc. 2009;2009:614-618.
54. USPSTF Recommendations. http://www.uspreventiveservicestaskforce.org/recommendations.htm . 2010. U.S. Preventive Services Task Force. 5-20-2013. Ref Type: Electronic Citation
55. ACIP Recommendations. http://www.cdc.gov/vaccines/pubs/ACIP-list.htm . 2011. Centers for Disease Control and Prevention Advisory Committee for Immunization Practices. 5-20-2013. Ref Type: Electronic Citation
56. Sittig DF, Classen DC. Safe electronic health record use requires a comprehensive monitoring and evaluation framework. JAMA. 2010;303:450-451.
57. Ash JS, Sittig DF, Wright A et al. Clinical decision support in small community practice settings: a case study. J Am Med Inform Assoc. 2011;18:879-882.
58. Ash JS, Sittig DF, Guappone KP et al. Recommended practices for computerized clinical decision support and knowledge management in community settings: a qualitative study. BMC Med Inform Decis Mak. 2012;12:6.
59. Sittig DF, Wright A, Simonaitis L et al. The state of the art in clinical knowledge management: an inventory of tools and techniques. Int J Med Inform. 2010;79:44-57.

60. Wright A, Phansalkar S, Bloomrosen M et al. Best Practices in Clinical Decision Support: the Case of Preventive Care Reminders. Appl Clin Inform. 2010;1:331-345.
61. Wright A, Sittig DF, Ash JS et al. Governance for clinical decision support: case studies and recommended practices from leading institutions. J Am Med Inform Assoc. 2011;18:187-194.
62. Horsky J, Schiff GD, Johnston D, Mercincavage L, Bell D, Middleton B. Interface design principles for usable decision support: A targeted review of best practices for clinical prescribing interventions. J Biomed Inform. 2012.
63. Osheroff J, Teich J, Levick D et al. Improving Outcomes with Clinical Decision Support: An Implementer's Guide. Second Edition ed. Healthcare Information and Management Systems Society; 2012.
64. Bennett JW, Glasziou PP. Computerised reminders and feedback in medication management: a systematic review of randomised controlled trials. Med J Aust. 2003;178:217-222.
65. Morris AH. Developing and implementing computerized protocols for standardization of clinical decisions. Ann Intern Med. 2000;132:373-383.
66. van der SH, Aarts J, Vulto A, Berg M. Overriding of drug safety alerts in computerized physician order entry. J Am Med Inform Assoc. 2006;13:138-147.
67. Carvalho CJ, Borycki EM, Kushniruk A. Ensuring the safety of health information systems: using heuristics for patient safety. Healthc Q. 2009;12 Spec No Patient:49-54.
68. Aarts J, Ash J, Berg M. Extending the understanding of computerized physician order entry: implications for professional collaboration, workflow and quality of care. Int J Med Inform. 2007;76 Suppl 1:S4-13.
69. Geissbuhler A, Grande JF, Bates RA, Miller RA, Stead WW. Design of a general clinical notification system based on the publish-subscribe paradigm. Proc AMIA Annu Fall Symp. 1997;126-130.
70. Kuperman GJ, Teich JM, Tanasijevic MJ et al. Improving response to critical laboratory results with automation: results of a randomized controlled trial. J Am Med Inform Assoc. 1999;6:512-522.
71. Sittig DF, Teich JM, Osheroff JA, Singh H. Improving clinical quality indicators through electronic health records: it takes more than just a reminder. Pediatrics. 2009;124:375-377.
72. Bates DW, Boyle DL, Rittenberg E et al. What proportion of common diagnostic tests appear redundant? Am J Med. 1998;104:361-368.
73. Zhou L, Maviglia SM, Mahoney LM et al. Supratherapeutic Dosing of Acetaminophen Among Hospitalized Patients. Arch Intern Med. 2012;1-8.
74. Overhage JM, Tierney WM, Zhou XH, McDonald CJ. A randomized trial of "corollary orders" to prevent errors of omission. J Am Med Inform Assoc. 1997;4:364-375.
75. Wright A, Bates DW, Middleton B et al. Creating and sharing clinical decision support content with Web 2.0: Issues and examples. J Biomed Inform. 2009;42:334-346.
76. Del FG, Huser V, Strasberg HR, Maviglia SM, Curtis C, Cimino JJ. Implementations of the HL7 Context-Aware Knowledge Retrieval ("Infobutton") Standard: Challenges, strengths, limitations, and uptake. J Biomed Inform. 2012;45:726-735.
77. Birkmeyer, JD and Dimick, JB. Leapfrog safety standards: potential benefits of universal adoption. 2004. Washington, DC, The Leapfrog Group. Ref Type: Report

78. Kilbridge PM, Welebob EM, Classen DC. Development of the Leapfrog methodology for evaluating hospital implemented inpatient computerized physician order entry systems. Qual Saf Health Care. 2006;15:81-84.
79. Metzger JB, Welebob E, Turisco F, Classen DC. The Leapfrog Group's CPOE Standard and Evaluation Tool. Patient Safety & Quality Healthcare. 2008.
80. McCoy AB, Waitman LR, Lewis JB et al. A framework for evaluating the appropriateness of clinical decision support alerts and responses. J Am Med Inform Assoc. 2012;19:346-352.
81. Sengstack P. CPOE configuration to reduce medication errors: a literature review on the safety of CPOE systems and design recommendations. Journal of Health Care Information Management. 2010;24:1-6.
82. Sittig DF, Singh H. Defining Health Information Technology-Related Errors: New Developments Since To Err Is Human. Arch Intern Med. 2011;171:1281-1284.
83. Schumacher, R. M. and Lowry, S. Z. NIST Guide to the Processes Approach for Improving the Usability of Electronic Health Records. http://www.nist.gov/itl/hit/upload/Guide_Final_Publication_Version.pdf . 2013. National Institute of Standards and Technology. 5-20-2013. Ref Type: Electronic Citation
84. Khajouei R, Jaspers MW. CPOE system design aspects and their qualitative effect on usability. Stud Health Technol Inform. 2008;136:309-314.
85. ISMP's List of Error-Prone Abbreviations, Symbols, and Dose Designations. http://www.ismp.org/tools/errorproneabbreviations.pdf . 2012. Institute of Safe Medication Practices. Ref Type: Electronic Citation
86. Joint Commission. Information Management Standards. IM.02.02.01., Elements of Performance 2 and 3. 2010. Ref Type: Bill/Resolution
87. Passiment E, Meisel JL, Fontanesi J, Fritsma G, Aleryani S, Marques M. Decoding laboratory test names: a major challenge to appropriate patient care. J Gen Intern Med. 2013;28:453-458.
88. ISMP's List of Confused Drug Names. http://www.ismp.org/Tools/confuseddrugnames.pdf . 2012. Institute for Safe Medical Practices. Ref Type: Electronic Citation
89. Joint Commission. Medication Management Standards. MM.01.02.01, Element of Performance 1. 2010. Ref Type: Bill/Resolution
90. Filik R, Purdy K, Gale A, Gerrett D. Drug name confusion: evaluating the effectiveness of capital ("Tall Man") letters using eye movement data. Soc Sci Med. 2004;59:2597-2601.
91. Weir CR, McCarthy CA. Using implementation safety indicators for CPOE implementation. Jt Comm J Qual Patient Saf. 2009;35:21-28.

CHAPTER 11

ASSESSMENT OF DIAGNOSTIC TEST RESULT REPORTING AND FOLLOW-UP

IMPROVING FOLLOW-UP OF ABNORMAL CANCER SCREENS USING ELECTRONIC HEALTH RECORDS: TRUST BUT VERIFY TEST RESULT COMMUNICATION

Hardeep Singh, Lindsey Wilson, Laura A. Petersen, Mona K. Sawhney, Brian Reis, Donna Espadas, and Dean F. Sittig

11.1.1 BACKGROUND

Fewer than 75% of patients with abnormal cancer screening examinations receive follow-up diagnostic care subsequent to the initial screening [1-5]. This inadequate follow-up of abnormal cancer screens compromises the benefits of population-based screening programs [6-9]. For instance, the rate of follow-up for positive fecal occult blood test (FOBT) results in

---

*Improving Follow-Up of Abnormal Cancer Screens Using Electronic Health Records: Trust but Verify Test Result Communication. © Singh H, Wilson L, Petersen LA, Sawhney MK, Reis B, Espadas D and Sittig DF; liensee BioMed Central Ltd. BMC Medical Informatics and Decision Making **9**,49 (2009), doi:10.1186/1472-6947-9-497. Licensed under Creative Commons Attribution 2.0 Generic License, http://creativecommons.org/licenses/by/2.0/.*

the Veterans Affairs health care system is low; more than 40% of veterans with positive FOBTs may not be receiving timely diagnostic colonoscopies [10,11]. Lack of timely follow-up has also been documented outside the VA system [12,13].

An important, largely preventable but relatively unexplored reason for lack of follow-up is a problem in communication of the positive test result from the laboratory to the clinician who ordered it [14,15]. The use of electronic health records, especially those that utilize such features as automated communication of abnormal results from laboratories to clinicians, can potentially improve follow-up of abnormal cancer screens [16-19]. Electronically "alerting" the ordering provider about an abnormal test result such as positive FOBT can improve the availability of vital information at the point of care [18]. As one of several multifaceted interventions to improve follow-up of positive FOBTs, our institution previously implemented standard operating procedures for the electronic health record's test result communication system [19], including the transmission of a mandatory alert to the patient's clinician for every positive FOBT result. This procedure was expected to reduce breakdowns in communication between the laboratory and clinicians.

A significant increase in timely responses to positive FOBT notifications (defined as a documented response within two weeks of the test) followed implementation of this and several related interventions. However, we found that 40% of automated notifications of FOBT results had no documented response by a treating clinician at two weeks even though all of the patients with these positive FOBT tests were eligible to receive a diagnostic colonoscopy. Our research question was to determine why a large number of FOBT alerts were not followed by clinician response at 2-weeks and to investigate if technical and workflow-related aspects of automated communication in the electronic health record were responsible. We also sought to implement and evaluate a potential solution to the issue(s) we identified.

11.1.2 METHODS

The study was conducted at the Michael E. DeBakey Veterans Affairs Medical Center and its satellite clinics and was approved by the local institutional review board. We used a mixed methods approach analogous to root cause analysis [20] to uncover potential workflow or technical reasons for lack of clinician response to positive FOBT results. We conducted eleven semi-structured interviews with key informants from the laboratory, primary care, and information technology sections to gather details related to FOBT alert generation, transmission, and receipt. Concurrently, we obtained quantitative data to track the alert receipt and follow-up actions by providers.

Clinicians in the VA health care system receive notifications of high-priority information such as abnormal test results in a "View Alert" window of the electronic health record. To understand the technical issues surrounding electronic communication, we analyzed and mapped the associated system-level processes involved. We discovered that the FOBT alert communication system is driven by an underlying component of the electronic health record that continually monitors test order and result entry. Alerts are automatically generated and recipients selected based on a set of predefined rules and parameters. For instance, entry of a test result such as positive FOBT (which was pre-determined to be a high-priority test result) will generate an automated notification to one or more clinicians. The proper recipients for this notification are chosen based on the setting of certain system parameters. After delivery to recipients, alerts stay active in the clinician's inbox up to two weeks, or until acknowledged.

Using the alert tracking system of the electronic health record, we identified all positive FOBT alerts transmitted daily during our study period. Approximately three weeks after alert generation, a trained physician reviewed the electronic health record for evidence of timely FOBT follow-up using a standardized data collection form that had been pilot tested in previous work [19]. Any documented response to the FOBT, such as colonoscopy referral, patient notification, or mention of exclusion criteria for colonoscopy, was considered timely follow-up. If no follow-up action was documented, an additional investigator confirmed the findings and called the ordering clinician (usually the primary care practitioner-PCP). If the clinician gave convincing information to support any undocumented actions, we considered this response as evidence of timely follow-up as well. We also recorded clinicians' comments and actions.

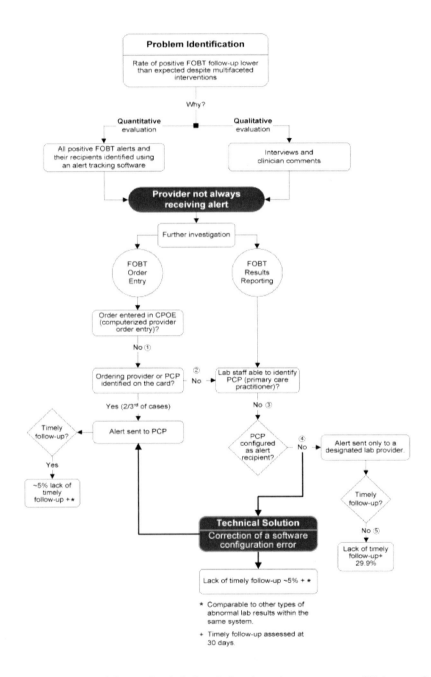

FIGURE 11.1.1: Workflow and technical analysis to investigate root cause of high rates of non-response to positive FOBT (fecal occult blood test). Contributing steps 1-5 identified.

Assessment of Diagnostic Test Result Reporting and Follow-Up 305

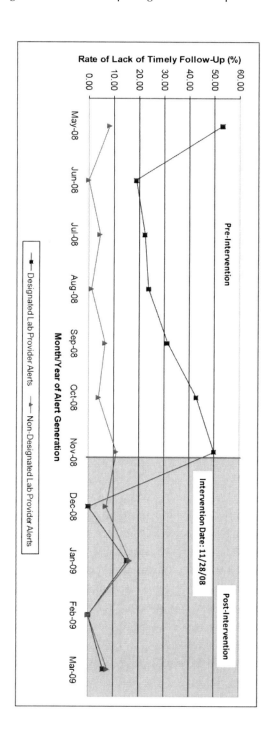

FIGURE 11.1.2: Follow Up of Positive Fecal Occult Blood Tests Pre- and Post-Intervention to Correct Software Configuration Error.

Following a trail of positive FOBTs that were found to have lack of timely follow-up, we used purposeful sampling and snowball techniques to identify our study subjects [21]. We initially purposefully sampled three PCPs whose FOBT results were found on chart review to have not received follow-up. Information from these PCPs led to further interviews with 1 additional provider (a subspecialist) and representatives that were involved with FOBT performance (laboratory personnel) and FOBT reporting (laboratory and Information Technology personnel). Additionally, 3 institutional representatives from leadership and administration that oversee workflow related to FOBT results were also interviewed.

We gathered data from interviews, clinicians' comments and FOBT tracking to uncover reasons for lack of timely follow-up (Figure 11.1.1). Using themes generated from this data, we found that five steps contributed to the problem, one of which was a software configuration error in the alert communication system. The latter step was the most significant one in the final common pathway and most amenable to a systems based intervention to improve communication and follow-up of positive FOBTs. To assess effect of the intervention we implemented, we compared rates of follow-up of positive FOBTs pre- and post-intervention using a Z test of two proportions.

11.1.3 RESULTS

11.1.3.1 PROBLEM IDENTIFICATION

Data from PCP interviews and reported comments suggested that PCPs were not receiving positive FOBT alerts consistently, leading us to further investigate the processes associated with FOBT alert generation. Workflow analysis revealed that a large number of patients who are given FOBT cards never return them to the lab for processing, and therefore an order for the test (through a computerized order-entry system) is only placed upon receipt of the card by the lab. However, in the absence of a provider-generated computerized order, the ordering provider is not easily identifiable unless written on the card. Because lab technicians use a different order-entry system, it is difficult for them to identify the ordering provider (and hence the primary recipient of the alert).

Further analysis of alert generation revealed that, regardless of an identifiable ordering provider in the system, the alert management software is designed to communicate all high priority alerts to the PCP as long as a primary alert recipient is identified. We discovered that in positive FOBTs where the ordering provider was not identified, a laboratory staff member served as the designated "ordering" provider, i.e. the primary recipient for the alert. This workaround (nonstandard procedures typically used because of deficiencies in system or workflow design) [22] was intended to enable the completion of the order and subsequent transmission of any alert generated to the patient's PCP; a fail-safe or safety-net mechanism designed to prevent loss of FOBT follow-up.

However, additional technical analysis of the alert tracking data revealed that in all cases where the designated lab provider was alerted as the "ordering" provider, there was no concomitant alert transmission to the PCP. Thus, only the lab provider was receiving the positive FOBT results and had no knowledge of this technical problem. We categorized such alerts as designated lab provider alerts and found that lack of timely follow-up was much more prevalent in this subgroup of alerts (29.9% vs. 4.5% in non-designated lab provider alerts).

11.1.3.2 INTERVENTION

We surmised that a lack of PCP awareness (in over a third of cases with positive FOBTs) contributed substantially to the prevalence of FOBT results with no documented follow-up, and that a software configuration error was the root of the problem. Once the electronic health record determines the need to generate an alert, proper recipients are selected based on their relationship to the patient (i.e., ordering provider, PCP, etc.). We found an improper configuration of the parameter that defines these default recipients, such that the PCP was not selected as a recipient for designated lab provider alerts (i.e. when PCPs were not listed as ordering providers). However, we could not determine when and how this error occurred in the system. Nevertheless, we posited that a problem-specific fix of this incorrect software configuration would reduce the risk of loss of follow-up for these alerts. The solution to this problem, an addition of a code to link patients to their PCP for tests ordered by others, was implemented on November 28, 2008 (date of intervention).

11.1.3.3 EVALUATION

We reviewed 360 alerts (117 designated lab provider alerts) pre-intervention and 130 alerts (55 designated lab provider alerts) post-intervention. Figure 11.1.2 shows the monthly prevalence of designated lab provider and non- designated lab provider alerts without timely follow-up pre- and post-intervention. Pre-intervention, lack of timely follow-up was observed for 29.9% of the designated lab provider alerts and 4.5% of non- designated lab provider alerts group. However, in the time period following the intervention, the percentage of designated lab provider alerts without timely follow-up decreased to 5.4% ($p < 0.01$) and was not statistically significantly different from that of non- designated lab provider alerts (6.6%; $p = 0.9$). This rate decrease occurred immediately following the intervention and remained stable (i.e. lower than pre-intervention levels) in the subsequent four months (Figure 11.1.2). Post-intervention tracking data confirmed that alerts assigned to the designated lab provider were now also being transmitted to the patient's PCP.

11.1.4 DISCUSSION

We investigated reasons why follow-up actions on a large proportion of positive FOBT results that needed a diagnostic colonoscopy were not documented by clinicians despite the use of a system to electronically communicate positive results. In addition to order-entry workarounds in the electronic health record, we discovered that the communication system intended to alert PCPs of positive FOBT results was not configured correctly, leading to certain situations in which PCPs never received the test result. Upon correction of the software configuration error, the percentage of positive FOBT results lacking follow-up were dramatically reduced. Although the rate did not drop to zero, it was comparable to the rate of lack of timely follow-up we found for other types of non-life threatening, high-priority lab notifications in the same system [23]. Our findings suggest that communication of cancer-related test results in the electronic health record must be monitored to avoid compromising the promise of cancer screening programs.

Of the over 800 patients each year who have positive FOBTs at our institution, about 10-15% of them are eventually diagnosed with some form of colon disease (including cancer). None of the patients in our study had any delay in cancer diagnosis or related harm. Although it is possible that follow-up may have occurred beyond our 30 day "timely response" window had we not intervened, previous work suggests that many of these findings would ultimately never be followed-up [10,11,19]. Thus, our seemingly small intervention could potentially have a large impact on decreasing time to referral for colonoscopy, thereby reducing the risk of a missed or delayed diagnosis of colorectal cancer, a common reason for ambulatory malpractice claims [24-26]. Previous literature has highlighted the need for system-based interventions to improve follow-up of positive cancer screens and our study is one of few that contributes to this body of knowledge [6].

Our findings also highlight how electronic health record use can have dramatic effects on follow-up care of patients. Electronic health records have potential to address the fragmented and discontinuous care that usually characterizes care in the outpatient setting. Critical information flow between different practitioners, settings and systems of care is essential to high quality care. Through good decision support systems, transmission of information to the right provider at the right time is within the reach of integrated electronic health records. However, as we find, electronic health record use must take into account the effect electronic communication will have on workflow and vice versa. Not doing this correctly would lead to circumstances that reduce the situational awareness of providers and perhaps other unintended adverse effects.

A limitation of our study was a lack of comparable data from other VA or non-VA facilities. However, our work illustrates how electronic test result communication systems are susceptible to errors that may limit their intended outcomes. Furthermore, it should be noted that other VA investigators [10,11] have demonstrated high rates of lack of positive FOBT follow-up, so it is possible that this problem exists at other VA sites. We are currently investigating whether this problem exists in other VA facilities or if this was an isolated event. Additionally, in this study we did not address many other systems issues that should be considered to address follow-up of abnormal test results in addition to provider, technology and work-flow. In our work, we are now using a socio-technical model that accounts for many other systems issues beyond the responsible provider, including the role of organizations and policies and procedures to address monitoring of abnormal test results [27]. For instance, an institutional policy that all FOBTs are ordered through computerized order entry would be another intervention to address this area. In our future work, we will propose multifaceted solutions to address the many complex issues related to abnormal test result follow-up.

Although electronic health records likely offer many benefits over paper-based systems for improving communication of abnormal cancer screening results [17], our findings highlight the need to account for inherent complexities of clinical practice. This complexity may introduce circumstances requiring special attention to EHR workflow to prevent loss of follow-up of important clinical information. In our setting, several workarounds of the FOBT ordering and reporting process resulted in disruption of the normal electronic health record workflow, creating a reliance on a secondary PCP notification system, which was not functioning as intended. The challenge of recognizing these complexities and their effects underscores the need for continuous monitoring of key electronic health record features that may impact safety. The work described here was a direct result of quality assurance work that is highly regarded in the VA health care system. Other institutions could use our methods to track the effectiveness of electronic communication. However, quality monitoring procedures such as used by the VA to ensure system safety must also be used to identify red flags that would lead to similar future investigations. Without the safeguards used by the VA, the problems related to test result communication may go undetected.

Health care systems should aim to achieve a high reliability for tracking delivery of abnormal cancer screening results. An example viewed as an ideal model for tracking systems is that of FedEx, which is considered to have 99.6% tracking reliability for its packages [28]. To achieve such high tracking reliability would not only require implementation of comprehensive technology-based systems for communication, but also formal policies and procedures regarding their use [28]. The test result communication system evaluated in our study addressed several criteria [28] for effective critical results reporting systems, such as computerized tracking and back-up procedures. However, to achieve tracking comparable to other industries, cancer screening programs should continuously monitor and oversee the timely delivery of positive cancer screening results to the right clinicians. For example, we recommend that cancer screening programs using electronic health record systems should develop and monitor multiple metrics of performance of automated communication processes. Failure to implement such monitoring systems could lead to sub-optimal screening success, which may otherwise be difficult if not impossible to trace.

11.1.5 CONCLUSION

In conclusion, we believe that electronic health records are beneficial in communicating abnormal cancer screening results to clinicians and will improve their follow-up care; however, we cannot assume that electronic communication is always working exactly as expected especially when workarounds are used. To achieve the most benefits of cancer screening programs, robust monitoring systems are necessary in electronic health record systems to ensure that abnormal cancer screening results are being delivered to the correct providers in a timely manner.

REFERENCES

1. Yabroff K, Washington KS, Leader A, Neilson E, Mandelblatt J: Is the Promise of Cancer-Screening Programs Being Compromised? Quality of Follow-Up Care after Abnormal Screening Results. Med Care Res Rev 2003, 60:294-331.
2. Baig N, Myers RE, Turner BJ, et al.: Physician-reported reasons for limited follow-up of patients with a positive fecal occult blood test screening result. The American Journal of Gastroenterology 2003, 98:2078-2081.
3. Levin B, Hess K, Johnson C: Screening for colorectal cancer. A comparison of 3 fecal occult blood tests. Arch Intern Med 1997, 157:970-976.
4. Morris JB, Stellato TA, Guy BB, Gordon NH, Berger NA: A critical analysis of the largest reported mass fecal occult blood screening program in the United States. Am J Surg 1991, 161:101-105.

5. Burack RC, Simon MS, Stano M, George J, Coombs J: Follow-up among women with an abnormal mammogram in an HMO: is it complete, timely, and efficient? Am J Manag Care 2000, 6:1102-1113.
6. Bastani R, Yabroff KR, Myers RE, Glenn B: Interventions to improve follow-up of abnormal findings in cancer screening. Cancer 2004, 101:1188-1200.
7. Mandel JS, Church TR, Bond JH, et al.: The effect of fecal occult-blood screening on the incidence of colorectal cancer. N Engl J Med 2000, 343:1603-1607.
8. Mandel JS, Bond JH, Church TR, et al.: Reducing mortality from colorectal cancer by screening for fecal occult blood. Minnesota Colon Cancer Control Study. N Engl J Med 1993, 328:1365-1371.
9. Kronborg O, Jorgensen OD, Fenger C, Rasmussen M: Randomized study of biennial screening with a faecal occult blood test: results after nine screening rounds. Scand J Gastroenterol 2004, 39:846-851.
10. Etzioni D, Yano E, Rubenstein L, et al.: Measuring the Quality of Colorectal Cancer Screening: The Importance of Follow-Up. Diseases of the Colon & Rectum 2006, 49:1002-1010.
11. Fisher DA, Jeffreys A, Coffman CJ, Fasanella K: Barriers to full colon evaluation for a positive fecal occult blood test. Cancer Epidemiol Biomarkers Prev 2006, 15:1232-1235.
12. Myers RE, Hyslop T, Gerrity M, et al.: Physician Intention to Recommend Complete Diagnostic Evaluation in Colorectal Cancer Screening. Cancer Epidemiol Biomarkers Prev 1999, 8:587-593.
13. Myers RE, Turner B, Weinberg D, et al.: Impact of a physician-oriented intervention on follow-up in colorectal cancer screening. Prev Med 2004, 38:375-381.
14. Jimbo M, Myers RE, Meyer B, et al.: Reasons Patients With a Positive Fecal Occult Blood Test Result Do Not Undergo Complete Diagnostic Evaluation. Ann Fam Med 2009, 7:11-16.
15. Wahls T: Diagnostic errors and abnormal diagnostic tests lost to follow-up: a source of needless waste and delay to treatment. J Ambul Care Manage 2007, 30:338-343.
16. Poon EG, Wang SJ, Gandhi TK, Bates DW, Kuperman GJ: Design and implementation of a comprehensive outpatient Results Manager. J Biomed Inform 2003, 36:80-91.
17. Singh H, Arora HS, Vij MS, Rao R, Khan M, Petersen LA: Communication outcomes of critical imaging results in a computerized notification system. J Am Med Inform Assoc 2007, 14:459-466.
18. Singh H, Naik A, Rao R, Petersen L: Reducing Diagnostic Errors Through Effective Communication: Harnessing the Power of Information Technology. Journal of General Internal Medicine 2008, 23:489-494.
19. Singh H, Kadiyala H, Bhagwath G, et al.: Using a multifaceted approach to improve the follow-up of positive fecal occult blood test results. Am J Gastroenterol 2009, 104:942-952.
20. Bagian JP, Gosbee J, Lee CZ, Williams L, McKnight SD, Mannos DM: The Veterans Affairs root cause analysis system in action. Jt Comm J Qual Improv 2002, 28:531-545.
21. Ash JS, Smith AC, Stavri PZ: Performing subjectivist studies in the qualitative traditions responsive to users. In Evaluation Methods in Biomedical Informatics. 2nd edition. Edited by Friedman CP, Wyatt JC. Springer New York; 2006:267-300.

22. Koppel R, Wetterneck T, Telles JL, Karsh BT: Workarounds to Barcode Medication Administration Systems: Their Occurrences, Causes, and Threats to Patient Safety. J Am Med Inform Assoc 2008, 15:408-423.
23. Singh H, Thomas E, Petersen LA: Automated Notification of Laboratory Test Results in an Electronic Health Record: Do Any Safety Concerns Remain? American Journal of Medicine, in press.
24. Gandhi TK, Kachalia A, Thomas EJ, et al.: Missed and delayed diagnoses in the ambulatory setting: A study of closed malpractice claims. Ann Intern Med 2006, 145:488-496.
25. Phillips RL Jr, Bartholomew LA, Dovey SM, Fryer GE Jr, Miyoshi TJ, Green LA: Learning from malpractice claims about negligent, adverse events in primary care in the United States. Qual Saf Health Care 2004, 13:121-126.
26. Singh H, Sethi S, Raber M, Petersen LA: Errors in cancer diagnosis: current understanding and future directions. J Clin Oncol 2007, 25:5009-5018.
27. Sittig DF, Singh H: Eight Rights of Safe Electronic Health Record Use. JAMA 2009, 302:1111-1113.
28. Bates DW, Leape LL: Doing better with critical test results. Jt Comm J Qual Patient Saf 2005, 31:66-67.

IMPROVING TEST RESULT FOLLOW-UP THROUGH ELECTRONIC HEALTH RECORDS REQUIRES MORE THAN JUST AN ALERT

Dean F. Sittig and Hardeep Singh

A recent American Medical Association report highlighted failures in communication of abnormal test results as an important but understudied facet of improving safety in ambulatory care. [1] Because many outpatient test results are not life-threatening and don't require verbal communication, health information technology (IT) has potential to reliably transmit result information in the fragmented outpatient setting. Thus, few will disagree

Springer and the Journal of General Internal Medicine, *27,10, 2012, pp 1235-1237, Improving Test Result Follow-up through Electronic Health Records Requires More than Just an Alert, Sittig DF and Singh H, © Society of General Internal Medicine 2012. With kind permission from Springer Science and Business Media.*

that communication of abnormal test results is an obvious context where advantages of health IT will be observed.

In this issue of JGIM, Callen et al. report the results of a timely systematic review of 19 studies that documented quantitative evidence of test results not followed up in ambulatory settings. [2] They found wide variation in abnormal results lacking follow-up: 7 % to 62 % for laboratory, and 1 % to 36 % for imaging tests. Although evidence of the effectiveness of electronic test management systems was limited, there was a general trend towards improved follow-up in electronic systems.

In another article in this issue, El-Kareh et al. discuss the results of a randomized controlled trial that put electronic communication to the test. The authors studied the effectiveness of sending microbiology test result alerts via a secure, internal e-mail system to clinicians when results were finalized post-discharge. [3] They found better documented evidence of appropriate follow-up within 3 days in the intervention group (28 % vs. 13 % in controls). Neither group's laboratory follow-up rate was particularly encouraging.

On the bright side, both studies used distinctly different research approaches to reach similar conclusions, i.e., application of information and communication technologies, such as electronic health records (EHRs) with alerting capability, can increase the likelihood of appropriate test result follow-up. In paper-based systems, evaluating evidence of follow-up is itself challenging. On the other hand, both articles remind us that using EHR-based technology by itself does not entirely solve the problem of failure to follow up test results. Callen et al., as well as others, have made a strong case for addressing these failures based on safety implications. Additionally, Stage 2 meaningful EHR use (slated for implementation in 2014) includes laboratory test result reporting criteria. Time is now ripe for novel approaches to understand and improve this complex problem.

The use of technology in the complex healthcare system must take into context the social environment where technology is embedded. For example, Callen et al. found lack of clear policies and procedures in relation to test result follow-up. We previously identified ambiguity of responsibility for test result follow-up to be a key factor in failure to follow up abnormal results. [4] Several EHRs now use asynchronous alert notifications to transmit results, but providers often receive many other types of notifications in their electronic in-box. We found that primary care providers (PCPs) receive a mean of 57 alerts a day in an integrated delivery system's EHR, all with new information they need to process and/or act upon. [5] Important information about abnormal results might get buried among other alerts.

To help understand the complexities involved with electronic communication of test results and facilitate progress in developing multifaceted solutions, a "sociotechnical" approach is needed. In our work, we use an eight-dimension sociotechnical model to study both problems and solutions related to safe and effective EHR implementation and use. [6] In the sections below, we illustrate the usefulness of this model by discussing each of its eight dimensions, as applied to issues raised by the two studies. We also take the liberty of making several recommendations that might be useful to reduce failures in test result follow-up in EHR-based systems.

1. *Hardware/Software:* To maintain superiority over paper, EHRs must be configured to ensure that results are reported to the correct provider in a timely fashion. Thus, all test orders should be placed via a Computer-based Provider Order Entry (CPOE) system. Orders should be transmitted in a coded format to the entity performing the test, and the transmission should occur via a two-way system-to-system interface that can send orders and receive results. Otherwise, results might not make it back into the EHR in a form that allows clinical decision support interventions (e.g., alert for abnormal creatinine will not fire while entering an order for metformin).

2. *Clinical Content:* Results should be stored as structured/coded data to facilitate reporting and tracking of results. This feature enabled El-Kareh and colleagues to extract results from the EHR, and can facilitate result-tracking functions. Institutions must also define standardized result categories and definitions (e.g., critical, normal, etc.) to facilitate prioritization and reporting. For instance, certain levels of abnormalities can be flagged in the EHR for more immediate action based on urgency. Care should be taken to avoid flagging borderline or clinically insignificant results as urgent.

3. *User Interface:* A poor user interface can lead providers to miss critical information. EHRs should have result review screens that ensure that all critical information is displayed on one screen (i.e., no scrolling is required) and all columns are sufficiently wide to allow users to see all pertinent information. In addition, users should be able to sort, or filter, results by date, type, patient, or urgency. [7]
4. *Personnel:* Providers should be trained to process their alerts in a timely manner and document follow-up and communication of results to patients in the EHR. Poor documentation is widely prevalent and might be one reason to explain the low follow-up rates reported in El-Kareh et al.'s study.
5. *Workflow/Communication:* Institutions must avoid partial use of EHRs for test result management (i.e., results or notes, but not both, available electronically [8]), because this leads to a higher risk of test result follow-up failures. Workflows related to certain high-risk areas (tests ordered by residents, part-time physicians, emergency deparment physicians; send-out tests; and post-discharge results) must be well-defined. This process should include creation of back-up procedures (including use of surrogates) and fail-safe escalation systems to safeguard against results "falling through the cracks". To what extent this was done, if at all, in the El-Kareh study is unclear, and thus a seemingly straightforward technological intervention might not have reached its full potential. Additionally, practices must create robust processes to send both normal and abnormal test results to patients. In the Veterans' Health Administration (VA), providers can generate letters though EHR templates, which are then sent to patients through centralized mailing facilities. Many institutions use web-based portals to make results accessible to patients, and some directly notify patients bypassing provider review. Whether the latter approach reduces follow-up failures is unclear. [9]
6. *Internal Organizational Policies, Procedures, Culture and Environment:* Responsibility for test result follow-up is an under-recognized and underemphasized contributory factor in follow-up failures. Responsibility should always be clear, and can be delegated to someone as long as that procedure is clear to both parties. It

is unclear how physicians in the El-Kareh study perceived their test result follow-up responsibilities; many that did not answer the survey or follow up appropriately might have attributed this responsibility to someone else (e.g., inpatient physician thought that the PCP was responsible post-discharge). We also recommend that all institutions/clinics should have an annually updated, written policy on all aspects of test result management (e.g., provider notification, patient notification, follow-up responsibilities). [10] This document should define processes and procedures for test result communication, including which results are critical and need verbal communication. Institutions should also maintain updated contact information for all providers and patients. Some of the e-mail alerts sent by the investigators might not have reached the study physicians.

7. *External Rules and Regulations:* In 2009, the VA released a policy directive requiring communication of all test results to patients within 14 calendar days after the test result is available to the ordering practitioner. To the best of our knowledge, there are no other federal or state policies giving guidance on definitions and measurement of timeliness of test result follow-up.

8. *Measurement and Monitoring:* The VA is now instituting a measurement system for test results follow-up, and we encourage other institutions to do the same. Logs of test result values, alerts, and provider acknowledgment of alert receipt (results review) could be used for this purpose. However, acknowledgment of a test result receipt does not guarantee that the follow-up action has taken place;4 alternative measurement systems should be in place to monitor test result follow-up.

11.2.1 CONCLUSIONS

Timely follow-up of test results remains a problem even in institutions that use state-of-the-art EHR systems to alert providers about abnormalities. We believe that solutions to these problems will require a comprehensive sociotechnical approach beyond just implementing alerts and other technologies to improve information transfer. Both research reports in this issue of the journal convincingly illustrate this point.

REFERENCES

1. Lorincz CY, Drazen E, Sokol PE, Neerukonda KV, Metzger J, Toepp MC, Maul L, Classen DC, Wynia MK. Research in Ambulatory Patient Safety 2000–2010: A 10-Year Review. American Medical Association, Chicago IL 2011. Available at: www.ama-assn.org/go/patientsafety.
2. Callen JL, Westbrook JI, Georgiou A, Li J. Failure to Follow-Up Test Results for Ambulatory Patients: A Systematic Review. J Gen Intern Med. 2012 doi:10.1007/s11606-011-1949-5.
3. El-Kareh R, Roy C, Williams DH, Poon EG. Impact of Automated Alerts on Follow-Up of Post-Discharge Microbiology Results: A Cluster Randomized Controlled Trial. J Gen Intern Med. 2012 doi:10.1007/s11606-012-1986-8.
4. Singh H, Thomas EJ, Mani S, Sittig D, Arora H, Espadas D, Khan MM, Petersen LA. Timely follow-up of abnormal diagnostic imaging test results in an outpatient setting: are electronic medical records achieving their potential? Arch Intern Med. 2009;169(17):1578–86. doi: 10.1001/archinternmed.2009.263.
5. Murphy DR, Reis B, Sittig DF, Singh H. Notifications received by primary care practitioners in electronic health records: a taxonomy and time analysis. Am J Med. 2012 Feb;125(2):209.e1-7.
6. Sittig DF, Singh H. A new sociotechnical model for studying health information technology in complex adaptive healthcare systems. Qual Saf Health Care. 2010;19(Suppl 3):i68–74. doi: 10.1136/qshc.2010.042085.
7. Singh H, Wilson L, Reis B, Sawhney MK, Espadas D, Sittig DF. Ten strategies to improve management of abnormal test result alerts in the electronic health record. J Patient Saf. 2010;6(2):121–3. doi: 10.1097/PTS.0b013e3181ddf652.
8. Casalino LP, Dunham D, Chin MH, Bielang R, Kistner EO, Karrison TG, Ong MK, Sarkar U, McLaughlin MA, Meltzer DO. Frequency of failure to inform patients of clinically significant outpatient test results. Arch Intern Med. 2009;169(12):1123–9. doi: 10.1001/archinternmed.2009.130.
9. Davis Giardina T, Singh H. Should patients get direct access to their laboratory test results? An answer with many questions. JAMA. 2011 Dec 14;306(22):2502-3.
10. Singh H, Vij MS. Eight recommendations for policies for communicating abnormal test results. Jt Comm J Qual Patient Saf. 2010;36(5):226–32.

TEN STRATEGIES TO IMPROVE MANAGEMENT OF ABNORMAL TEST RESULT ALERTS IN THE ELECTRONIC HEALTH RECORD

Hardeep Singh, Lindsey Wilson, Brian Reis, Mona K. Sawhney, Donna Espadas, and Dean F. Sittig

Missed abnormal test results are a significant patient safety problem, especially in the outpatient setting. Failure to communicate and follow up on abnormal diagnostic test results can lead to diagnostic errors, adverse events, and liability claims. [1–4] Automated alert notification systems integrated within electronic health records (EHRs) offer a potential solution. [5,6] For instance, communication of abnormal clinical information through "alerts" (computerized notifications of significantly abnormal or critical test results) can potentially facilitate rapid review of patient information. [7] The Computerized Patient Record System (CPRS), an integrated EHR used at all Veterans Affairs (VA) facilities, uses an automated notification system (the View Alert system) to communicate abnormal diagnostic test results (Figure 11.3.1). Despite this automated notification system, we recently found that 7% of abnormal outpatient laboratory results and 8% of abnormal imaging results lacked follow-up within 30 days. [8, 9] Therefore, electronic alerts do not eliminate the problem of missed results. We also found that clinicians did not acknowledge 18% of diagnostic imaging alerts and 10% of diagnostic lab alerts. Some clinicians received an overwhelming number of alerts (e.g., > 50 per day), some of which they never reviewed. Many clinicians had inconsistent knowledge of specific features in the EHR to help manage alerts.

Improving critical test result reporting is a national patient safety goal of the Joint Commission. [10] Additionally, the VA recently released a directive emphasizing timeliness of test result communication to practitioners and patients and further recommended that each VA facility address

---

Ten Strategies to Improve Management of Abnormal Test Result Alerts in the Electronic Health Record. Singh H, Wilson L, Reis B, Sawhney MK, Espadas D, and Sittig DF. Journal of Patient Safety *6,2 (2011). Reprinted with permission from Wolters Kluwer.*

ordering and reporting test results. [11] Based on our ongoing quantitative and qualitative evaluation work, we have identified ten strategies that clinicians can use immediately to improve their management of automated notifications related to abnormal test results. We identified these strategies on the basis of two chart review studies, [9,12] a focus group study, [13] and in-depth task analysis sessions [14] that we conducted over the course of a 2-year project funded by the VA National Center for Patient Safety. Subsequently, we obtained informal feedback from numerous primary care physicians who agreed that adoption of these strategies could help them manage alerts more reliably and effectively. Consistent with our recently proposed model for safe EHR use, [15] the strategies are divided into three groups: clinician (user) centered, human-computer interface centered and communication and workflow centered.

11.3.1 CLINICIAN (USER) CENTERED

Clinician centered strategies generally require additional user training. They include the following.

11.3.1.1 ADJUSTING VOLUME OF NOTIFICATIONS ACCORDING TO CLINICIAN PREFERENCES

Clinicians receive several types of notifications, not just abnormal test results. Many clinicians do not realize that certain non-mandatory notifications can be turned on or off. For instance, if they don't believe particular types of notifications are useful in their practice, they may decide to turn them off to reduce information overload and alert fatigue. Similarly, new users may expect to receive certain types of notifications, but these may have been turned off in the EHR by default. Therefore, clinicians must use the notification menu in the EHR to customize their notifications according to their preferences. Conversely, the EHR should allow system administrators to identify some alerts as "mandatory" so that they cannot be turned off.

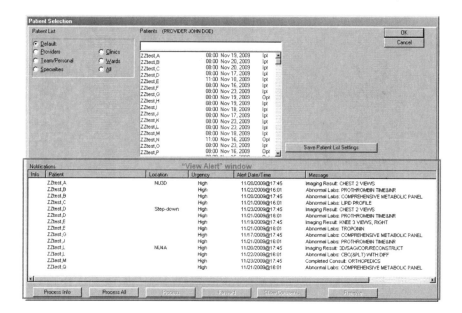

FIGURE 11.3.1: The View Alert Notification window of the VA's electronic health record

11.3.1.2 PREVENT LOSING TRACK OF NOTIFICATIONS

Once acknowledged, notifications may not always stay in the clinician's inbox. Thus, if the clinician is interrupted while processing an alert, the information may be lost. CPRS offers a "Renew Alert" feature to prevent the alert from disappearing (if, for instance, the clinician is called out of the office while processing an alert). Even though this feature currently exists in the software, and clinicians would like to have the ability to "save" alerts, we found most clinicians had no knowledge of this feature.

11.3.1.3 MAKE THE PROCESSING SOFTWARE ("PROCESS ALL") FEATURE WORK

In CPRS, the "Process All" feature allows clinicians to process alerts one after the other without returning to the View Alert window. This may increase the efficiency of processing. Using this feature does not imply clinicians have to process all alerts at once; it may still be done piecemeal.

11.3.1.4 CREATE A STRATEGY TO PRIORITIZE

Clinicians who are short on time should prioritize alerts based on urgency level. For instance, we recommend processing high priority and critical (abnormal imaging and laboratory) alerts first. We also recommend that clinicians avoid processing alerts altogether when they are particularly rushed. Setting aside a specific time of day best suited to the clinician's workflow may be required to manage the increasing number of alerts being generated.

11.3.2 HUMAN-COMPUTER INTERFACE CENTERED

The human-computer interface centered strategies require optimal use of the existing display features of the EHR screen. They include the following.

11.3.2.1 SORTING ALERT NOTIFICATIONS FOR EASIER PROCESSING

Depending on personal preferences, clinicians can sort alerts by any heading in the View Alert window (Figure 11.3.1). They may use the sorting feature to view higher priority alerts at the top of the list or to process similar kinds of alerts at the same time. For instance, sorting by location would generate a view of all inpatient alerts categorized by ward location, followed by all outpatient alerts. This technique can improve the efficiency of alert processing.

11.3.2.2 RESIZE THE NOTIFICATION WINDOW TO SEE MORE ALERTS

The size of the notification window can be adjusted to see more alerts on the screen. This may help to decrease the risk of missing an important alert. Particular care should be taken to ensure that the default window size does not force users to use a horizontal scroll bar to see all alert details. Horizontal scrolling could be a difficult skill for some users to master.

11.3.2.3 DON'T MISS CRITICAL INFORMATION IN THE EHR DUE TO SMALL COLUMN SIZE

When viewing information about patient diagnostic tests or labs in the EHR, clinicians may miss supplemental information due to narrow column width. For instance, in the CPRS imaging reports menu, abnormal reports are often hidden with only the "A" visible after the report. Resizing data columns will show the entire word "Abnormal." As described above in (6), default column widths should be set wide enough to see all text in the longest string.

11.3.3 COMMUNICATION AND WORKFLOW CENTERED

Patients are often seen by multiple providers, and test results must continue to be monitored and acted upon when the primary or ordering clinician is unavailable. Strategies that focus on communication and workflow aspects include:

11.3.3.1 USING THE "SURROGATE CLINICIAN" FEATURE IN THE EHR

While clinicians are away from their offices, especially for extended periods, they must identify another clinician to receive their alert notifications. Before designating surrogates, we recommend that clinicians minimize the volume of alerts that will be transmitted to their covering partner by temporarily customizing their notifications (e.g., turning off non-urgent alerts).

11.3.3.2 ENSURING ANOTHER CLINICIAN ALSO GETS A NOTIFICATION ABOUT A PARTICULAR TEST

A CPRS feature "Alert When Results" allows clinicians to notify an additional clinician when the results of an order are available. This is most useful when a specific clinician needs to be notified of a particular test on a one-time basis, or when residents want to hand off a potentially important test to their supervisors or alternates. When clinicians need to track a particular test order to ensure proper follow-up action (regardless of result), this is the most appropriate feature to use. This feature is of special significance because the Joint Commission has recently recommended that organizations distinguish "critical tests" and "critical results." [16] "Critical tests" always require rapid communication of the results, even if normal.

11.3.3.3 REMAIN "ALERT" ABOUT RESPONSIBILITY

When multiple clinicians receive notification of the same test, responsibility for follow-up may be ambiguous. For instance, a sub-specialist and a primary care provider who receive the same alert may both assume that the other will provide follow-up. While there is a need for other reliable procedures to assign message responsibility in the EHR, currently it is best to communicate verbally and clarify responsibility with the other clinician in cases where no follow-up actions have been documented.

Although our strategies have face validity, we recommend them with the caveat that we have no systematic evidence linking these strategies to improved outcomes. We plan to conduct such validation studies in the future. Another potential limitation is that we identified these strategies through research on the particular EHR used in VA health care facilities. However, because other EHR systems have similar notification capabilities and features, we believe that many of these strategies can be utilized by providers outside the VA. For example, some of the interface suggestions would be applicable to any EHR that uses basic Microsoft Windows user interface features (e.g., sorting a column by clicking on the heading; resizing a column or a window by moving the pointer to the border and dragging the column border to the desired width). Other strategies involve setting user preferences, which most EHRs allow in some form or another. Finally, several strategies describe key features or functions that have proven useful to healthcare providers. Users should work with their EHR training and support personnel or EHR vendors to propose additional desired functions in future versions of their applications.

In conclusion, we propose ten strategies to help providers better manage alert notifications related to abnormal test results in the EHR. Once these strategies have been implemented, health care organizations using automated EHR-based notification systems could potentially see fewer communication failures and improvements in test result follow-up.

REFERENCES

1. Bates DW, Leape LL. Doing better with critical test results. Jt Comm J Qual Patient Saf. 2005 February;31(2):66–67.
2. Gandhi TK. Fumbled handoffs: one dropped ball after another. Ann Intern Med. 2005 March 1;142(5):352–358.
3. Schiff GD. Introduction: Communicating critical test results. Jt Comm J Qual Patient Saf. 2005 February;31(2):63–65.
4. Wahls T. Diagnostic errors and abnormal diagnostic tests lost to follow-up: a source of needless waste and delay to treatment. J Ambul Care Manage. 2007 October;30(4):338–343.
5. Poon EG, Wang SJ, Gandhi TK, Bates DW, Kuperman GJ. Design and implementation of a comprehensive outpatient Results Manager. J Biomed Inform. 2003 February;36(1–2):80–91.
6. Singh H, Naik A, Rao R, Petersen L. Reducing Diagnostic Errors Through Effective Communication: Harnessing the Power of Information Technology. Journal of General Internal Medicine. 2008 April;23(4):489–494.
7. Singh H, Arora HS, Vij MS, Rao R, Khan M, Petersen LA. Communication outcomes of critical imaging results in a computerized notification system. J Am Med Inform Assoc. 2007;14(4):459–466.
8. Singh H, Thomas EJ, Mani S, et al. Timely Follow-Up of Abnormal Diagnostic Test Results: Are Electronic Medical Records Achieving Their Potential? 2009 Ref Type: Unpublished Work.
9. Singh H, Thomas E, Petersen LA. Automated Notification of Laboratory Test Results in an Electronic Health Record: Do Any Safety Concerns Remain? American Journal of Medicine. 2009 In press.
10. The Joint Commission announces the 2009 National Patient Safety Goals and requirements. Jt Comm Perspect. 2008 July;28(7):1–11. 1–15.
11. VHA Directive 2009-019: Ordering and Reporting Test Results. Veterans Health Administration. 2009. Mar 24, Available at: URL: http://www1.va.gov/vhapublications/ViewPublication.asp?pub_ID=1864.

12. Singh H, Thomas EJ, Mani S, et al. Timely follow-up of abnormal diagnostic imaging test results in an outpatient setting: are electronic medical records achieving their potential? Arch Intern Med. 2009 September 28;169(17):1578–1586.
13. Singh H, Hysong S, Esquivel A, Sawheny M, Wilson L, Sittig DF. A Human Factors Engineering Approach to Improve Safety of Test Result Management. Human Factors Engineering in Health Informatics Symposium; 11-12-2009; Sonoma, CA. Ref Type: Abstract.
14. Hysong S, Sawheny M, Wilson L, et al. Provider Management Strategies of Abnormal Test Result Alerts: A Cognitive Task Analysis. JAMIA. 2009 In press.
15. Sittig DF, Singh H. Eight Rights of Safe Electronic Health Record Use. JAMA. 2009 September 9;302(10):1111–1113.
16. 2009 Standards FAQs NPSG.02.03.01 Critical tests, results and values. The Joint Commission. 2008. Dec 9 [Accessed May 5, 2009]. Available at: URL: http://www.jointcommission.org/AccreditationPrograms/LaboratoryServices/Standards/09_FAQs/NPSG/Communication/NPSG.02.03.01/Critical_tests_results_values.htm.

SAFER SELF-ASSESSMENT GUIDE: TEST RESULT REPORTING AND FOLLOW-UP

SAFER Guides

OVERVIEW

Test results reporting practices, which include communication of test results from diagnostic services (e.g. radiology and laboratory) to referring clinical practitioners, are complex and vulnerable to breakdown. In the EHR-enabled healthcare environment, we rely upon technology to support and manage these processes. EHRs can incorporate standardized and automated features to improve the safety and effectiveness of how test results information is communicated. However, best practices for EHR-based results reporting are not well defined yet. This self-assessment guide is intended to increase awareness of practices to improve the safety of EHR-based results reporting and support proactive evaluation of selected high-risk areas. It helps you identify and evaluate where test result reporting and follow-up breakdowns may occur in your healthcare delivery system.

The guide focuses primarily on processes of getting test results back to the providers, i.e., when providers are notified electronically of the results and are then responsible for reviewing the results and follow-up with patients. Use of this assessment guide is intended to stimulate implementation of the recommended practices, as well as sustain those already present. When assessing EHRs at repeated intervals (e.g., initially, annually, and when changes are made), the guide can be used to establish a baseline for measuring the effect of interventions designed to improve the safety of test result communication. The guide is applicable to ambulatory physician practices and other outpatient settings as well as hospitals.

EXPECTATIONS

Healthcare professionals should use this assessment to identify and prioritize patient safety issues related to EHR-based test results reporting and appropriate patient follow-up. Prioritization could consider both the frequency and severity of a safety event that might result in absence of a specific practice. We anticipate this to be a useful tool in ongoing safety and risk management programs, allowing you to address new risks that arise in EHR-enabled healthcare settings and helping you take advantage of the safety benefits of EHRs. Please refer to the guide for additional information, including the specific risks and rationale addressed by the recommended practices, and examples of potential strategies to support the recommended practices.

INSTRUCTIONS

A multidisciplinary team should work together to complete this assessment and evaluate patient safety risks addressed by the recommended practices within the context of your healthcare delivery system. Different team members will be needed for input depending on what aspect of test results reporting is being assessed (see Practice Table). Input will be needed from a number of individuals, which could include IT managers (e.g., IT service provider, EHR vendor, CIO, or CMIO), risk managers, practice managers, patient safety and quality personnel, as well as key clinical stakeholders that are involved in ensuring the safety of diagnostic service reporting practices and patient follow-up (nurses, laboratorians, pathologists, radiologists, etc.). The following table can be used as a guide to facilitate multidisciplinary input and collaboration to achieve practice implementation:

Legend	Key Multidisciplinary Facilitators of Practice Implementation
C	Clinicians, support staff, and/or clinical administration (e.g., Medical Records and Risk Managers)
Dx	Diagnostic services, such as laboratory or radiology—could be local or remote
Ev	EHR vendor
IT	IT support staff, could be local or contracted. Responsible for maintaining the EHR and infrastructure
Rx	Pharmacy – could be local or remote

TEST RESULTS REPORTING AND FOLLOW-UP

Recommended Practices	Rationale for Practice or Risk Addressed	Examples of Potentially Useful Practices/Scenarios
Phase 1 – Make Health IT Safer		
Principle: Data Availability (EHRs and the data contained within them are available to authorized individuals where and when required to support healthcare delivery and business operations.)		
1. Test names, values, and interpretations for laboratory results are stored in the EHR as structured data using standardized nomenclature. [6,11,13-17] Dx, Ev, IT	Structured laboratory results facilitate EHR-based result reporting and tracking functions. [4] Structured data enable use of clinical decision support (CDS) that can avoid errors and optimize patient safety.	• Test result IDs (e.g., sodium, potassium) that are sent with LOINC codes are stored as coded data. [18] • Abnormal test result values and interpretations are defined and stored in a standardized, coded format (e.g., high/low sodium; critical potassium; positive/negative fecal occult blood test, etc.). [9]

Assessment of Diagnostic Test Result Reporting and Follow-Up

Recommended Practices	Rationale for Practice or Risk Addressed	Examples of Potentially Useful Practices/Scenarios
		• There is a process to handle paper-based test results that includes, at a minimum, the entry of a coded value into the EHR to indicate whether the result was normal or abnormal along with a scanned copy of the report in the EHR.
2. Predominantly text-based test reports (e.g., radiology or pathology reports) have a coded (e.g., abnormal/normal at a minimum) interpretation associated with them. Dx, Ev, IT	Coded results in structured fields facilitate EHR-based result reporting and tracking functions. [4]	• Imaging results are coded as abnormal using a structured code if there is a new or unexpected abnormality that requires follow-up. [19,20]
		• Mammography results are stored according to BI-RADS® criteria

Phase 1 – Make Health IT Safer

Principle: Data Integrity (Data are accurate, consistent and not lost, altered or created inappropriately.)

3. Functionality for ordering and reporting results is tested pre-and post-go live. C, Dx, Ev, IT	Problems related to system configuration errors leading to results routing logic errors are inevitable. With testing, many such unforeseen problems can be identified and addressed before they result in patient harm. Errors related to closed loop test order entry and results delivery are difficult to detect and can lead to delays in care.	• Efforts are made to proactively identify failure points related to EHR-enabled test results delivery.
		• Specifically designed testing scripts are used to identify points remediable points of vulnerability [21] in order to build systems that are more fault-tolerant.
		• Specific testing of routing logic, provider recipients, and configuration is performed to ensure accurate results delivery.
4. After system changes in components or applications related to CPOE and diagnostic services, the data and data presentation are reviewed to ensure accuracy and completeness. Dx, Ev, IT	System changes can unexpectedly affect the integrity of the data as it moves through organizations in ways that may not be recognized without proactive review.	• Organizations identify specific types of EHR system changes that impact CPOE and diagnostic services, such as application upgrades or changes to interfaces, and carefully review data integrity at all points where data is used.

Recommended Practices	Rationale for Practice or Risk Addressed	Examples of Potentially Useful Practices/Scenarios
		• Problems related to tables out of sync are identified with thorough testing
		• Error queues are used to monitor for proper system performance; results that cannot be automatically delivered are manually delivered.
		• Order entry and result reporting interfaces are tested after every change to the laboratory/imaging ordering catalog.

Phase 2 – Safer Application and Use of IT

Principle: Complete/Correct EHR Use (Correct system usage [i.e., features and functions used as designed, implemented, and tested] is required for mission-critical clinical and administrative processes throughout the organization.)

5. Orders for diagnostic tests are placed using CPOE and electronically transmitted to the diagnostic service provider (e.g., laboratory or radiology). [6,22,23] (MU) Dx, Ev, IT	A hybrid paper and electronic environment for test ordering is hazardous. CPOE can facilitate closed loop communication and results accessibility via the EHR, but only if the results are available in the system. Test results can be lost or missed if on paper, when clinicians have come to rely on the EHR.	• For common tests, there is a two-way system-to-system interface (i.e., for ordering, resulting, acknowledging, and cancelling orders) between the clinic/institution and the testing facility. [24] • Diagnostic tests that are not orderable through CPOE for any reason are promptly added to the system.
6. EHR is able to track the status of all orders and related procedures (e.g., specimen received and collected; test completed, reported, and acknowledged). [4] Dx, Ev, IT	• Tracking orders facilitates closed loop communication. This enables detection of problems regarding processing and delivery of test results.	• EHR can track whether the specimen was received, collected, test completed, results reported, and acknowledged. • Clinical practices where test result information is not yet fully integrated into the EHR use additional tracking strategies to enable follow-up. [25]
7. The ordering clinician is identifiable on all ordered tests and test reports, and, if another clinician is responsible for follow-up, that clinician is also identified in the EHR. [8] C, Ev, IT	• Clear identification of the ordering clinician facilitates closed loop communication. • Ambiguous responsibility increases the risk of follow-up failure.[4]	• Result routing systems supports delivery of results to the ordering provider. [5,9,11] • EHR supports assignment/transfer of responsibility for test order follow-up.

Assessment of Diagnostic Test Result Reporting and Follow-Up

Recommended Practices	Rationale for Practice or Risk Addressed	Examples of Potentially Useful Practices/Scenarios
8. When test results are amended, the change is clearly visible in the EHR and printed reports. [9] Dx, Ev, IT	Results that are subsequently changed carry a significant potential for delayed or wrong treatment based on outdated, incorrect results.	• Changed results are clearly flagged as such in the EHR (such as marked as "amended").
9. When test results are changed or amended, the ordering clinician and other clinicians responsible for follow-up are notified electronically. For clinically significant changes, the clinicians are also contacted directly. [26] C, Dx, Ev, IT	Results that are subsequently changed carry a significant potential for delayed or wrong treatment based on old (incorrect) result/interpretation.	The individual changing the results is responsible for notifying appropriate clinicians of those changes. Since electronic systems do not always ensure that a critical communication will be received and reviewed promptly, for clinically important changes to results appropriate clinicians are also contacted directly. • Policies and procedures ensure that changes in test results (and accompanying documentation) are effectively communicated to the appropriate clinicians responsible for patient care, including after the patient has transitioned to another setting of care
10. Send-out (or reference lab) tests are electronically tracked, and their results are incorporated into the EHR, with a coded test name, result value and interpretation. C, Dx, Ev, IT	Send-out tests are vulnerable to loss of follow-up.	• The EHR facilitates the tracking of "send-out" tests and provides a mechanism to allow clinicians or organizations to incorporate these results into the EHR and assign them to the correct patient. • Procedures exist to ensure that all test results, including those received from outside the institution through fax or mail, are properly incorporated into the EHR.
11. Written policies specify unambiguous responsibility for test result follow-up with a shared understanding among all involved in providing follow-up care [4,6,9,13,14,27,28] C, Dx	New workflows resulting from the introduction of EHRs can introduce new hazards related to miscommunication of responsibility for follow-up. Ambiguous responsibility increases the risk of follow-up failure.	• In the outpatient setting, ordering provider is responsible for follow-up unless he or she delegates this (e.g., covering provider). Delegation should be documented and accepted by the delegate.

Recommended Practices	Rationale for Practice or Risk Addressed	Examples of Potentially Useful Practices/Scenarios
		• Ordering clinicians in any setting assume responsibility for follow-up care, unless that responsibility is unambiguously transferred to another clinician, who accepts responsibility.
12. Workflows that are particularly vulnerable to mishandling of test results, especially critical ones, are identified, [29] and back-up procedures ensure test results are received by someone responsible for the affected patient's care. [6,26] C, Dx, Ev, IT	Lost or mishandled test results, especially critical ones, are a significant risk to patients, especially in situations with workflows particularly vulnerable to such failures, such as shift changes or transitions of care. [30]	• Situations that are vulnerable to test results follow-up failures are identified. These include handoffs between clinicians (such as between residents, part-time physicians, ER physicians, and hospitalists), and care transitions between clinical settings (such as between different units of a hospital, and between the hospital and home or a post-acute facility). In these situations, processes should be in place to ensure that test results are communicated to a clinician responsible for follow-up care
		• Life threatening results are notified through verbal means to ensure positive confirmation of receipt. [9]
		• Notifications that remain unacknowledged after a pre-specified number of days are forwarded (or escalated) to an alternate responsible provider. [31]
		• Diagnostic services should ensure that test results are communicated to a back-up provider in a timely fashion in the event that the primary provider is not available. The necessary timeliness is dependent on the significance of the test result. [32]
		• Institution maintains an updated contact list of all providers that practice in it and this list includes their coverage schedules. [8]

Assessment of Diagnostic Test Result Reporting and Follow-Up

Recommended Practices	Rationale for Practice or Risk Addressed	Examples of Potentially Useful Practices/Scenarios
		• Institution maintains a patient-provider link (e.g., patient's PCP is identified).
Phase 2 – Safer Application and Use of IT		
Principle: System Usability (All EHR features and functions required to manage the treatment, payment, and operations of the healthcare system are designed, developed, and implemented in such a way to minimize the potential for errors. In addition all information in the system must be clearly visible, understandable, and actionable to authorized users.)		
13. Results outside normal reference ranges (or determined to be abnormal) are flagged (presented in a visually distinct way). [6,9] Dx, Ev, IT	Although absence of flags does not necessarily mean the result is normal, flagging can reduce likelihood of missing abnormal or critical results.	• Abnormal results are flagged (e.g., bolded font, asterisk beside values, use of "H" or "L," different colors, etc.) or marked for better visualization in the EHR. • Color is not used as the only visual indicator of clinical significance. • Critical values are flagged in a distinct way from simply abnormal values
14. Display of results (e.g., numeric, text, graphic, image) should be easily accessible, clearly visible (and not easily overlooked), and understandable. Dx, Ev, IT	Missed or misunderstood test results as the result of a poorly designed human-computer interface are as dangerous to patients as lost or wrong results. Results visualization and display should maximize safety in order to ensure critical information isn't missed.	• Displays of test results undergo usability testing for the intended clinical users. • Information is displayed in columns that are sufficiently wide to allow review of all pertinent information (i.e., providers do not need to drag columns on the user interface to detect abnormalities). [1]1 • Multicomponent results are reported in one place (e.g., lupus anticoagulant has 2-3 subcomponents that may be individually positive or negative but should be reported together). • Result details are reported on one screen, eliminating the need for horizontal scrolling. For example, providers should not have to use additional scrolling (e.g., on to the "next page"), [6,11] to access critical information.

Recommended Practices	Rationale for Practice or Risk Addressed	Examples of Potentially Useful Practices/Scenarios
		• If the screen is not displaying the full message, there are salient indicators directing the user to the non-displayed remainder of the message (e.g., obvious scroll bars).
		• Most recent test results should be displayed first (e.g., either at the top of a row-based display or at the left-side on a columnar display) to ensure that clinicians are always aware of current data. [33]
15. Automated non-interruptive results notifications (also called "in-basket alerts" or flags) are limited to those that are clinically relevant in order to minimize "alert fatigue." [4,11,14,27,28,34,35] Dx, Ev, IT	• Information overload from too many alerts is associated with more missed test results. [36] • Results that are poorly displayed increase risk of misinterpretation or being overlooked completely.	• A multidisciplinary committee that includes frontline clinician decides which abnormal result alerts the providers are required (i.e., mandated) to receive and which ones clinicians can choose to suppress.
		• Outpatient clinicians have the option to receive results from their patients in the their electronic inboxes.
		• Notifications of a patient's results are batched (aggregated) by type and/or date to minimize the number of notifications.
		• Institution/clinic monitors providers' inbox, i.e., the total number of alert notifications sent to providers.
		• The institution/clinic provides workflow support to help a provider when the number of unread notifications in his or her inbox grows large.
16. Results notifications remain in the clinician inbox until a clinician action occurs to address them. [4,11,37] C, Ev, IT	If notifications drop off, providers can miss results.	Notifications remain in the inbox until a clinician signs them.

Assessment of Diagnostic Test Result Reporting and Follow-Up

Recommended Practices	Rationale for Practice or Risk Addressed	Examples of Potentially Useful Practices/Scenarios
17. There is an EHR-based process for clinicians to either assign surrogates [6,8,38] for reviewing notifications or to enable surrogates to look at the principal clinicians' inboxes. C, Ev, IT	Not using surrogate features and functions appropriately increases risk of loss of test result follow-up.	• If providers plan to be away, they assign a covering provider to whom the system can automatically forward test results. • Organizations have policies and procedures that establish expectations for timely review of test results and specifically address planned and unplanned absences
18. There are mechanisms to forward results and results notifications from one clinician to another. [11,27] C, Ev, IT	Notifications sometimes are sent to incorrect providers, and in this situation, this functionality allows providers to forward alerts to the correct person.	• In addition to automatic forwarding, such as when a clinician is on vacation, forwarding can be done under clinician control (e.g., when the notification is transmitted to the incorrect clinician). • Mechanisms are in place for tracking acknowledgment and acceptance of forwarded notifications
19. Summarization tools to trend and graph laboratory data are available in the EHR. Ev, IT	Displaying certain laboratory test results over time helps identify clinically relevant anomalies or trends. Summarization tools in the EHR improve visualization, interpretation, and accessibility of results.	• The EHR incorporates automated tools and reports that enable selected lab results to be easily graphed and displayed over time to view trends.
20. Test results can be sorted in the clinician's EHR inbox according to clinically relevant criteria (e.g., date/time, severity, hospital location, or patient). [6,11,26,28] Ev, IT	Clinicians need ways to prioritize results review so they can address the most pressing issues first and cope with information overload.39 Sorting also improves visualization and accessibility of results.	Results can be sorted according to important parameters such as date, type, urgency, patient, and location.
21. The EHR has the capability for the clinician to set reminders for future tasks to facilitate test result follow-up. [28,40] Ev, IT	The EHR can help clinicians' follow-up with patients regarding test results. Unless they set reminders for themselves, clinicians may forget about follow-up tasks they need to do.	Functionality to record a follow-up action due at a future date exists in the EHR.

Recommended Practices	Rationale for Practice or Risk Addressed	Examples of Potentially Useful Practices/Scenarios
Phase 3 – Leverage IT to Facilitate Oversight and Improvement of Patient Safety		
Principle: Safety Surveillance and Optimization (Monitor, detect and report on safety-critical clinical and administrative aspects of EHRs and healthcare processes and make iterative refinements to optimize safety.)		
22. As part of quality assurance activities, organizations monitor selected practices related to test result reporting and follow-up. Monitored practices include clinician use of the EHR for test results review and clinician follow-up on abnormal test results. [4-6,13,26,41-44] C, Ev, IT	Effective quality assurance patient safety programs include monitoring of core clinical metrics. Errors related to missed or delayed follow-up of test results are a significant cause of adverse events that harm patients.	• The organization has in place processes to monitor and report alert responses (e.g., acknowledged or not; time to acknowledgement)8 and test result follow-up with patients. [5] • Clinicians document communication of test results to patients in their EHR. [45] • Organizational QA activities select and measure test results-related benchmarks for ongoing monitoring, starting in areas of identified concern and high risk. A measurement system for test results reporting exists with the following potential measures: 1. Percentage of all active clinicians who have reviewed at least one laboratory test result within the last month. If greater than 95%, this measure could indicate if the EHR is the "source of truth" for laboratory test results (vs. dependence on paper-based communication). 2. Test results with the lowest follow-up rate are investigated to understand root causes of the problem. [6,43] 3. Percentage of all test results reviewed by the ordering provider within 4 days. This should be greater than 90%. 4. Results not reviewed for more than a week. This should be minimal.

Recommended Practices	Rationale for Practice or Risk Addressed	Examples of Potentially Useful Practices/Scenarios
23. There is a process to monitor results related to certain high-risk areas: patients undergoing transitions (e.g., pending test results of discharged patients) or providers undergoing transitions (e.g., tests ordered by residents that routinely rotate to new services, clinics, or locations). [6,26,29,44] C, Dx, Ev, IT	Test results are missed in EHR systems despite advanced systems for notification.	Test results with the lowest follow-up rate are investigated to understand root causes of the problem. [6,43]
24. As part of quality assurance, the organization monitors and addresses test results sent to the wrong clinician or never transmitted to any clinician (e.g., due to an interface problem or patient/provider misidentification). [21] C, Dx, Ev, IT	When test results are "lost in the system," there is a danger that there will be no follow-up, posing a significant risk of patient harm.	• Error logs are used to detect results such as those that were never delivered, results without any ordering providers, results with unidentifiable providers, etc. • National Provider ID (NPI) is used for provider attribution of orders. • Monitor provider master files to ensure that they are synchronized to avoid scenarios in which the ordering provider's contact information is outdated or unknown.

REFERENCES

1. Singh H, Naik A, Rao R, Petersen L. Reducing Diagnostic Errors Through Effective Communication: Harnessing the Power of Information Technology. Journal of General Internal Medicine. 2008;23:489-494.
2. Hickner J, Fernald D, Harris D, Poon E, Elder N, Mold J. Issues and initiatives in the testing process in primary care physician offices. Jt Comm J Qual Patient Saf. 2005;31:81-89.
3. Schiff GD. Medical error: a 60-year-old man with delayed care for a renal mass. JAMA. 2011;305:1890-1898.
4. Singh H, Thomas EJ, Mani S et al. Timely follow-up of abnormal diagnostic imaging test results in an outpatient setting: are electronic medical records achieving their potential? Arch Intern Med. 2009;169:1578-1586.

5. Singh H, Thomas EJ, Sittig DF et al. Notification of abnormal lab test results in an electronic medical record: do any safety concerns remain? Am J Med. 2010;123:238-244.
6. Sittig DF, Singh H. Improving Test Result Follow-up through Electronic Health Records Requires More than Just an Alert. J Gen Intern Med. 2012;27:1235-1237.
7. Laxmisan A, Sittig DF, Pietz K, Espadas D, Krishnan B, Singh H. Effectiveness of an Electronic Health Record-Based Intervention to Improve Follow-up of Abnormal Pathology Results: a Retrospective Record Analysis. Medical Care. 2012.
8. Lab Communication Checklist Validation - Geisinger Health System. 2012. 2012. Ref Type: Personal Communication
9. Singh H, Vij MS. Eight recommendations for policies for communicating abnormal test results. Jt Comm J Qual Patient Saf. 2010;36:226-232.
10. Singh H, Kadiyala H, Bhagwath G et al. Using a multifaceted approach to improve the follow-up of positive fecal occult blood test results. Am J Gastroenterol. 2009;104:942-952.
11. Singh H, Wilson L, Reis B, Sawhney MK, Espadas D, Sittig DF. Ten strategies to improve management of abnormal test result alerts in the electronic health record. J Patient Saf. 2010;6:121-123.
12. Sittig DF, Singh H. Electronic health records and national patient-safety goals. N Engl J Med. 2012;367:1854-1860.
13. Callen JL, Westbrook JI, Georgiou A, Li J. Failure to follow-up test results for ambulatory patients: a systematic review. J Gen Intern Med. 2012;27:1334-1348.
14. Dalal AK, Poon EG, Karson AS, Gandhi TK, Roy CL. Lessons learned from implementation of a computerized application for pending tests at hospital discharge. J Hosp Med. 2011;6:16-21.
15. El-Kareh R, Roy C, Williams DH, Poon EG. Impact of automated alerts on follow-up of post-discharge microbiology results: a cluster randomized controlled trial. J Gen Intern Med. 2012;27:1243-1250.
16. Elder NC, McEwen TR, Flach J, Gallimore J, Pallerla H. The management of test results in primary care: does an electronic medical record make a difference? Fam Med. 2010;42:327-333.
17. Murphy DR, Laxmisan A, Reis B et al. Electronic Health Record-Based Triggers to Detect Potential Delays in Cancer Diagnosis. BMJ Quality and Safety. In press.
18. Vreeman DJ, McDonald CJ, Huff SM. LOINC(R) - A Universal Catalog of Individual Clinical Observations and Uniform Representation of Enumerated Collections. Int J Funct Inform Personal Med. 2010;3:273-291.
19. Burnside ES, Sickles EA, Bassett LW et al. The ACR BI-RADS experience: learning from history. J Am Coll Radiol. 2009;6:851-860.
20. Russ G, Bigorgne C, Royer B, Rouxel A, Bienvenu-Perrard M. [The Thyroid Imaging Reporting and Data System (TIRADS) for ultrasound of the thyroid]. J Radiol. 2011;92:701-713.
21. Yackel TR, Embi PJ. Unintended errors with EHR-based result management: a case series. J Am Med Inform Assoc. 2010;17:104-107.
22. Callen J, Paoloni R, Georgiou A, Prgomet M, Westbrook J. The rate of missed test results in an emergency department: an evaluation using an electronic test order and results viewing system. Methods Inf Med. 2010;49:37-43.

23. Passiment E, Meisel JL, Fontanesi J, Fritsma G, Aleryani S, Marques M. Decoding laboratory test names: a major challenge to appropriate patient care. J Gen Intern Med. 2013;28:453-458.
24. Georgiou A, Prgomet M, Toouli G, Callen J, Westbrook J. What do physicians tell laboratories when requesting tests? A multi-method examination of information supplied to the microbiology laboratory before and after the introduction of electronic ordering. Int J Med Inform. 2011;80:646-654.
25. Improving your office testing process: A toolkit for rapid-cycle patient safety and quality improvement. 2012. Agency for Healthcare Research and Quality. Ref Type: Generic
26. Poon EG, Wang SJ, Gandhi TK, Bates DW, Kuperman GJ. Design and implementation of a comprehensive outpatient Results Manager. J Biomed Inform. 2003;36:80-91.
27. Dalal AK, Schnipper JL, Poon EG et al. Design and implementation of an automated email notification system for results of tests pending at discharge. J Am Med Inform Assoc. 2012;19:523-528.
28. Hysong SJ, Sawhney MK, Wilson L et al. Understanding the management of electronic test result notifications in the outpatient setting. BMC Med Inform Decis Mak. 2011;11:22.
29. Roy CL, Poon EG, Karson AS et al. Patient safety concerns arising from test results that return after hospital discharge. Ann Intern Med. 2005;143:121-128.
30. Beckwith, B. A., Aller, R. D., Brassel, J. H., Brodsky, V. B., and deBaca, M. F. Laboratory interoperability best practices: Ten mistakes to avoid. http://www.cap.org/apps/docs/committees/informatics/cap_dihit_lab_interop_final_march_2013.pdf . 13. College of American Pathologists. 5-31-2013. Ref Type: Electronic Citation
31. Litvin C, Cavanaugh JS, Callanan M, Tenner CT. To err is human continued: a failure of follow-up. J Clin Outcomes Manag. 2008;15:21-23.
32. Kuperman GJ, Teich JM, Bates DW et al. Detecting alerts, notifying the physician, and offering action items: a comprehensive alerting system. Proc AMIA Annu Fall Symp. 1996;704-708.
33. Horsky J, Kuperman GJ, Patel VL. Comprehensive analysis of a medication dosing error related to CPOE. J Am Med Inform Assoc. 2005;12:377-382.
34. Hysong SJ, Sawhney MK, Wilson L et al. Provider management strategies of abnormal test result alerts: a cognitive task analysis. J Am Med Inform Assoc. 2010;17:71-77.
35. Murphy DR, Reis B, Sittig DF, Singh H. Notifications received by primary care practitioners in electronic health records: a taxonomy and time analysis. Am J Med. 2012;125:209-7.
36. Murphy DR, Reis B, Kadiyala H et al. Electronic health record-based messages to primary care providers: valuable information or just noise? Arch Intern Med. 2012;172:283-285.
37. Singh H, Spitzmueller C, Petersen NJ et al. Primary care practitioners views on test result management in EHR-enabled health systems: a national survey. J Am Med Inform Assoc. 2012.
38. Singh H, Spitzmueller C, Petersen NJ, Sawhney MK, Sittig DF. Information Overload and Missed Test Results in Electronic Record-Based Settings. JAMA Internal Medicine. 2013;173:702-703.

39. Woods DD, Patterson E, Roth EM. Can we ever escape from data overload? A cognitive systems diagnosis. Cognition, Technology & Work. 2002;4:22-36.
40. Poon EG, Kuperman GJ, Fiskio J, Bates DW. Real-time notification of laboratory data requested by users through alphanumeric pagers. J Am Med Inform Assoc. 2002;9:217-222.
41. Boohaker EA, Ward RE, Uman JE, McCarthy BD. Patient notification and follow-up of abnormal test results. A physician survey. Arch Intern Med. 1996;156:327-331.
42. Greenes DS, Fleisher GR, Kohane I. Potential impact of a computerized system to report late-arriving laboratory results in the emergency department. Pediatr Emerg Care. 2000;16:313-315.
43. Singh H, Wilson L, Petersen LA et al. Improving follow-up of abnormal cancer screens using electronic health records: trust but verify test result communication. BMC Med Inform Decis Mak. 2009;9:49.
44. Smith, M. W., Murphy, D. R., Laxmisan, A., Sittig, D. F., Reis, B., Esquivel, A., and Singh, H. A multifaceted approach to development of a software aid for delayed follow-up. Applied Clinical Informatics . 2013. Ref Type: Unpublished Work
45. VHA Directive 2009-019: Ordering and Reporting Test Results. Veterans Health Administration . 3-24-2009. Washington DC, Department of Veterans Affairs. Ref Type: Electronic Citation

CHAPTER 12

ASSESSMENT OF CLINICIAN-TO-CLINICIAN E-COMMUNICATION

IMPROVING THE EFFECTIVENESS OF ELECTRONIC HEALTH RECORD-BASED REFERRAL PROCESSES

Adol Esquivel, Dean F. Sittig, Daniel R. Murphy, and Hardeep Singh

12.1.1 INTRODUCTION

Outpatient referrals, defined as processes that include a transfer of responsibility for some aspect of patient's care from a referring provider to a secondary service or provider, [1] are an important but challenging aspect of primary care practice. Successful coordination of referrals hinges upon effective and timely communication to facilitate information sharing and transfer of patient care responsibilities between outpatient providers [2-10]. However, referral communication related to both provider-provider and provider-patient interactions [3,11-14] is prone to breakdown [2,14-

---

Improving the Effectiveness of Electronic Health Record-Based Referral Processes. © Esquivel A, Sittig DF, Murphy DR, and Singh H; liensee BioMed Central Ltd. BMC Medical Informatics and Decision Making *12*,107 (2012) doi:10.1186/1472-6947-12-107. Licensed under Creative Commons Attribution 2.0 Generic License, http://creativecommons.org/licenses/by/2.0/.

22]. The growing use of referral care [23] suggests the need for improving reliability and efficiency of the referral process to create a greater impact on health care quality.

In accordance with the 2009 Health Information Technology for Economic and Clinical Health Act (HITECH) and its Meaningful Use goals for effective use of electronic health records (EHRs), healthcare institutions are increasingly adopting technology to support patient care. By 2015, hospitals are expected to demonstrate, among other things, the capability to exchange key clinical information among providers of care and other patient-authorized entities electronically [24]. This increasing adoption of health information technology holds promise for improving referral communication in health care [25-28]. However, early adopters of these technologies, mostly large integrated systems, have encountered novel communication challenges and unintended consequences that are important to understand in order to reduce future care delays [18,29-35].

Many referrals between primary care providers (PCPs) and specialists do not take place within the same practice or institution; and in general, providers don't have access to the same EHR. However, efforts to address communication challenges using EHRs will be essential given the emphasis on coordination of care and exchange of relevant clinical information by the Patient Protection and Affordable Care Act of 2010 [36]. Recent reform initiatives call for healthcare institutions to become Accountable Care Organizations (ACOs) [37] and demonstrate the use of evidence-based medicine and the application of evolving technologies to support a strong foundation for coordinated primary care. They also create an expectation of continuous process improvement based on measurement of clinical quality and outcomes [38]. EHR-based referrals thus would be an essential component of patient care through ACOs. Even when supported by technology, referral communication between PCPs and specialists is often unsatisfactory [39]. This might be partially due to lack of attention on how communication technology fits with the social environment in which it is implemented [40,41]. Addressing these key challenges in making electronic referral communication effective [11,12,42] requires a multifaceted "socio-technical" approach [43].

Although efforts have been made to improve and standardize overall EHR usability, [44,45] there are presently no standards that specifically

Assessment of Clinician-to-Clinician E-Communication 343

address the design or use of electronic systems in outpatient referral communication, and best practices in this area are limited [6,19,39,46-48]. In fact, no available turn-key EHR system can fully support the complexities of most referral processes. Furthermore, referral processes are highly variable across health care settings, and EHRs that support referrals are often heavily customized to reflect unique organizational requirements [19,49,50]. Although complete standardization of referral practices is neither possible nor desirable, several aspects of referral communication are amenable to strategies to reduce the risk of unintended consequences and delays in patient care.

TABLE 12.1.1: Recommendations Summary and their relation to Socio-Technical dimensions

	Recommendation	Primary Socio-Technical Dimension*
1	Include real-time clinician-to-clinician communication features as part of the referral system.	Hardware & Software
2	Design and use electronic standardized referral templates that include both structured and free-text fields.	Human-Computer Interface
3	Enforce electronic capture of the reason for the referral.	Clinical Content
4	Bring PCPs and specialists together to collaboratively develop referral guidelines for inclusion into the electronic referral system	People
5	Integrate patient communication into the electronic referral process	People
6	Use automation to pre-populate electronic referral requests with patient-specific data	Workflow & Communication
7	Include the capability of electronic consultations (information-only referrals).	Workflow & Communication
8	Close the communication loop by providing referral status tracking and feedback capabilities and integrating these tools into providers' workflows	Workflow & Communication
9	Standardize and maintain up-to-date institutional policies and procedures for electronic referrals.	Organization Policies & Procedures
10	Monitor electronic referral communication performance.	Measurement & Monitoring

Although recommendation may be associated with more than one dimension of the socio-technical mode, this table identifies the dimension each recommendation most directly relates to.

This article describes ten recommendations that represent potential best practices to design, develop, implement, improve, and monitor electronic outpatient referral communication. Recommendations are grounded in a socio-technical model for health information technology [43]. This model uses 8 interrelated dimensions to identify challenges related to developing, implementing, and using information technology within health care (hardware & software, clinical content, human-computer interface, people, workflow & communication, organizational features, external rules and regulations, and measurement & monitoring). The recommendations are also based on current literature, sound clinical practice, our previous work, and a systems-based approach to understanding and implementing health information technology solutions. We also categorized recommendations according to the dimensions of the socio-technical model with which they are most closely related (Table 12.1.1). Some recommendations have an established evidence-base and others are based on our experiences or perspectives, but most are not widely adopted by institutions and/or current EHRs. Thus, we believe these recommendations are relevant to all system designers, practicing clinicians, and other stakeholders considering the use of EHRs to support referral communication.

12.1.2 RECOMMENDATION #1: INCLUDE REAL-TIME CLINICIAN-TO-CLINICIAN COMMUNICATION FEATURES AS PART OF THE REFERRAL SYSTEM

Providers often prefer traditional face-to-face or synchronous communication, such as telephone conversations. While excessive reliance on the EHR and other health information technology may diminish the use of real-time communication, certain critical situations require the interactivity afforded by direct conversation. In fact, some estimates propose that up to 60% of providers' time in clinic is devoted to synchronous conversation [51]. In some cases, such as when a referral is urgent, real-time communication may be required to expedite the referral process [52,53]. Specialists may also want to speak directly to referring providers if there is any doubt about a referral's appropriateness or urgency, even when PCPs and specialists share access to the patient's record. EHRs can facilitate real-time phone conversations or internet-based audio-, video-, or text-based conferencing interactions by providing easily accessible and updated contact information for specialists and PCPs (or their clinics) on the referral interface [54-57]. This flexibility should be specified in any policies and procedures governing outpatient referrals [58].

12.1.3 RECOMMENDATION #2: DESIGN AND USE STANDARDIZED ELECTRONIC REFERRAL TEMPLATES THAT INCLUDE BOTH STRUCTURED AND FREE-TEXT FIELDS

The content, form, and style of referral letters influences the referral process [5,16,59-61]. Several studies have shown increased provider satisfaction and more consistent and timely feedback from specialists when referral templates are used to standardize referral communication [16,62,63]. Electronic systems provide an excellent opportunity to create, maintain, and disseminate the use of standardized templates [64,65]. However, the interface of electronic referral templates should be designed to avoid excessive constraints that can limit providers' ability to explain and document relevant findings [4]. Thus, when designing electronic referral templates, human-computer interface designers must maintain a delicate balance between structured fields to capture required essential information and free-text fields to allow providers to qualify and expand on their findings freely.

12.1.4 RECOMMENDATION #3: ENFORCE ELECTRONIC CAPTURE OF THE REASON FOR THE REFERRAL

More than fifty years ago, Williams et al. determined that providing a clear reason for a referral was an essential step in the outpatient referral process [13]. Since then, multiple studies have shown that providers' failure to clearly state the reason for referral (a problem identified in 20-88% of referrals [7,8,21,50]) remains a major barrier to effective referral communication [20,66]. The inclusion of a clear reason to justify a referral is not only regarded as good professional practice but it has also been shown to expedite the referral process [2,7,22,67]. Therefore, electronic systems should be designed to prevent referrals from being transmitted unless they have a clearly defined reason to justify them. In addition to a standard set of generic choices, such as those proposed by Forrest et al. (to seek advice, to request a technical procedure, and to request co-management of the patient), electronic systems should give providers the option to expand and elaborate on their selection when needed [68].

12.1.5 RECOMMENDATION #4: BRING PCPS AND SPECIALISTS TOGETHER TO COLLABORATIVELY DEVELOP REFERRAL GUIDELINES FOR INCLUSION INTO THE ELECTRONIC REFERRAL SYSTEM

EHRs offer a robust platform for integrating referral guidelines into providers' workflows at the point of care, and referral guidelines can improve the referral process in several ways. For instance, they can help providers determine the appropriateness of a referral prior to initiating the request [42,47] or allow a provider to anticipate the specialist's referral information and patient work-up needs, improving efficiency and quality. People comprise one of the key dimensions of the socio-technical model. While EHRs are valuable delivery vehicles for referral guidelines, effective outcomes will only be achieved by collaborative efforts between referring providers and specialists to facilitate communication, decrease referral denials, and clarify referral expectations. While collaboration across different practice settings and institutions will be challenging to operationalize, it must also be encouraged keeping in line with the national focus on reducing health care costs and overutilization [19,69]. For instance, solo practitioners and small independent practices lacking formal organizational structures can leverage their existing networks of specialists to develop mutually agreed-upon referral guidelines. Additionally, third parties involved in regulatory, reimbursement, or quality improvement activities (e.g., regional extension centers, payers, or medical societies) can facilitate the development and dissemination of a basic set of guidelines as a starting point. Service agreements between PCPs and specialists that include referral guidelines can facilitate provider access to specialists and reduce inappropriate referrals by suggesting evidence-based pathways or alternatives to referrals [70-73]. However, given the complexity of some referrals, systems should remain sufficiently flexible to allow providers to bypass guidelines and submit a referral request that may not appear to adhere to guideline criteria by appropriately justifying its urgency and clinical need.

12.1.6 RECOMMENDATION #5: INTEGRATE PATIENT COMMUNICATION INTO THE ELECTRONIC REFERRAL PROCESS

As early as 1971, researchers pointed out that the success of outpatient referrals was related in part to patient-related variables, [2] such as patient's illness and socioeconomic background. However, subsequent work has paid little attention to the patient's role in outpatient referral communication. In recent years, the growth of personal health records and other consumer electronic communication tools have modernized and fundamentally transformed patient-provider communication [74]. Nevertheless, patient-related communication remains vulnerable to breakdowns. For instance, these communication failures can account for a substantial number of incomplete referrals resulting in missed appointments and delays in care [53,75]. Attributes similar to those expected of provider-to-provider electronic communication (i.e., secure, timely, reliable, and actionable) [76] must also be used to inform tools to enhance patient-centered communication [77]. These attributes should be the hallmark of effective electronic communication within the patient-centered medical home model [78-80]. Hence, EHRs aimed at supporting referral communication should include functionality to allow the patient to provide additional information if and when needed, and to permit patients to become an active decision-maker during the referral process (i.e. allow them to schedule and cancel appointments, select providers, ask questions). Given the low adoption and use of existing patient communication tools [81,82], novel methods beyond traditional web-based portals are needed. System developers and administrators should explore how to leverage technologies such as smart phone apps, social media portals, and electronic outreach programs [83,84] as well as consider alternative forms of patient access or outreach in order to make patient communication more reliable. This will enable patients to have secure and timely access to relevant information such as referral status updates, reminders to increase patient compliance, and tools to facilitate communication with their physician.

12.1.7 RECOMMENDATION #6: USE AUTOMATION TO PRE-POPULATE ELECTRONIC REFERRAL REQUESTS WITH PATIENT-SPECIFIC DATA

If used appropriately, electronic referrals have the potential to enhance provider workflow by automating certain tedious or repetitive steps where manual effort is unnecessary. The cognitive load imposed by the use of structured templates, referral guidelines, and use of computerized interfaces increases the time commitment and complexity of initiating and managing referrals [85]. In a recent study, referring PCPs and specialists both suggested the use of automation to pre-populate electronic referral requests in order to decrease both workload and cognitive load [9]. In a separate study, auto-population was commended by providers as a mechanism to improve the efficiency of the consultation process [86]. Electronic referrals should harness the benefits of EHR data and use it to automatically pre-populate fields in the referral template whenever possible (e.g., demographic data, current medication list, recent relevant laboratory test results [18]). Ultimately, more advanced EHRs could even use rule-based pre-population to supply additional relevant information based on the patient's diagnosis or age group.

12.1.8 RECOMMENDATION #7: INCLUDE THE CAPABILITY OF ELECTRONIC CONSULTATIONS (INFORMATION-ONLY REFERRALS)

The conceptual definition of "referral" implies an actual transfer of responsibility for some aspect of the patient's care and an encounter with another provider. In contrast, a strict consultation involves seeking a colleague's opinion about a particular aspect of the care of the patient, but at no time is the patient under the direct care of the consultant [1,87].

For example, certain referral questions are addressed more efficiently through consultation or information exchanges between the referring PCP and the specialist, which does not necessarily require a physical encounter between the patient and the specialist [9]. Workflow efficiency might be improved if electronic consultations are effectively used. Electronic health records can facilitate these consultations through more flexible and efficient electronic consultation processes that minimize delays (i.e., "information-only" referrals that do not require a patient visit). A successful example of this practice is the established telemedicine modality known as "store-and-forward" in which the provider exchanges relevant patient information with the consultant asynchronously and requests his or her opinion electronically [88,89]. These strategies, if implemented appropriately, can also minimize delays and inefficiencies in care related to unnecessary referrals [48].

12.1.9 RECOMMENDATION #8: CLOSE THE COMMUNICATION LOOP BY PROVIDING AND INTEGRATING REFERRAL STATUS TRACKING AND FEEDBACK CAPABILITIES INTO PROVIDERS' WORKFLOWS

Coordination of care is more effective when all interested parties are aware of the status of the referral request. Referring providers should receive timely feedback from the specialists upon denial, approval, or completion of each referral [9]. However, studies suggest that specialists fail to provide feedback in 15-45% of referrals [4,7,22,61]. Similarly, specialists may need to discuss requests with the referring providers before or after approving them. In engineering, a closed-loop control system is one in which feedback is needed to control the states or outputs of a dynamic system [90]. Often used in decision support systems, [91,92] closed-loop control can improve electronic referrals by ensuring that communication is coupled with timely and appropriate feedback. Effectively closing the loop on all outpatient referral communication requires considerable resources and efforts from all stakeholders; however, EHRs can help to close the referral communication loop in multiple ways. For example allowing providers to document and access each other's notes about encounters, orders, and other relevant information, or by automatically notifying providers of changes in the status of the referral as it progresses through the referral stages. Additionally, the EHR can notify the referring provider when the specialist has reviewed, approved, or denied a referral request or has asked for additional information [86]. These tools must integrate into providers' workflow in order to leverage improvements in reliability and efficiency. Nevertheless, as with other types of electronic communication in healthcare, it is important not to overload providers with excessive notifications about status updates [93]. Thus, while electronic referral communication must be comprehensive, it should be implemented in a non-intrusive manner so that information remains available to providers and patients on demand.

12.1.10 RECOMMENDATION #9: STANDARDIZE AND MAINTAIN UP-TO-DATE INSTITUTIONAL POLICIES AND PROCEDURES FOR ELECTRONIC REFERRALS

Within institutions, lack of clear policies and procedures can result in unnecessary heterogeneity across referral processes causing inefficiencies in patient care, provider dissatisfaction, and potential for delays in diagnosis and treatment [9]. Even when organizations develop policies and procedures governing referrals, the adoption of health information technology often translates into profound changes in performance and culture [94,95]. Organizations must carefully review and continuously update policies and procedures related to referrals to ensure they reflect appropriate use of electronic tools [40]. Referral policies and procedures should provide detailed guidance with respect to every facet of the use of technology supporting the referral process. For example, to assure compliance and effective use of health information technology for referrals, organizations need to have clearly documented roles and responsibilities for PCPs, specialists, and supporting staff during key stages of the referral process. Additionally, referral policies and procedures should outline the minimum information PCPs should include in the electronic referral request, as well as expected turnaround times for specialists to respond to the referral. They should also incorporate details about the tools available to providers to monitor timeliness and effectiveness of electronic referral communication [19]. Finally, they should allow the flexibility to account for different levels of urgency and importance across clinical problems and specialties, permitting providers to expedite a particular referral when necessary [40,52,96,97]. A clear and common understanding of referral processes with documented policies and procedures of how the technology should be used by PCPs, specialists, and supporting staff is essential for success.

12.1.11 RECOMMENDATION #10: MONITOR ELECTRONIC REFERRAL COMMUNICATION PERFORMANCE

Recent literature has revealed several serious health information technology-related errors that arose from faulty system design, configuration, or implementation processes [98-101]. Organizations must continuously monitor and evaluate the usability, performance, benefits, and drawbacks of their electronic referral systems [40]. As with any health information technology-related process, referral communication should be monitored and revised, as needed, [43] to ensure that all stakeholders' needs are being met in a safe and efficient manner. For instance, in our previous work we found that about 7% of electronic referrals at our institution had no follow-up action by specialists at 30 days [29]. Continuous monitoring and frequent assessments of several process measurements (e.g., completed referrals, no-shows/missed appointments, and denied or cancelled referrals) should be part of the organization's ongoing efforts to ensure the effectiveness of their electronic referral communication practices.

12.1.12 CONCLUSION

EHR-based referrals offer the possibility of greatly improving existing outpatient referral processes. However, technology-facilitated referral processes have not yet reached their potential and will soon be put to the test given the rapid adoption of EHRs. Our proposed recommendations highlight the need to consider the socio-technical context in which information technology-based tools are implemented. Allowing for some flexibility in the referral process and monitoring communication outcomes are vital to effective implementation. As healthcare organizations continue to adopt and use EHRs, the success of technology-enabled referral processes will depend on their ability to remain patient-centered and responsive to providers' needs. The recommendations presented address key areas within seven of the eight socio-technical dimensions, all of which must be performed while adhering to external rules and regulations (e.g., HIPAA or HITECH act), as suggested by the model's eighth dimension. We envision that these recommendations will be useful for several types of stakeholders as they move forward in designing, implementing, and improving their electronic referral systems.

REFERENCES

1. McWhinney IR: A textbook of family medicine. USA: Oxford University Press; 1997.
2. Shortell SM, Anderson OW: The physician referral process: a theoretical perspective. Health Serv Res 1971, 6:39-48.
3. Byrd JC, Moskowitz MA: Outpatient consultation: interaction between the general internist and the specialist. J Gen Intern Med 1987, 2:93-98.
4. Newton J, Eccles M, Hutchinson A: Communication between general practitioners and consultants: what should their letters contain? BMJ 1992, 304:821-824.
5. Westerman RF, Hull FM, Bezemer PD, Gort G: A study of communication between general practitioners and specialists. Br J Gen Pract 1990, 40:445-449.
6. Chen AHM, Yee HF Jr: Improving the primary care-specialty care interface: getting from here to there. Arch Intern Med 2009, 169:1024-1026.
7. McPhee SJ, Lo B, Saika GY, Meltzer R: How good is communication between primary care physicians and subspecialty consultants? Arch Intern Med 1984, 144:1265-1268.
8. Gandhi TK, Sittig DF, Franklin M, Sussman AJ, Fairchild DG, Bates DW: Communication breakdown in the outpatient referral process. J Gen Intern Med 2000, 15:626-631.
9. Hysong SJ, Esquivel A, Sittig DF, Paul LA, Espadas D, Singh S, Singh H: Towards successful coordination of electronic health record based-referrals: a qualitative analysis. Implement Sci 2011, 6:84.
10. O'Malley AS, Reschovsky JD: Referral and consultation communication between primary care and specialist physicians: Finding common ground. Arch Intern Med 2011, 171:56-65.
11. Forrest CB, Majeed A, Weiner JP, Carroll K, Bindman AB: Comparison of specialty referral rates in the United Kingdom and the United States: retrospective cohort analysis. BMJ 2002, 325:370-371.

12. Roland M: General practitioner referral rates. BMJ. 1988, 297:437-438.
13. Williams TF, White KL, Fleming WL, Greenberg BG: The referral process in medical care and the university clinic's role. J Med Educ 1961, 36:899-907.
14. Deckard GJ, Borkowski N, Diaz D, Sanchez C, Boisette SA: Improving timeliness and efficiency in the referral process for safety net providers: application of the Lean Six Sigma methodology. J Ambul Care Manage 2010, 33:124-130.
15. Javalgi R, Joseph WB, Gombeski WR Jr, Lester JA: How physicians make referrals. J Health Care Mark 1993, 13:6-17.
16. Jenkins S, Arroll B, Hawken S, Nicholson R: Referral letters: are form letters better? Br J Gen Pract 1997, 47:107-108.
17. Munro C: Referral of Patients-A Neglected Aspect of Medical Practice. Hong Kong Prac 1989, 11:523-6.
18. Sittig DF, Gandhi TK, Franklin M, Turetsky M, Sussman AJ, Fairchild DG, Bates DW, Komaroff AL, Teich JM: A computer-based outpatient clinical referral system. Int J Med Inform 1999, 55:149-158.
19. Kim Y, Chen AH, Keith E, Yee HF Jr, Kushel MB: Not perfect, but better: primary care providers' experiences with electronic referrals in a safety net health system. J Gen Intern Med 2009, 24:614-619.
20. Lee T, Pappius EM, Goldman L: Impact of inter-physician communication on the effectiveness of medical consultations. Am J Med 1983, 74:106-112.
21. Conley J, Jordan M, Ghali WA: Audit of the consultation process on general internal medicine services. Qual Saf Health Care 2009, 18:59-62.
22. Cummins RO, Smith RW, Inui TS: Communication failure in primary care. Failure of consultants to provide follow-up information. JAMA 1980, 243:1650-1652.
23. Barnett ML, Song Z, Landon BE: Trends in Physician Referrals in the United States, 1999–2009. Arch Intern Med 2012, 172:163-170.
24. Public Inspection: Medicare and Medicaid Programs: Electronic Health Record Incentive Program -Stage 2 [https://www.federalregister.gov/articles/2012/03/07/2012-04443/electronic-health-record-incentive-program--stage-2-medicare-and-medicaid-programs webcite]
25. Kalogriopoulos NA, Baran J, Nimunkar AJ, Webster JG: Electronic medical record systems for developing countries: review. Conf Proc IEEE Eng Med Biol Soc 2009, 2009:1730-1733.
26. McCullough JS, Casey M, Moscovice I, Prasad S: The effect of health information technology on quality in U.S. hospitals. Health Aff (Millwood) 2010, 29:647-654.
27. Roberts J: Personal electronic health records: from biomedical research to people's health. Inform Prim Care 2009, 17:255-260.
28. Blumenthal D: Launching HITECH. N Engl J Med 2010, 362:382-385.
29. Singh H, Esquivel A, Sittig DF, Murphy D, Kadiyala H, Schiesser R, Espadas D, Petersen LA: Follow-up actions on electronic referral communication in a multispecialty outpatient setting. J Gen Intern Med 2011, 26:64-69.
30. Novak LL: Improving health IT through understanding the cultural production of safety in clinical settings. Stud Health Technol Inform 2010, 157:175-180.
31. Callahan D: Medical progress: unintended consequences. Hastings Cent Rep 2009, Suppl:13-14.

32. Bernstam EV, Hersh WR, Sim I, Eichmann D, Silverstein JC, Smith JW, Becich MJ: Unintended consequences of health information technology: a need for biomedical informatics. J Biomed Inform 2010, 43:828-830.
33. Weiner M, El Hoyek G, Wang L, Dexter PR, Zerr AD, Perkins AJ, James F, Juneja R: A web-based generalist-specialist system to improve scheduling of outpatient specialty consultations in an academic center. J Gen Intern Med 2009, 24:710-715.
34. Shaw LJ, de Berker DAR: Strengths and weaknesses of electronic referral: comparison of data content and clinical value of electronic and paper referrals in dermatology. Br J Gen Pract 2007, 57:223-224.
35. Campbell EM, Sittig DF, Guappone KP, Dykstra RH, Ash JS: Overdependence on technology: an unintended adverse consequence of computerized provider order entry. AMIA Annu Symp Proc 2007, 94-98.
36. U.S. Congress: Patient Protection and Affordable Care Act. 2010.
37. Fisher ES, Shortell SM: Accountable Care Organizations. JAMA 2010, 304:1715-1716.
38. Mountford J, Davie C: Toward an Outcomes-Based Health Care System. JAMA 2010, 304:2407-2408.
39. Chen AH, Kushel MB, Grumbach K, Yee HF Jr: Practice profile. A safety-net system gains efficiencies through "eReferrals" to specialists. Health Aff (Millwood) 2010, 29:969-971.
40. Sittig DF, Singh H: Eight rights of safe electronic health record use. JAMA 2009, 302:1111-1113.
41. Berg M, Aarts J, van der Lei J: ICT in health care: sociotechnical approaches. Methods Inf Med 2003, 42:297-301.
42. Grimshaw JM, Winkens RAG, Shirran L, Cunningham C, Mayhew A, Thomas R, Fraser C: Interventions to improve outpatient referrals from primary care to secondary care. Cochrane Database Syst Rev 2005. CD005471
43. Sittig DF, Singh H: A new sociotechnical model for studying health information technology in complex adaptive healthcare systems. Qual Saf Health Care 2010, 19(Suppl 3):i68-74.
44. Armijo D, McDonnell C, Werner K: Electronic Health Record Usability: Interface Design Considerations. Rockville, MD: Agency for Healthcare Research and Quality; 2009. AHRQ Publication No. 09(10)-0091-2-EF
45. Schumacher RM, Lowry SZ: NIST Guide to the Processes Approach for Improving the Usability of Electronic Health Records. Gaithersburg, MD: National Institute of Standards and Technology; 2010:5-10.
46. Chen AH, Yee HF Jr: Improving primary care-specialty care communication: lessons from San Francisco's safety net: comment on "Referral and consultation communication between primary care and specialist physicians. Arch Intern Med 2011, 171:65-67.
47. Kim-Hwang JE, Chen AH, Bell DS, Guzman D, Yee HF Jr, Kushel MB: Evaluating electronic referrals for specialty care at a public hospital. J Gen Intern Med 2010, 25:1123-1128.
48. Katz MH: How can we know so little about physician referrals? Arch. Intern. Med. 2012, 172:100.

49. Augestad KM, Revhaug A, Vonen B, Johnsen R, Lindsetmo R-O: The one-stop trial: does electronic referral and booking by the general practitioner (GPs) to outpatient day case surgery reduce waiting time and costs? A randomized controlled trial protocol. BMC Surg 2008, 8:14.
50. Gandhi TK, Keating NL, Ditmore M, Kiernan D, Johnson R, Burdick E, Hamann C: Improving referral communication using a referral tool within an electronic medical record. In Advances in Patient Safety: New Directions and Alternative Approaches Edited by Henriksen K, Battles JB, Keyes MA, Grady ML Rockville MD. 2008, 4. [Agency for Healthcare Research and Quality]
51. Tang PC, Jaworski MA, Fellencer CA, Kreider N, LaRosa MP, Marquardt WC: Clinician information activities in diverse ambulatory care practices. Proc AMIA Annu Fall Symp 1996, 12-16.
52. Coiera E: Communication systems in healthcare. Clin Biochem Rev 2006, 27:89-98.
53. Singh H, Petersen LA, Daci K, Collins C, Khan M, El-Serag HB: Reducing referral delays in colorectal cancer diagnosis: is it about how you ask? Qual Saf Health Care 2010, 19:e27.
54. Robertson KJ: Diabetes and the Internet. Horm Res 2002, 57:110-112.
55. Saxena S, Kumar V, Giri V: Telecardiology for effective healthcare services. J Med Eng Technol 2003, 27:149.
56. Forti S, Galvagni M, Galligioni E, Eccher C: A real time teleconsultation system for sharing an oncologic web-based electronic medical record. AMIA Annu Symp Proc 2005, 2005:959.
57. Gwozdek AE, Klausner CP, Kerschbaum WE: The utilization of Computer Mediated Communication for case study collaboration. J Dent Hyg 2008, 82:8.
58. Coiera E: When conversation is better than computation. J Am Med Inform Assoc 2000, 7:277-286.
59. Esquivel A, Dunn K, McLane S, Te'eni D, Zhang J, Turley JP: When your words count: a discriminative model to predict approval of referrals. Inform Prim Care 2009, 17:201-207.
60. Graham PH: Improving communication with specialists. The case of an oncology clinic. Med J Aust 1994, 160:625-627.
61. Epstein RM: Communication between primary care physicians and consultants. Arch Fam Med 1995, 4:403-409.
62. Tan GB, Cohen H, Taylor FC, Gabbay J: Referral of patients to an anticoagulant clinic: implications for better management. Qual Health Care 1993, 2:96-99.
63. Elcuaz Viscarret R, Beorlegui Aznárez J, Cortés Ugalde F, Goñi Murillo C, Espelosín Betelu G, Sagredo Arce T: Analysis of emergency referrals to dermatology. Aten Primaria 1998, 21:131-136.
64. Cameron JR, Ahmed S, Curry P, Forrest G, Sanders R: Impact of direct electronic optometric referral with ocular imaging to a hospital eye service. Eye (Lond) 2009, 23:1134-1140.
65. Scott K: The Swansea electronic referrals project. J Telemed Telecare 2009, 15:156-158.
66. Piterman L, Koritsas S: Part II. General practitioner-specialist referral process. Intern Med J 2005, 35:491-496.

67. Goldman L, Lee T, Rudd P: Ten commandments for effective consultations. Arch Intern Med 1983, 143:1753-1755.
68. Forrest CB: A typology of specialists' clinical roles. Arch Intern Med 2009, 169:1062-1068.
69. Salerno SM, Hurst FP, Halvorson S, Mercado DL: Principles of effective consultation: an update for the 21st-century consultant. Arch Intern Med 2007, 167:271-275.
70. Mitus AJ: The birth of InterQual: evidence-based decision support criteria that helped change healthcare. Prof Case Manag 2008, 13:228-233.
71. CM protocol results in decreased denials Healthcare Benchmarks Qual Improv 2009, 16:20-22.
72. Lucassen A, Watson E, Harcourt J, Rose P, O'Grady J: Guidelines for referral to a regional genetics service: GPs respond by referring more appropriate cases. Fam Pract 2001, 18:135-140.
73. Fertig A, Roland M, King H, Moore T: Understanding variation in rates of referral among general practitioners: are inappropriate referrals important and would guid lines help to reduce rates? BMJ 1993, 307:1467-1470.
74. Reti SR, Feldman HJ, Ross SE, Safran C: Improving personal health records for patient-centered care. J Am Med Inform Assoc 2010, 17:192-195.
75. Singh H, Hirani K, Kadiyala H, Rudomiotov O, Davis T, Khan MM, Wahls TL: Characteristics and Predictors of Missed Opportunities in Lung Cancer Diagnosis: An Electronic Health Record–Based Study. J Clin Oncol 2010, 28:3307-3315.
76. de Meyer F, Lundgren PA, de Moor G, Fiers T: Determination of user requirements for the secure communication of electronic medical record information. Int J Med Inform 1998, 49:125-130.
77. Tang PC, Ash JS, Bates DW, Overhage JM, Sands DZ: Personal health records: definitions, benefits, and strategies for overcoming barriers to adoption. J Am Med Inform Assoc 2006, 13:121-126.
78. Davis K, Schoenbaum SC, Audet A-M: A 2020 vision of patient-centered primary care. J Gen Intern Med 2005, 20:953-957.
79. Nutting PA, Miller WL, Crabtree BF, Jaen CR, Stewart EE, Stange KC: Initial lessons from the first national demonstration project on practice transformation to a patient-centered medical home. Ann Fam Med 2009, 7:254-260.
80. Reid RJ, Fishman PA, Yu O, Ross TR, Tufano JT, Soman MP, Larson EB: Patient-centered medical home demonstration: a prospective, quasi-experimental, before and after evaluation. Am J Manag Care 2009, 15:e71-87.
81. Carrell D, Ralston JD: Variation in Adoption Rates of a Patient Web Portal with a Shared Medical Record by Age, Gender, and Morbidity Level. AMIA Annual Symposium Proceedings 2006, 2006:871.
82. Kaelber DC, Jha AK, Johnston D, Middleton B, Bates DW: A Research Agenda for Personal Health Records (PHRs). Journal of the American Medical Informatics Association 2008, 15:729-736.
83. Eysenbach G: Medicine 2.0: Social Networking, Collaboration, Participation, Apomediation, and Openness. Journal of Medical Internet Research 2008, 10(3):e22.
84. Gibbons MC: Use of Health Information Technology among Racial and Ethnic Underserved Communities. Perspectives in Health Information Management / AHIMA, American Health Information Management Association; 2011:8.

85. Patel VL, Kushniruk AW: Interface design for health care environments: the role of cognitive science. Proc AMIA Symp 1998, 29-37.
86. Warren J, White S, Day KJ, Gu Y, Pollock M: Introduction of Electronic Referral from Community Associated with More Timely Review by Secondary Services. Applied Clinical Informatics 2011, 2:546-564.
87. Palen TE, Price D, Shetterly S, Wallace KB: Comparing virtual consults to traditional consults using an electronic health record: an observational case¿control study. BMC Medical Informatics and Decision Making 2012, 12:65.
88. Hersh W, Helfand M, Wallace J, Kraemer D, Patterson P, Shapiro S, Greenlick M: A systematic review of the efficacy of telemedicine for making diagnostic and management decisions. J Telemed Telecare 2002, 8:197-209.
89. Callahan CW, Malone F, Estroff D, Person DA: Effectiveness of an Internet-based store-and-forward telemedicine system for pediatric subspecialty consultation. Arch Pediatr Adolesc Med 2005, 159:389-393.
90. The control handbook. New York\: CRC Press; 1996.
91. Gardner RM: Clinical decision support systems: the fascination with closed-loop control. Yearb Med Inform 2009, 17-21.
92. Gaudinat A: Closing the loops in biomedical informatics from theory to daily practice. Yearb Med Inform 2009, 37-39.
93. Murphy DR, Reis B, Sittig DF, Singh H: Notifications received by primary care practitioners in electronic health records: a taxonomy and time analysis. Am J Med 2012, 125(209):e1-7.
94. Brynjolfsson E, Hitt LM: Beyond computation: Information technology, organizational transformation and business performance. J Econ Perspect 2000, 14:23-48.
95. Southon FC, Sauer C, Grant CN: Information technology in complex health services: organizational impediments to successful technology transfer and diffusion. J Am Med Inform Assoc 1997, 4:112-124.
96. Toussaint PJ, Coiera E: Supporting communication in health care. Int J Med Inform 2005, 74:779.
97. Ash JS, Berg M, Coiera E: Some Unintended Consequences of Information Technology in Health Care: The Nature of Patient Care Information System-related Errors. J Am Med Inform Assoc 2004, 11:104-112.
98. Magrabi F, Ong M-S, Runciman W, Coiera E: An analysis of computer-related patient safety incidents to inform the development of a classification. J Am Med Inform Assoc 2010, 17:663-670.
99. Magrabi F, Ong M-S, Runciman W, Coiera E: Using FDA reports to inform a classification for health information technology safety problems. J Am Med Inform Assoc 2012, 19:45-53.
100. Sittig DF, Singh H: Defining health information technology-related errors: new developments since to err is human. Arch Intern Med 2011, 171:1281-1284.
101. Sittig DF, Ash JS, Zhang J, Osheroff JA, Shabot MM: Lessons from "Unexpected increased mortality after implementation of a commercially sold computerized physician order entry system.". Pediatrics 2006, 118:797-801.

SAFER SELF-ASSESSMENT GUIDE: CLINICIAN COMMUNICATION

SAFER Guides

OVERVIEW

Processes relating to clinician communication are complex and vulnerable to breakdown. In the EHR-enabled healthcare environment, we rely upon technology to support and manage these complex communication processes. If implemented and used correctly, EHRs have potential to improve the safety and effectiveness of how information is communicated between clinicians. This self-assessment guide is intended to increase awareness of practices to improve the safety of EHR-based communication and support the proactive evaluation of select risk areas. It helps you identify and evaluate where communication breakdowns may occur in your healthcare delivery system and focuses on processes relating to electronic communication between clinicians. While the guide is broadly applicable, it is focused on three high-risk processes: consultations or referrals, discharge-related communication messages and patient-related messaging between clinicians. Thoughtful use of this assessment guide by EHR users is intended to stimulate implementation of the recommended practices, as well as sustain those that are already present. When assessing EHRs at repeated intervals, (such as initially, annually and when changes are made), the guide can be used to establish a baseline for measuring the effect of interventions designed to improve the safety of clinician communication. The guide works for ambulatory physician practices and other outpatient settings as well as for hospitals.

EXPECTATIONS

Healthcare professionals should use this assessment to aid in identifying and prioritizing patient safety issues related to EHR-enabled clinician communication. For example, you could consider both the frequency and severity of a safety event that might result in absence of these practices. We anticipate this to be a useful tool in ongoing safety and risk management programs, allowing you to address new risks that arise in EHR-enabled healthcare settings and helping you take advantage of the safety benefits of EHR-enabled healthcare settings. Please refer to the guide for additional information, including the specific risks and rationales addressed by the recommended practices and example strategies implemented in other clinical settings to support the recommended practices.

Legend	Key Facilitators of Practice Implementation
C	Clinicians, support staff, and/or clinical administration (e.g., Medical Records and Risk Managers)
Dx	Diagnostic services, such as laboratory or radiology—could be local or remote
Ev	EHR vendor
IT	IT support staff, could be local or contracted. Responsible for maintaining the EHR and infrastructure
Rx	Pharmacy – could be local or remote

CLINICIAN COMMUNICATION

Communication is a key aspect of nearly all processes of patient care and has an enormous potential to impact patient safety. [1-6] Communication breakdowns between clinicians are one of the most common causes of medical errors and patient harm. Several attributes of electronic health record-based communication can result in a disconnect between the sender and the receiver of clinical information, including:

- It is generally asynchronous, and often the sender cannot be sure when or if the message has been received.
- It is structured mostly around a single patient record, whereas work and relationships happen across patients.

Communication processes have increasingly become integrated into the electronic health record. [7,8] These include sending and receiving referral and consult communication, transitioning the patient from the inpatient to outpatient setting (peri-discharge period), and communicating clinical messages in the EHR. In general, there are a number of ways that communication in these processes can fail:

- Failure to include all the necessary information within the message
- Failure of the information to reach the correct person at the correct time (e.g., to an alternate clinician when primary clinician is unavailable)
- Failure to support situational awareness by overloading the user by presenting too much unstructured or irrelevant information (e.g., too many messages or alerts) [5,9]

Throughout this guide, the term "electronic communication" will primarily refer to electronic communication related to three broad activities: (1) Referral and consultation-related communication (2) Clinician-to-clinician messages, and (3) Communication during the peri-discharge period; although many of the recommended practices apply to other forms of electronic communication.

Recommended Practices	Rationale for Practice or Risk Addressed	Examples of Potentially Useful Practices/Scenarios
Phase 1 – Make Health IT Safer		
Principle: Data Availability (EHRs and the data contained within them are available to authorized individuals where and when required to support healthcare delivery and business operations.)		
1. Urgent clinical information is delivered to clinicians in a timely manner, and delivery is recorded in the EHR. C, Ev, IT	• If active efforts are not taken to inform clinicians of the presence of critical information, this information may be missed by clinicians resulting in delays in care. [11,12] • If primary care physicians (PCPs) do not receive a timely discharge summary, they may incorrectly restart or change medications for which contraindications have been identified during hospitalization.	• The organization has a policy for verbal delivery of critical information that supplements use of the EHR. • Hospitals have policies and procedures to address timely electronic delivery of important clinical information. For example, hospital discharge summaries are delivered to clinicians responsible for follow-up within two business days. • Messages are automatically forwarded to an alternate clinician if not responded to within certain time period appropriate to the time-urgency of the message. • The EHR allows automatic forwarding of messages to a surrogate clinician during a specific time period or circumstance, such as when the clinician is absent.

Assessment of Clinician-to-Clinician E-Communication

Recommended Practices	Rationale for Practice or Risk Addressed	Examples of Potentially Useful Practices/Scenarios
		• Messages are delivered to a "pool" that several clinicians are held accountable for and the responsibility of which clinician has to follow-up and when is clear.
		• When a patient transitions to another setting, a clinician provides a summary of care record to the receiving hospital or clinician in a timely manner. The summary record should include at a minimum, the Common Meaningful Use Data Set. [13]
2. Policies and training facilitate appropriate use of messaging systems and limit unnecessary messaging. C	Information overload is a significant problem in EHR systems. When a large amount of information that is not clinically relevant is transmitted through the same channels as information with high urgency, the latter may be missed leading to potential patient harm. [5,9]	• The organization has a policy on secure messaging that specifies what should and should not be transmitted, and users are trained on it • Messages are sent only to persons who may need to act upon them. 'Reply to all' is used only when necessary. • Mechanisms are in place to allow communication of non-clinical information (e.g., appointment request) in a way that does not impact communication of clinical information (e.g., abnormal laboratory results).
3. The EHR includes the capability for clinicians to look up the status of their electronic communications (e.g., delivered, opened, acknowledged).1 Ev, IT	Delays in care may result from referrals, consults, and clinician-to-clinician messages that do not receive timely action. [1,14,15]	• A real-time tracking system allows referring clinicians to determine the status of all their referrals and consults transmitted and allows specialists to identify all referrals and consults that are pending. • Clinicians and specialists are able to print a report of all their referrals and consults with the respective status of each. • Clinicians are able to identify whether their messages have been opened (read receipt). • The EHR automatically notifies the ordering clinician or team when referrals or consults are canceled or completed.

Recommended Practices	Rationale for Practice or Risk Addressed	Examples of Potentially Useful Practices/Scenarios
		• Clinicians are notified if a message they sent has not been opened within a pre-specified number of days.
		• The EHR can track whether a message was received or not.
		• Outpatient practices where messaging systems are not yet fully integrated into the EHR use additional tracking strategies to enable follow-up.

Phase 1 – Make Health IT Safer

Principle: Data Integrity (Data are accurate, consistent and not lost, altered or created inappropriately.)

4. Messages clearly display the individual who initiated the message and the time and date it was sent. Ev	In order to make informed and appropriate decisions, clinicians need to know the source and timing of a message.	• The EHR message interface prominently shows the date, time, and sender

Phase 2 – Safer Application and Use of IT

Principle: Complete/Correct EHR Use (Correct system usage [i.e., features and functions used as designed, implemented, and tested] is required for mission-critical clinical and administrative processes throughout the organization.)

5. The EHR facilitates provision of all necessary information for referral and consult request orders prior to transmission. [1,16] C, Ev, IT	• Referral and consult processing and routing may be delayed if information provided with the request is inadequate, resulting in care delays. • Referral and consultation request without certain fields filled, such as "Specialty" or "reason for referral" might be delayed	• Templates are used to facilitate completion of electronic referrals and consults to meet the specialist's requirements. • Clinicians are prompted when certain key fields, such as the "reason for referral" or "specialty" field, are left blank. • Referral requests should include, at a minimum, the Common MU Data Set. [13]
6. The EHR facilitates accurate routing of clinician-to-clinician messages and enables forwarding of messages to other clinicians. Ev, IT	Delays in patient care may results when important information is inadvertently transmitted to an incorrect recipient and cannot be redirected to the correct one.	• In the EHR, "To:" and "From:" fields are visible on message inbox and at the top of message content. • The EHR supports forwarding of incorrectly routed messages to other clinicians.

Assessment of Clinician-to-Clinician E-Communication

Recommended Practices	Rationale for Practice or Risk Addressed	Examples of Potentially Useful Practices/Scenarios
		• Clinicians can forward messages they received incorrectly to the correct recipients
		• Additional mechanisms exist for tracking acknowledgment and acceptance of forwarded notifications
7. Clinicians are able to electronically access up to date patient and clinician contact information (e.g., email address, telephone and fax numbers, etc.) and identify clinicians currently involved in a patient's care. [17] C, Ev	Patient care delays result from time spent searching for correct clinician contact information, a patient's treating clinician, or provider's care team members. Care delays may also result from incorrect message routing based on inaccurate contact information.	• The EHR system is updated at least monthly with a contact list of all practicing clinicians, and, for hospitals, includes clinician coverage schedules
		• The EHR automatically addresses internal messages between clinicians, so that email address or fax numbers need not be typed
8. Electronic message systems include the capability to indicate the urgency of messages. Ev	Communicating the urgency of a message, such as a referral or consult, is necessary to facilitate triaging, and to ensure timely follow-up	• The EHR has functionality to allow clinicians to flag referrals or consults as urgent when needed.
		• Specialists are given immediate access to all referral and consult requests, and can triage patients and schedule appointments based on urgency.
		• Messages that are administrative in nature are clearly differentiated from clinical alerts.
9. The EHR contains a copy of clinician-to-clinician communications. Ev	• Clinicians may miss important information related to a particular patient because it is "hidden" in secondary data repositories or in paper-based record storage.	• Written clinician-to-clinician communication is documented into or scanned into the EHR.
		• The EHR includes a secure messaging module with external access (i.e., to facilitate electronic communication with patients or providers not using the EHR) that does not require separate, external software.
	• Delays in care may result when specialist recommendations (such as to order further testing) are not received by the ordering clinician.	
		• If clinical messaging systems external to the EHR are used, a copy of every message is stored in the EHR.

Recommended Practices	Rationale for Practice or Risk Addressed	Examples of Potentially Useful Practices/Scenarios
Phase 2 – Safer Application and Use of IT		
Principle: System Usability (All EHR features and functions required to manage the treatment, payment, and operations of the healthcare system are designed, developed, and implemented in such a way to minimize the potential for errors. In addition all information in the system must be clearly visible, understandable, and actionable to authorized users.)		
10. The EHR displays time-sensitive and time-critical information more prominently than less urgent information. Ev	• Clinicians may miss urgent information when commingled with other less urgent messages, resulting in delayed care. • A clinician may miss a small section of relevant and important information within several pages of a referral or consults note sent to him or her.	• Messages with critical or urgent information are made visually distinct (e.g., visually highlighted). • The EHR allows sorting of clinician-to-clinician messages by urgency. • When sending notes/documentation to other clinicians (such as for co-signing), the EHR allows the sender to add recipient-specific explanatory messages, highlighting, or markups.
11. Both EHR design and organizational policy facilitate clear identification of clinicians who are responsible for action or follow-up in response to a message. [1] C, Ev	On messages addressed to multiple recipients, each recipient may incorrectly assume that the other recipient(s) will take follow-up action, leading to no action being taken at all.	• Message screens display a "responsible clinician" indicator. • The system supports forwarding and accepting responsibility for follow-up. • The EHR is able to display when responsibility for follow-up action is accepted by a clinician. • A comprehensive policy exists outlining responsibility for follow up action for certain situations (e.g., no-shows).
Phase 3 – Leverage IT to Facilitate Oversight and Improvement of Patient Safety		
Principle: Safety Surveillance and Optimization (Monitor, detect and report on safety-critical clinical and administrative aspects of EHRs and healthcare processes and make iterative refinements to optimize safety.)		
12. Mechanisms exist to monitor the timeliness of acknowledgment and response to messages. [1,18] C, Ev, IT	System problems related to delayed acknowledgment of clinician-to-clinician messages may go unnoticed if monitoring systems are not in place and checked regularly.	• Referring clinicians, specialists, and/or leadership receive an alert when no action is taken on a referral or consult request or a clinician-to-clinician message within 14 days. • Referrals and consult response times are tracked by organization leadership.

Recommended Practices	Rationale for Practice or Risk Addressed	Examples of Potentially Useful Practices/Scenarios
		• Messaging is periodically monitored to understand and improve quality of communication.
		• Policies and procedures are in place to prevent messages "lost" in the system, such as messages sent to clinicians no longer employed by the organization

REFERENCES

1. Esquivel A, Sittig DF, Murphy DR, Singh H. Improving the effectiveness of electronic health record-based referral processes. BMC Med Inform Decis Mak. 2012;12:107.
2. Gandhi TK, Sittig DF, Franklin M, Sussman AJ, Fairchild DG, Bates DW. Communication breakdown in the outpatient referral process. J Gen Intern Med. 2000;15:626-631.
3. Saxena K, Lung BR, Becker JR. Improving patient safety by modifying provider ordering behavior using alerts (CDSS) in CPOE system. AMIA Annu Symp Proc. 2011;2011:1207-1216.
4. McDonald CJ. Protocol-based computer reminders, the quality of care and the non-perfectability of man. N Engl J Med. 1976;295:1351-1355.
5. Murphy DR, Reis B, Kadiyala H et al. Electronic health record-based messages to primary care providers: valuable information or just noise? Arch Intern Med. 2012;172:283-285.
6. Sittig DF, Singh H. Eight rights of safe electronic health record use. JAMA. 2009;302:1111-1113.
7. Chaudhry B, Wang J, Wu S et al. Systematic review: impact of health information technology on quality, efficiency, and costs of medical care. Ann Intern Med. 2006;144:742-752.
8. Saleem JJ, Russ AL, Neddo A, Blades PT, Doebbeling BN, Foresman BH. Paper persistence, workarounds, and communication breakdowns in computerized consultation management. Int J Med Inform. 2011;80:466-479.
9. Murphy DR, Reis B, Sittig DF, Singh H. Notifications received by primary care practitioners in electronic health records: a taxonomy and time analysis. Am J Med. 2012;125:209-7.
10. Sittig DF, Singh H. Electronic health records and national patient-safety goals. N Engl J Med. 2012;367:1854-1860.
11. El-Kareh R, Roy C, Williams DH, Poon EG. Impact of automated alerts on follow-up of post-discharge microbiology results: a cluster randomized controlled trial. J Gen Intern Med. 2012;27:1243-1250.

12. Sittig DF, Singh H. Improving Test Result Follow-up through Electronic Health Records Requires More than Just an Alert. J Gen Intern Med. 2012;27:1235-1237.
13. Table of Meaningful Use Stage 2 Criteria. http://www.healthit.gov/sites/default/files/meaningfulusetablesseries2_110112.pdf . 2013. Center for Medicare and Medicaid Services. 4-26-2013. Ref Type: Electronic Citation
14. Hysong SJ, Esquivel A, Sittig DF et al. Towards successful coordination of electronic health record based-referrals: a qualitative analysis. Implement Sci. 2011;6:84.
15. Singh H, Espadas D, Schiesser R, Petersen L. Follow-up of electronic referrals in a multispecialty outpatient clinic. Society of General Internal Medicine 32nd Annual Meeting. 2009.
16. Sittig DF, Gandhi TK, Franklin M et al. A computer-based outpatient clinical referral system. Int J Med Inform. 1999;55:149-158.
17. Hiltz FL, Teich JM. Coverage List: a provider-patient database supporting advanced hospital information services. Proc Annu Symp Comput Appl Med Care. 1994;809-813.
18. 2014 Clinical Quality Measures (CQMs): Adult recommended core measures. 2012. Centers for Medicare and Medicaid Services. Ref Type: Pamphlet

CHAPTER 13

ASSESSMENT OF HANDHELD COMPUTING DEVICES

SOCIOTECHNICAL EVALUATION OF THE SAFETY AND EFFECTIVENESS OF POINT-OF-CARE MOBILE COMPUTING DEVICES: A CASE STUDY CONDUCTED IN INDIA

Dean F. Sittig, Kanav Kahol, and Hardeep Singh

13.1.1 INTRODUCTION

Several countries are aiming to transform their health care delivery systems with unprecedented economic investments in their health information technology (IT) infrastructures. Con-currently, policy initiatives in India and elsewhere have called for technology implementation to enhance health care quality and access [1]. Despite this momentum and commitment of resources, the pace of health IT adoption initiatives has been slower and more variable than expected [2]. Globally, only a few organizations have achieved successful transformation of their systems. [3,4]

Sociotechnical evaluation of the safety and effectiveness of point-of-care mobile computing devices: a case study conducted in India. Sittig DF, Kahol K, and Singh H. Studies in Health Technology and Informatics *192 (2013). Reprinted with permission.*

Many health care settings are just now beginning their health IT journey while others are using health IT partially and still modifying their work processes to make health IT fit. [5,6] The unexpected slower pace of health IT adoption could partially be explained by challenges to successful health IT implementation within the workflow of a complex health care system. For example, a number of unanticipated problems, including issues with patient safety and provider productivity [7,8,9,10] have occurred with IT adoption.

In view of the challenges that clinicians and organizations face with implementation of health IT, we previously developed an 8-dimension, socio-technical model of safe and effective IT use [11]. This model (see Figure 13.1.1) offers a comprehensive framework for evaluating the design, development, implementation, use, and monitoring of health IT within complex health care systems and was recently applied to electronic communication [12]. We are also using this model as a guide to proactively identify risks and opportunities to improve new and existing health IT systems [13]. Using this model to guide our current project, we sought to evaluate a mobile computing device in rural Indian healthcare settings.

13.1.2 BACKGROUND

To reform India's highly fragmented healthcare system, one essential prerequisite is a safe and effective "health IT-enabled clinical work system" that has potential to reach and im-prove the health of over one billion patients. In October 2010, the Planning Commission of India convened a High-Level Expert Group (HLEG) on Universal Health Coverage (UHC) and charged this Group to develop a framework for providing easily accessible and affordable health care to all Indians. [1] One of HLEG's recommendations was to develop a national health information technology network based on uniform standards to ensure inter-operability between all healthcare stakeholders. More than two-thirds of the population in India lives in rural areas where health care access is limited, technology penetration is low and physicians are scarce. Never-theless, one possible method of outreach is through front-line non-physician health care workers who are technology-enabled and use mobile devices to collect/interpret basic clinical data. Globally, informaticians and clinicians have always antici-

Assessment of Handheld Computing Devices

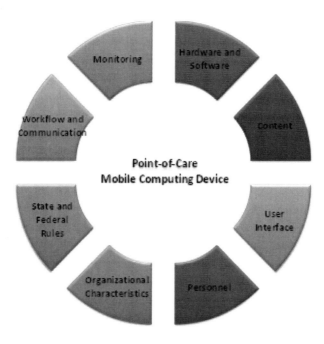

FIGURE 13.1.1: 8-dimension socio-technical model used to identify and categorize the items in the guide.

pated a small, inexpensive portable device that is capable of collecting, interpreting, storing, and transmitting patient data from the point-of-care (POC) for clinical and administrative functions. The widespread availability of tablet computers with Bluetooth and 3G/4G networking capabilities has brought such tools within closer reach.

With this vision, the Public Health Foundation of India, Division of Health Technologies, developed the "Swasthya Slate," [15] a state-of-the-art Android-based tablet computer that is designed to collect and process administrative, demographic, and physiologic data relevant to all aspects of primary care, including maternal and child care [16]. The Swasthya Slate system was designed primarily to empower frontline health workers to deliver high quality care. This system provides a seamless interface to the electronic medical record, which, when combined with cloud computing technologies, can automate data reporting to central authorities, reducing the burden of secondary data entry. Additionally, using global positioning satellites (GPS) and images, it is possible to validate and authenticate care delivery. For example, a supervisor who oversees 5-6 providers at different primary health centers can review the GPS locations at which visit data were entered and review pictures for authentication.

13.1.2.1 SWASTHYA SLATE FUNCTIONALITY

In addition to supporting manually entered information, Swasthya Slate enables digitization of test data and point-of-care-diagnostics. For example, Bluetooth-enabled blood pressure monitors and blood sugar monitors can transmit data directly to the device [17]. The tablet can also image and analyze reactive test strips to diagnose, for instance, high levels of blood glucose and anemia.

The system further facilitates the provision of high-quality healthcare by including clinical decision support (CDS) systems as part of the tablet. CDS systems use artificial intelligence algorithms or basic logical flowcharts that encode guidelines from governmental health agencies to provide healthcare workers with on-the-spot help in delivering care. These systems can also provide logistical support to the healthcare workers, for example, by enabling them to easily access daily plans, plot their care delivery routes, get reminders, and access emergency services and learn about resources at nearest care facilities to properly guide the patient.

The peripherals used with the Swasthya Slate are equivalent to those used in standard practice in the Indian public health system and include: (1) stethoscope; (2) water quality meter; (3) 3-lead ECG; (4) digital thermometer; (5) heart rate and Sp02 sensor (for oxygen saturation); (6) blood pressure monitor; (7) hemoglobin color scale; (8) urinalysis test strips; (9) blood glucose monitor; (10) digital weight scale; (11) flashlight; and (12) measuring tape. Thus, the device could enable a non-physician health care worker to collect many of the basic parameters for a medical assessment of common conditions.

Assessment of Handheld Computing Devices

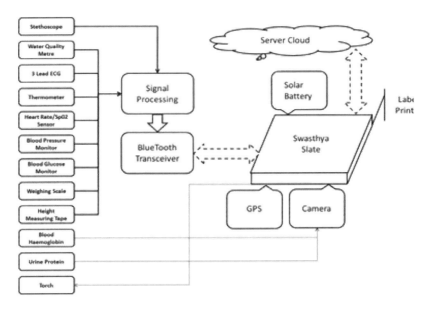

FIGURE 13.1.2: Swasthya Slate Hardware Block Diagram and system (see www.swasthyaslate.org)

To ensure that the robust functionality of this system fits within the social context of India's health system, we developed a sociotechnical assessment tool for its formative evaluation. Our study objective was to apply our sociotechnical model to develop a comprehensive evaluation strategy for the Swasthya Slate and use this evaluation to address both technical and non-technical areas of improvement during the all-important design, development, and usability testing phases of user-centered design. Our ultimate goal was to ensure the device's safe and effective, large-scale use in rural India.

13.1.3 MATERIALS AND METHODS

13.1.3.1 DEVELOPMENT OF A "SOCIOTECHNICAL" ASSESSMENT GUIDE

First, we developed an itemized assessment guide to identify potential risks or challenges to safe and effective use of the tablet under realistic clinical practice conditions. Item content was derived from several sources: 1) an extensive review of the literature, 2) interviews with experts in clinical care and health IT implementation, 3) surveys of challenges and oppor-tunities to user acceptance of these types of devices, and 4) field observations of primary care workers with various levels of clinical and computing expertise working with these and similar devices.

1. *Literature reviews:* We reviewed the literature relevant to each of the eight dimensions of our model to identify items that were applicable to safe and effective use of IT, particularly those which were directly applicable to tablet devices.
2. *Interviews:* We conducted interviews in both the US and India with experts in public health, medicine, and health IT. We focused to a large degree on frontline health workers, who bear the burden of delivering most clinical care in rural India. Interviews with health workers were important to identify potential improvements to the system in terms of usability (user interface), training requirements, compliance with local, regional and national laws and reporting re-

quirements for specific clinical conditions (e.g. pregnancy), workflow and communication, and supervision and monitoring by physicians. We interviewed administrators to further understand legal, monitoring, and workflow issues which could pose as barriers and facilitators to implementation of such a device. Finally, we interviewed physicians regarding issues of clinical content (knowledge, rules and logic embedded into the device) and whether the communication and reporting channels under development and supervision mechanisms of front-line personnel collecting data would be aligned with their expectations.

3. *Documenting/observing user acceptance:* We administered the IsoMetrics Usability Inventory [19] to frontline healthcare workers to document usability in the following 7 domains: 1) suitability for the task; 2) self-descriptiveness of the system (e.g. functions of the system are self-explanatory); 3) controllability of the system; 4) conformity with user expectations; 5) error tolerance; 6) suitability for individualization; and 7) suitability for learning. We also administered a custom developed questionnaire to evaluate how well the system fulfilled the reporting requirements of selected conditions such as pregnancy.

4. *Field observations:* Field observations were used to examine the effectiveness of training of trainers and evaluate the durability of the tablet. It also helped us study how environmental factors (e.g., temperature, rain, direct sunlight) affect the usability of the system.

13.1.3.2 MOBILE COMPUTING DEVICE EVALUATION GUIDE

An initial set of evaluation items from each of the eight sociotechnical dimensions was created from the results of the literature search and interviews. The items were then refined based on additional expert opinion, user acceptance testing, and observations. The following items (under each dimension) were determined to be most relevant to the safety and effectiveness of the device (and potentially other similar devices) and were included in the final draft of the guide.

13.1.3.2.1 HARDWARE AND SOFTWARE

Reliable hardware and software is essential for any mobile POC mobile device. The following items were found to be most relevant for safety and effectiveness of POC devices.

- The tablet will run the required software applications for at least 4 hours on battery power.
- The device has a protective case to reduce breakage or damage and prevent entry of dust into the system.
- The device is water-resistant; the screen can be cleaned with liquid disinfectant.
- The device has up-to-date virus protection software.
- The device's hardware interfaces have been tested with all external, ancillary devices (i.e., thermometer, water quality gauge, blood glucose monitor, etc.).
- The device is password protected.
- The device's hard drive is encrypted and can be erased by remote command in the event the device is lost or stolen.
- The device can connect to the Internet through a variety of means (e.g., either a wireless LAN or 3G/4G connection) and can store data locally and then upload it at a later time in the event that Internet connections are not available.

13.1.3.2.2 CLINICAL CONTENT

Up-to-date clinical content (i.e., data, information, and knowledge) is required to encode the user entered information as well as provide clinicians with reference information at the point of care.

- Required clinical content has been loaded on the device.
- Clinical content can be updated remotely.
- Clinical guidelines and CDS content are up to date.
- Clinical content is available in one of the native languages of the user. (An example of the system in Hindi appears in Figure 13.1.3 below.)

13.1.3.2.3 HUMAN-COMPUTER USER INTERFACE

The user interface enables users to interact with the data, information, and knowledge required to understand the patient's physiologic state and document their findings and intended actions.

- Users can see the information on the screen in direct sunlight.
- The fonts are large enough for middle-aged and older health care workers to read without difficulty.
- The touch screen is properly registered (i.e., when the user touches an item on the screen, the device recognizes that object has been touched).
- The device cannot be used with gloves on.
- The required software applications can be used with a finger or a stylus.
- The application allows both freehand and keyboard-based data entry.
- The device and key software applications provide multi-language support.
- Using the applications on the device requires limited text interface with audio support.
- The applications are easy to learn and text-based, audio, or video support is readily available.
- The software does not create tasks that are superfluous to the user's normal daily routine.
- The software adds value to the user's daily life.
- The software automatically produces reports and letters of discharge and referrals to minimize administrative work.

13.1.3.2.4 PERSONNEL

People are required to design, develop, implement, use, and manage all aspects of the IT-enabled healthcare system.

- All health care workers have had at least 2 hours of training on how to use the tablet in their native language.
- Centralized IT support personnel are accessible via cellphone or Voice-over-IP to health care workers.
- Health care workers are able to answer healthcare questions that are frequently asked by patients in rural areas who are unfamiliar with similar types of data collection instruments.

13.1.3.2.5 WORKFLOW AND COMMUNICATION

Modern healthcare requires extensive collaboration between disparate members of the healthcare team. Meeting the needs of various healthcare workers continues to be a challenge.

- Workflow observations are conducted and recorded prior to local implementation of the tablet.

FIGURE 13.1.3: Health maintenance reminders for maternal and child health are installed and working.

- Indications for referral are clearly specified and sent to the referring provider either via paper, fax, email, etc.

13.1.2.3.6 ORGANIZATIONAL POLICY, PROCEDURE, CULTURE, & ENVIRONMENT

Organizations that are involved with implementing and using the mobile device, policies and procedures and the culture and physical environment should empower workers and not burden them with constraints. Items that address this include:

- Standard operating procedure documents specify the scope and indications for use of the tablet.
- Procedures for maintenance and technical problem-solving are clearly delineated.

13.1.2.3.7 EXTERNAL RULES AND REGULATIONS

Local, regional and federal rules and regulations (i.e. those that originate outside of the organization) also have a significant impact on the safe and efficient functioning of the organization. This was addressed by the following items:

- Laws and provisions created by the government are adequate to protect the use of the tablet for its intended pur-poses and to prevent fraud and theft.
- Regulations create mechanisms to strictly reinforce the delivery of expedited clinical care and referrals for patients who are found to need urgent medical attention.

13.1.2.3.8 MEASUREMENT AND MONITORING

The key to improving the safety and efficiency of the IT-enabled healthcare system is to measure and monitor important details. This was addressed by the following items:

- The demographics interface is able to validate patient identity through a legitimate source such as user identification (UID), ration card, etc.
- Calibration of all physiologic or chemical sensors is per-formed every 3 months.
- 5% of data collected are validated for accuracy (e.g., 5% of automated EKG interpretations should be verified by a clinician).
- Outcome assessments are conducted using random samples of 5% of patients should be conducted to ensure that the tablet is serving its intended purpose (e.g., a positive diabetes screening should consistently prompt a referral or treatment).

13.1.3 RESULTS

Use of the guide for formative evaluation of the Swasthya Slate system resulted in several product enhancements and considerations of how the

FIGURE 13.1.4: Earlier box design

device fit within the larger social context of the health system. For instance, the tablet case was redesigned in response to feedback generated from these items (see Figure 13.1.4). The initial design emphasized the technology focus, but the final design aims to provide a more robust look with better protection against environmental factors.

Software reliability was significantly improved as well. The user interface was also improved by focusing on both affect (i.e., making it look more "sophisticated") and functionality. We utilized Microsoft's new "metro interface" design language emphasizing typography and large text on large buttons to catch the user's eye. This allowed users with limited education to use the tablet easily. We developed the reporting system to be in line with the reporting requirements of the government. For example, one of the requirements was that health workers complete a registry with a list of mothers. We interfaced the Slate with a label printer to automatically generate stickers for applicable cases, which the health worker could in turn simply stick on the register to save time and reduce omission or transcription errors. Following the ethnographic observations, the workflow was modified so that the upfront diagnostics were performed before the checkup which fit the user's workflow better as well as minimizing the time the kit needed to be turned on, maximizing the battery life. To comply with legal directives (external rules and regulations), our decision support system (content) was designed to limit interventions by frontline health workers to those that are non-pharmacological, i.e. so they didn't receive specific CDS interventions about prescribing medications beyond their expertise. We also identified skill sets of the types of personnel that would be using the device.

Quantitative data were also collected and analyzed with a specific focus on improving the usability of the tablet. To date, we have surveyed 100 community health workers, 50 nurse midwives, and 50 equivalent health workers in the private sector for our usability study. A composite scoring system was developed for each of the 7 usability domains. The mean usability rating across all of the domains was 8.9/10 (SD = 0.6). The lowest domain score was for user customization (mean 7.8/10, SD 1.1), although this was not unexpected because, by design, customization was limited to avoid potential interference with best practices. The highest domain score was suitability for the task (mean 9.2/10, SD 0.6).

Average learning time to first correct execution of the software was 10 minutes, and by 45 minutes users were able to use the apps with less than 1% "slip" errors (e.g., accidental pressing of buttons, etc.). Our training, which lasts 1 day, has been very successful in ensuring the full use of the system.

As the device is implemented more widely, we will continue to conduct additional iterative evaluation to inform device use as well as add additional items to the guide if needed for its subsequent use in other types of settings.

13.1.4 DISCUSSION

We developed a "sociotechnical" assessment guide for safe and effective use of a mobile computing health care device in India. A sociotechnical assessment can be used to help prevent unintended consequences of using mobile IT and for helping proactively detect, mitigate, and ameliorate unintended consequences and potential failures associated with the use of such devices. Our evaluation was grounded in our previously used multifaceted socio-technical model of health IT implementation and use. Based upon the work we conducted, others planning to collect and interpret data at the point of care in rural settings could consider similar formative evaluation methods to ensure successful design, development, implementation and use of these devices.

Health information technology is changing the way we deliver health care and can be used in reforming health care and improving health care access especially in developing countries. In India, there is a large deficit of physicians in rural settings, and thus point of care mobile devices can be used by trained non-physician health care workers to collect data and assist with primary health care needs. However, there might be little benefit of data collection and point of care devices unless the data is used successfully to improve clinical care in terms of improving quality, safety and efficiency. Thus, these devices must be integrated within the social context of the health system where they are implemented and used. We envision that stakeholders planning to use such devices would assemble multidisciplinary assessment teams to conduct such a comprehensive evaluation which will ensure that the device fits within the broader context of health care delivery and improvement.

Our study limitations include absence of outcome data on how Swasthya Slate impacts care processes or outcomes of clinical conditions. Nevertheless, the Slate is being pilot tested in several rural settings in India and data on impact will be available in the future. In addition, our evaluation might only be gener-alizable to certain types of rural healthcare settings.

13.1.5 CONCLUSION

To better leverage health IT, a sociotechnical approach is necessary to avoid unexpected challenges and failures. This includes both technical and non-technical formative and summative evaluations of health IT devices to ensure that they fit within the social context. Our evaluation strategy facilitated a comprehensive sociotechnical assessment and improvement of a promising point of care computing device in India. Our assessment revealed and addressed both technical (functionality, content, usability, user interface) and non-technical (work-flow, processes and policies etc.) areas of improvement.

REFERENCES

1. Thakur J. Key recommendations of high-level expert group report on universal health coverage for India. Indian Community Med. 2011 Dec;36(Suppl 1):S84-5. http://www.ncbi.nlm.nih.gov/pmc/articles/PMC3354908/.
2. Sittig D, Ash J. Clinical information Systems: Overcoming adverse consequences. Sudbury, MA: Jones and Bartlett Publishers, LLC; 2009.
3. Chaudhry B, Wang J, Wu S et al. Systematic review: im-pact of health information technology on quality, efficien-cy, and costs of medical care. Ann Intern Med. 2006;144:742-752.
4. Protti D. Comparison of information technology in general practice in 10 countries. Healthc Q. 2007;10:107-116.
5. Sittig DF, Ash JS, Zhang J, et al. Lessons from "Unex-pected increased mortality after implementation of a com-mercially sold computerized physician order entry sys-tem". Pediatrics. 2006;118:797-801.

6. Blumenthal D, Tavenner M. The "Meaningful Use" Regu-lation for Electronic Health Records. New England Journal of Medicine. 2010;363:501-504.
7. Campbell EM, Sittig DF, Ash JS, et al. Types of Unin-tended Consequences Related to Computerized Provider Order Entry. J Am Med Inform Assoc. 2006;13:547-556.
8. Metzger J, Welebob E, Bates DW, et al. Mixed results in the safety performance of computerized physician order entry. Health Aff (Millwood). 2010;29:655-663.
9. Magrabi F, Ong MS, Runciman W, et al. Using FDA re-ports to inform a classification for health information tech-nology safety problems. J Am Med Inform Assoc. 2011.
10. Harrington L, Kennerly D, Johnson C. Safety issues re-lated to the electronic medical record (EMR): synthesis of the literature from the last decade, 2000-2009. J Healthc Manag. 2011;56:31-43.
11. Sittig DF, Singh H. A New Socio-technical Model for Studying Health Information Technology in Complex Adaptive Healthcare Systems. Quality & Safety in Health-care, 2010 Oct;19 Suppl 3:i68-74.
12. Singh H, Spitzmueller C, Petersen NJ, et al. Primary care practitioners' views on test result management in EHR-enabled health systems: a national survey. J Am Med In-form Assoc. 2012 doi:10.1136/amiajnl-2012-001267
13. Singh H, Ash JS, Sittig DF. Safety Assurance Factors for Electronic Health Record Resilience (SAFER): study pro-tocol. BMC Med Inform Decis Mak, 2013; (in press).
14. Sittig DF, Singh H. Electronic health records and national patient-safety goals. N Engl J Med. 2012 Nov 8;367(19):1854-60. doi: 10.1056/NEJMsb1205420.
15. Swasthya Slate Website: http://swasthyaslate.org/usermanual.php
16. Demonstration of the Swasthya Slate Rev 2. Available at: http://www.youtube.com/watch?v=oTe_5IFgc7A
17. Contec Medical Systems Available from: http://www.contecmed.com/main/Default.asp. 2012
18. Loh B, Vuong N, Chan S, Lau C. Automated Mobile pH Reader on a Camera Phone. IAENG Intern. J Computer Science. 2011; 38(3): Advance Online Publication.
19. Gediga, Hamborg & Düntsch (1999). The IsoMetrics Usability Inventory: An opera-tionalisation of ISO 9241-10, Behaviour and Information Technology, 18, 151 - 164.
20. [ ]Esquivel A, Sittig DF, Murphy DR, et al. Improving the effectiveness of electronic health record based referral pro-cesses. BMC Med Inform Decis Mak. 2012 Sep 13;12:107.
21. Singh H, Sittig DF. A Socio-technical Model to Guide Safe and Effective Health Information Technology Use in India. Indian Journal of Medical Informatics; 6(1); 2012. http://ijmi.org/index.php/ijmi/article/view/189/74

CHAPTER 14

INCREASING RESILIENCE IN AN EHR-ENABLED HEALTHCARE ORGANIZATION

RESILIENT PRACTICES IN MAINTAINING SAFETY OF HEALTH INFORMATION TECHNOLOGIES

Michael W. Smith, Joan S. Ash, Dean F. Sittig, and Hardeep Singh

14.1.1 INTRODUCTION

The role of automation in supervisory control systems can facilitate and/or disrupt performance in many ways (e.g., Bainbridge, 1983). Likewise, the process of implementing changes in the delivery of health care can succeed or fail depending on numerous factors (Grol & Grimshaw, 2003). The volatile combination of the two can be seen in the current political, commercial, and scientific activities surrounding the adoption of electronic health records (EHRs) in the United States (Wright et al., 2013),

---

Resilient Practices in Maintaining Safety of Health Information Technologies. © *Smith MW, Ash JS, Sittig DF, and Singh H.* Journal of Cognitive Engineering and Decision Making **8**,3 *(2014), doi: 10.1177/1555343414534242. Reprinted with permission.*

especially the explorations and debates about EHR safety (Institute of Medicine, 2012).

Greenhalgh, Potts, Wong, Bark, and Swinglehurst (2009) describe several conceptualizations or "meta-narratives" about EHRs in the literature. The "health information systems" meta-narrative frames EHRs as tools for systematically managing clinical information, relieving health care workers of this burden, and thereby protecting patients from the associated safety risks. Technology is viewed primarily as a means of preventing mistakes by constraining human performance. This approach corresponds to a view of human error and safety that sees accidents as products of a faulty (usually human) element in proximity to the accident. Accordingly, strategies to increase safety focus primarily on the use of barriers (Dekker, 2006; Qureshi, Ashraf, & Amer, 2007).

In contrast, the "critical sociology" meta-narrative (Greenhalgh et al., 2009) conceptualizes EHRs as tools for systematically managing the work of clinical providers, imposing a model that presumes optimal efficiency and consistency while constraining the ability of clinical providers to respond to situated needs. This approach considers the underlying structure of the organization as a possible risk factor for errors. Similarly, some models of safety, such as the "Swiss cheese" model (Reason, 1997), view accidents not simply as the fault of front-line workers but as the result of multiple failures in a series of barriers, including latent failures in the organizational environment (Dekker, 2006; Hollnagel, 2008a; Qureshi et al., 2007).

Though these two meta-narratives reflect opposing perspectives, both contain valid insights on how EHRs affect clinical work. EHRs offer many potential safety enhancements and other benefits to providers and other clinical staff (Jha & Classen, 2011). At the same time, EHRs serve in some ways as supervisory control systems for the process of care delivery. For example, EHRs may shape provider choices by presenting certain options for tests and/or treatments for selection, or they may discourage providers from ordering potentially dangerous drug combinations (Teich et al., 2000). In some systems, deviations from computer-recommended treatments require a justification (Hsieh et al., 2004).

Although the two aforementioned perspectives highlight the impact of EHRs on clinical work, both are oversimplifications. First, an EHR is not a homogenous entity. At its simplest, an EHR is a database containing the health information of patients under the care of a facility, but in practice, EHRs are sophisticated software applications that contain and/or interact with other applications, including systems for computerized provider order entry, clinical decision support, test results management, pharmacy databases, and medication administration systems (i.e., bar-coding systems; Committee on Data Standards for Patient Safety, 2003). These software applications require networked hardware and clinical knowledge structures, such as decision rules and vocabularies, to operate (Sittig & Singh, 2010).

Second, EHRs are not static; rather, they are subject to change from other elements in the sociotechnical system, including providers (Hunte, Wears, & Schubert, 2013). For instance, EHRs evolve in response to changes in clinical knowledge structures and how providers encode medical problems (Aarts, 2011). Thus, the front-line providers at the "sharp end" of patient care (Cook & Woods, 1994) are not simply passive recipients but active agents in the ongoing development of the EHR. Being a technical system in people-centric health care delivery organizations, the EHR is a part of a multilevel network of pressures and influences. Patients, providers, managers, regulators, and many sociopolitical factors influence, and are influenced by, EHRs (Leveson, 2004; Rasmussen, 1997; Vicente, 2002).

Third, as dynamic systems embedded in organizations with quality, production, and resource pressures, EHRs need to be effectively managed, especially to maintain safety. Active maintenance and oversight activities are needed for adaptations in response to changes to health care systems (Reason, 1997; Reiman, 2011). Otherwise, changes may lead health care systems into unsafe states of "brittleness" in which additional stress may lead to sudden failure instead of smooth adaptation (Cook & Rasmussen, 2005; Woods & Wreathall, 2008).

The "systems approaches to risk management and integration" meta-narrative (Greenhalgh et al., 2009) acknowledges the role of EHRs as components of complex and dynamic sociotechnical systems, from whose interactions can emerge new modes of safety and risk. This conceptualization views safety as the product of complex interactions at multiple levels and the management of safety as involving awareness of risk and ongoing use of control processes. Thus it relates to a newer approach to safety where the focus is to ensure that the complex system can detect and adapt to new risks (Hale, Heming, Carthey, & Kirwan, 1997; Rankin, Lundberg, Woltjer, Rollenhagen, & Hollnagel, 2013; Saleh, Marais, Bakolas, & Cowlagi, 2010; Woods, Dekker, Cook, Johannesen, & Sarter, 2010). Resilience is a term that refers the capability of a complex system to maintain safe operations and the ability to fulfill its objectives despite new pressures or constraints. The resilience engineering approach to safety therefore emphasizes the system's ability to respond to changes in risk (Cook & Nemeth, 2006; Woods & Wreathall, 2008).

Functions identified as fundamental to resilience include monitoring for changes and threats, anticipating changes and being proactive, ensuring the capability to respond to disruptions, learning from past experiences (Hollnagel, 2009), management commitment, flexibility, buffering capacity, awareness of risk (Carthey, De Leval, & Reason, 2001; Costella, Saurin, & de Macedo Guimarães, 2009; Woods, 2006), and using control systems to maintain functioning in dynamic conditions (Hollnagel, 2008b; Leveson, 2012).

Much of the empirical research on resilience has occurred in the energy, aerospace, petrochemical, and transportation industries (Costella et al., 2009; Hollnagel, Woods, & Leveson, 2006). Compared to these domains, the system dynamics in health care can be considered to be more influenced by the intentions of social actors (Pejtersen & Rasmussen, 1997). Another difference is that preventable bad outcomes are relatively common and often undetected in health care, unlike in these other domains (Amalberti, 2006; Wears, 2012). While there is growing literature on resilience in health care (e.g., Wears, Hollnagel, & Braithwaite, 2013), including health information technology (HIT; Nemeth & Cook, 2007; Skorve, 2010), there are no empirical studies yet of resilience in management of EHRs.

Understanding successful practices in the management of EHR-related safety is critical given the inherent safety risks associated with EHRs (Ash, Sittig, Campbell, Guappone, & Dykstra, 2007; Karsh, Weinger, Abbott, & Wears, 2010; Sittig & Singh, 2009; Skorve, 2010; Walker et al., 2008) and the rapid adoption of HIT in the United States (Coiera, Aarts, & Kulikowski, 2012). In view of the dynamic complexity of EHR-enabled health care delivery systems (Carayon et al., 2006; Kannampallil, Schauer, Cohen, & Patel, 2011; Sittig & Singh, 2010), we propose that successful safety management of EHRs entails the use of resilience-related practices.

The primary goal of this study was to identify the role of resilient safety practices in the management of EHR safety. While the specific practices are situated in the domain of health care, the patterns in how the practitioners cope with complexity may reflect general strategies used in other domains that, like health care, are also trying to maintain resilience while introducing automation into complex sociotechnical systems where boundaries between safe and unsafe can often get fuzzy. Thus, a secondary goal of the study was to go beyond the specific domain and see how these practices relate to more general, domain-independent patterns of dealing with challenges in complex systems (Roth et al., 2013; Woods & Hollnagel, 2006). To collect evidence on practices to successfully manage EHRs within complex sociotechnical systems, we focused on safety practices used in large health care systems that have had many years of experience successfully managing HIT quality and safety.

14.1.2 METHOD

This study was part of a larger project on EHR safety (Singh, Ash, & Sittig, 2013) that involved numerous interviews, each one focused on one of the facets of HIT identified as key risk areas (Magrabi, Ong, Runciman, & Coiera, 2012; Myers, Jones, & Sittig, 2011). These key risk areas were computerized physician order entry, clinical decision support, test results reporting, communication between providers, patient identification, EHR downtime events, EHR customization and configuration, system-system interface data transfer, and HIT safety-related human skills. Our settings were two very large private health care systems in the United States regarded as successful pioneers in EHR implementation, each with over 20 years of experience using clinical IT systems. These two systems are Partners HealthCare (Teich et al., 1999) and Geisinger Health Systems (Paulus, Davis, & Steele, 2008).

We conducted interviews with 56 key informants (36 from Partners, 20 from Geisinger). The informants were identified by leadership contacts at each facility, based on each informant's expertise in one or more of the key risk areas listed above. Thus, the interviews were broadly focused on EHR safety as it related to the expertise of the key informant rather than only on the use of resilient safety practices. Informants' roles included chief medical information officer, director of nursing informatics, director of pharmacy informatics, director of IT optimization/innovation, risk manager, and physician project specialist. The interviews were semistructured, consisting of questions pertaining to that informant's unique expertise and responsibilities, and included open-ended questions inviting the informant to raise issues of his or her own. Hence a different set of questions was used for each participant. All interviews were recorded and transcribed. Interview transcript lengths ranged from 2,000 words to 16,000 words, averaging approximately 6,500 words.

Conceptually, we approached the analysis of interview data using a systems resilience engineering framework (Costella et al., 2009; Hollnagel, 2009; Woods, 2006). We analyzed interview transcripts using framework analysis (Ritchie & Spencer, 2002), a qualitative methodology that allows for both a "top-down" analysis using an existing framework and a "bottom-up" analysis for emergent themes or patterns. Thus, our analysis accounted for both anticipated resilience-related concepts, such as monitoring, anticipation, and sensitivity to risks, as well as new information that did not readily fit within our framework.

Using an iterative process, the transcripts were reviewed and statements relating to safety practices and any resilience-related concepts were coded as such. Then the coded items were reviewed to more specifically identify in what ways they reflected aspects of resilient safety practices. Afterward, these codes were reviewed and used in a process of iterative categorization in which coded items were grouped according to how they reflected concepts related to resilience. As various categorizations illuminated different patterns, items were recategorized accordingly. From this process, the final set of categories emerged. Then the literature was searched to confirm that each of the category concepts had been identified independently in other research on safety in complex domains (see Discussion). The Atlas.ti software package (ATLAS.ti Scientific Software Development GmbH, Berlin, Germany) was used for coding passages of transcripts. As part of the emphasis on domain-independent patterns, the coding was performed by a research team member with a background in cognitive systems engineering rather than health care operations or health information systems. To enhance credibility of the analysis (Patton, 2002), corroboration was performed by looking for conflicts between our initial coding and categorization results versus coding on general HIT safety issues and detailed interview summaries, all generated independently by other research team members as part of the parent project (Singh et al., 2013).

14.1.3 RESULTS

14.1.3.1 INFORMANTS AND STATEMENTS

From the interview transcripts, we identified 156 statements or references regarding resilient practices from 41 different informants. The informants represented a diverse range of roles and included physicians, IT personnel, and quality and safety personnel (Table 14.1.1).

TABLE 14.1.1: Facility and Roles of Key Informants

	Interviewed	Mentioned Resilient Practices
Facility		
Partners	36	25
Geisinger	20	16
Role		
IT	14	11
Informatics	15	11
Physicians	12	8
Other clinical operations	8	5
Safety, quality, and security	7	6
Total	56	41

14.1.3.2 CATEGORIZATION

Our analysis generated five main categories, each with two or more subcategories (Table 14.1.2). The categories reflect the resilience functions of sensitivity to risks, monitoring, control systems, responding, anticipation, and mindfulness. Examples and details for each category level and subcategory are in Table 14.1.2.

TABLE 14.1.2: Levels of Resilient Practices

	LEVEL	SUMMARY OF PRACTICES
1.	**Sensitivity to fundamental risks**	The informants recognized the dynamic nature of the HIT systems and how they are used, and the interdependencies between parts of the HIT systems and the larger healthcare system, and how these can affect patient safety risks.
a.	Awareness of need for monitoring	
b.	Sensitivity to dynamics and interdependencies	
2.	**Basic monitoring and responding practices**	They used a very wide range of approaches to monitor and evaluate the performance of the systems, including indicators of risk. Responses to problems involved work on software, but also on other facets of the sociotechnical system as well (e.g., software enhanced workarounds to mitigate risks due to poor system integration).
a.	Processes of testing and tracking	
b.	Processes for responding	
3.	**Management of monitoring and responding practices**	They had practices to ensure continued capability to effectively monitor and respond to risks. They used their understanding of dynamics and interdependencies to use resources more efficiently.
a.	Maintaining capability for testing and tracking	
b.	Maintaining capability for responding	
c.	Enabling safety and quality control	
4.	**Sensitivity to risks beyond the horizon**	Practices were in place to proactively assess for risks, and to deliberately avoid installing software that would unduly increase risk.
a.	Processes and mechanisms for being proactive	
b.	Controlling risk at governance level	
5.	**Reflecting on risks with the safety and quality control process itself**	Many of the informants were aware of limitations with the methods used in detecting and managing risk. Furthermore, some were aware of limitations and overly narrow system boundaries in the conceptual model of the system used to guide the quality control process.
a.	Limitations of monitoring methods	
b.	Sensitivity to failure in the quality control process	

14.1.3.3 SENSITIVITY TO FUNDAMENTAL RISKS

Statements in this category referred to the organization's awareness of basic threats to the safe operation of HIT systems.

> *We absolutely have to test every one of them [monthly software patches from vendor]. . . . We may report an issue . . . and they [the vendor] work on it, and they have a fix for it ready. . . . They say, "OK, yeah, we have this ready. We can send you the fix." But when we do that [implement the fix], that fix also touches all these other things . . . and all those other things also had fixes. So now this one fix you want, you have to bring in 50 others.*

14.1.3.3.1 AWARENESS OF THE NEED FOR ONGOING SAFETY ANALYSIS, INCLUDING REVIEW AND DOCUMENTATION

Informants described the need for testing and post-implementation monitoring, even when software was off-the-shelf without any add-ons ("No one ever just puts it in place and lets it go"). Informants also mentioned periodic reviews of policies and order sets and revisiting decisions to disable particular EHR safety features. They mentioned the importance of documenting the processes by which EHR components are implemented and monitored, and documenting workflow because workflows change. Two informants expressed sensitivity to the risks faced by health care facilities that are new to EHRs: "They don't know what they don't know."

14.1.3.3.2 SENSITIVITY TO DYNAMICS AND INTERDEPENDENCIES IN THESE SOCIOTECHNICAL SYSTEMS

Some comments emphasized the volatility of both the EHR system and the larger context of the health care industry. Informants mentioned several factors that changed the risk profile of the system over time, including the increasing length of patient notes ("note bloat"), providers' alert fatigue, and customization of the EHR. One informant pointed out how risks related to patient identification could affect a broad range of functions, from medication administration to food delivery.

Several comments referred to the interface between the EHR and other system components. For instance, one informant recounted a problem that resulted when another application vendor made changes to its EHR software without informing the facility, which led to problems with the integration of the updated software with other software in place. Another informant described the complexity of adding a new drug into the order entry system; in order for the drug to be dispensed correctly, it also needed to be added to other parts of the system (pharmacy and bar-coding administration systems). These are examples of the informants' sensitivity to the problem of asynchronous evolution of subsystems within integrated systems, whereby changes in one subsystem lead it to become incompatible with the other subsystems that lag behind (Leveson et al., 2006).

Finally, informants were aware of how changes in government policies, such as the Health Insurance Portability and Accountability Act of 1996 (HIPAA; Department of Health and Human Services, n.d.) and "meaningful use" criteria (Blumenthal & Tavenner, 2010), affect workflow and how these changes in workflow in turn affect software configuration and use (Campbell, Guappone, Sittig, Dykstra, & Ash, 2009). The interdependency between software and workflow was also acknowledged in the inclusion of workflows into the software testing process and in a vendor-led process of requirements and specifications validation.

14.1.3.4 BASIC MONITORING AND RESPONDING PRACTICES

Statements in this category referred to ways in which facilities ensured that their EHRs continued to operate as designed and fulfill their basic functions.

> *We do a tremendous amount of testing before any release upgrade. The team reads all of the release notes. We do integrative testing, unit testing. We do a dry run. We do testing where we have two weeks where almost all hands are on deck just the first two weeks that the new software is loaded into a test system to see what we're breaking.*

14.1.3.4.1 PROCESSES OF TESTING AND TRACKING

Informants used various testing practices and other strategies to assess the functioning of EHR systems. These included testing in development platforms and in the live platforms using test patients. One facility used the Leapfrog Group's safety assessment tool for computerized physician order entry systems (Classen, Avery, & Bates, 2007).

Software upgrades were often mentioned as an impetus for testing. The testing required by upgrades encompasses not only the upgrade itself but also the ancillary systems that are integrated with the upgraded software and the post-upgrade patches that are developed by the vendor as problems are discovered. System configurations were tested to ensure compatibility with clinical workflows. In addition, several informants mentioned testing the integration between components. One informant specifically described a test that included not only sending information to the pharmacy but also calling the pharmacy to ask how the information was displayed on the screen.

Informants mentioned practices designed to encourage incident reporting and regular review of IT-related safety concerns. Other sources of information about risks included announcements from the vendors and networking with other facilities that use the same EHR systems.

14.1.3.4.2 PROCESSES FOR RESPONDING AS PART OF QUALITY AND SAFETY CONTROL FEEDBACK

Informants presented many different ways they continually adjusted and corrected EHR systems. There were processes and tools that facilitated rapid response to EHR issues by monitoring for acute problems and alerting staff. For problems with a vendor's software, responses could involve informing the vendor of problems discovered and sharing information about problems and solutions with other facilities that use that vendor's EHR.

Reports of problems and incidents were reviewed with a focus on identifying potential areas for improvement and proposing solutions. Responses could be quite thorough in order to maintain safety. In one case, when it was discovered that many allergies had been entered into records not as coded data but as free text, which the drug allergy checking algorithm could not detect, the facility manually recoded the allergies in all the affected records.

The types of responses mentioned also included work-arounds. For example, because a software system for managing dosing of anticoagulant medication did not integrate well with the order entry system, there was a risk that patients would be prescribed the wrong dosage. In response, the hospital developed software that prompted providers to double-check the dosages when discharging a patient on anticoagulants. A similar solution was developed to prompt providers to double-check certain high-risk medications during the medication reconciliation process performed when patients are discharged from the hospital. In addition to exploring work-arounds as possible short-term solutions for risks, facilities assessed the proposed work-arounds for any risks they might introduce if they were implemented.

14.1.3.5 MANAGEMENT OF MONITORING AND RESPONDING PRACTICES

Statements in this category referred to what the facilities did to ensure they were able to continue to oversee and correct HIT operations.

> *One thing we try to stress, from a production support standpoint, is that when you do your project life cycles, there should be a line item there for an error [management] process.*

14.1.3.5.1 MAINTAINING CAPABILITY FOR TESTING AND TRACKING

Informants stressed the importance of maintaining resources for ongoing monitoring and review. They described staffing resources made available for such functions, including people to monitor logs of errors in information transmission and people to perform various manual testing operations.

Organizational structures were changed to use staffing resources more effectively and efficiently. For example, groups of informaticians and risk management specialists were moved to integrate better with front-line care staff.

Informants mentioned resources dedicated to help people report safety risks or incidents. These include incentives for physicians, systems to track the status of the facility's response to the issue, and specific tools to enable providers to easily comment on the appropriateness of decision support recommendations.

To use testing resources efficiently, they were applied based on the likelihood of uncovering something important. The degree of testing was adjusted to the estimated impact of the implementation on clinical processes. For example, informants mentioned how a forthcoming major update to the EHR was undergoing rigorous testing but that smaller changes to the system received less intensive testing. They also mentioned how most problems were usually discovered early on, in the pilot testing, which allowed for corrections to be made before the software was fully implemented.

Software tools to support monitoring were also mentioned. These included tools for automatically monitoring the interfaces between subsystems and a tool for tracking safety issues (Walker, Hassol, Bradshaw, & Rezaee, 2012). One facility replaced periodic error reports with continuous error log monitoring. Another use of technology was the automation of a multistep software testing process. Scripts were used to efficiently run a battery of standard tests. One informant mentioned testing the accuracy and potential impact of decision support algorithms in a way that did not impact providers. The algorithms were run in "stealth mode," without showing the alert messages to clinicians but storing results in a log for review. In an additional example of the use of tools to detect problems early on, a thermal scanner was used to detect unusually hot electrical connections between devices in the data center serving the IT system.

As the use and scale of the EHR evolved, tools were created in response to the accompanying risks. One evolving risk was from bad information accidently entered into charts due to imprecise copying and pasting from older notes in the EHR, instead of clinicians manually entering in all the text documentation (Hammond, Helbig, Benson, & Brathwaite-Sketoe, 2003). One facility developed a tool to monitor the extent of copying and pasting in their charts. Another tool was developed to identify instances of potential mismapping between a patient and a record. It detected unusual modifications made to patient identification information (e.g., updating birthday or full name), as would occur when Person B's record was being altered under the false assumption that it belonged to Person A. Another example of software developed in response to a problem was a tool that automatically checked for addressing and routing problems that could prevent correct delivery of pathology reports to the ordering providers.

14.1.3.5.2 MAINTAINING CAPABILITY FOR RESPONDING

Informants also stressed the importance of maintaining the resources necessary for responding to problems that arise. Furthermore, one mentioned the value of a process for ensuring resolution of issues by escalating unresolved ones to relevant leadership. Another informant emphasized preemptively validating particular error management processes to ensure that the team can fix any known problems with software should they occur after the software is implemented on the live platform. In other words, a team may set up an error on a test platform and see if it is possible to repair it with the constraints present on the live production platform.

Informants mentioned a few specific examples of maintaining response capabilities. As instructed by the supervisor, a junior IT person performed a maintenance operation on a live in-use (but redundant) part of the IT infrastructure in order to become more comfortable with working on live in-use IT systems should the need arise. To be available in case of problems with a significant upgrade being installed on the live system, a large IT team remained on site the entire night. In order to implement changes more rapidly, some informaticians were dedicated to and co-located with specific front-line clinical operations.

14.1.3.5.3 ENABLING SAFETY AND QUALITY CONTROL IN AN EFFECTIVE AND EFFICIENT WAY

Some informants mentioned practices for securing resources for quality improvement. One informant used data from assessments and monitoring

Increasing Resilience in an EHR-Enabled Healthcare Organization   397

to facilitate budget negotiations for resources for safety. Other practices mentioned were to use the resources more effectively and efficiently. One example was embedding informatics staff with clinical groups in order to speed up the cycle of problem detection and response. Many practices involved using IT tools to support the quality improvement process, such as providing information to stakeholders via databases and reporting engines. Informants mentioned specific examples of using IT to improve the process of monitoring performance and implementing adaptations. In one example concerning the tool used by clinicians to view information on patients who had been transferred into that facility, software tracked the way clinicians customized the display of information fields. That data were then used to identify which fields to prioritize in the redesigns of the patient transfer forms. Another example was a practice aimed at reducing alert fatigue. Pop-up warnings for drug-drug interactions were monitored to see which ones were overridden by providers on a consistent basis; those warnings for which the providers had found no value were removed in order to reduce alert fatigue (Phansalkar et al., 2010).

14.1.3.6 SENSITIVITY TO RISKS BEYOND THE HORIZON

Statements in this category referred to facilities' strategies of predicting and proactively managing problems that could occur in future HIT systems and in the larger sociotechnical system beyond the HIT system itself.

> *We have to be proactive here. . . . You know why? Because if you're up at 2 o'clock in the morning and trying to deal with a production problem where someone's life may be in danger, you want to be proactive. You want to have the work done ahead.*

14.1.3.6.1 PROCESSES AND MECHANISMS FOR BEING PROACTIVE

Several informants mentioned the role of testing and evaluation to proactively identify problems. One informant stressed the close evaluation of software upgrades to facilitate responding to potential problems during the upgrade process. Another mentioned that "most of the bigger issues get found in pilot phase.... We can ... make corrections before it even gets rolled out anywhere else."

Informants mentioned the use of reviews to identify potential hazards with new EHR implementations. In preparation for going live with a new inpatient system, one facility conducted a thorough prospective risk assessment, involving a wide range of clinical staff (Hundt et al., 2013). Another type of prospective risk assessment mentioned was validation sessions with the vendor of an EHR, to assess fit with workflows and identify gaps.

Additionally, one informant stressed the need for resources and processes to make sure issues were detected proactively and were acted upon. This included having effective communication channels with leadership to ensure that issues get addressed. Informants at both facilities mentioned having governance committees that reviewed proposed additions or modifications to the clinical IT systems and changes to clinical knowledge structures (e.g., decision rules, templates). These committees included people from clinical, IT, operations, and risk management. Also mentioned were efforts to ensure that representatives of various groups, including end users, were involved in HIT decisions.

Informants mentioned processes to keep leadership and other stakeholders informed about the risks in the system. This included reports to HIT and clinical steering committees, reviews of significant safety issues with the management team and the executive committee, and notifications to senior leadership (Belmont et al., 2013). Another practice was a daily teleconference involving representatives from various locations in the system, in which open issues including safety and EHR concerns were discussed.

14.1.3.6.2 CONTROLLING RISK AT GOVERNANCE LEVEL

Informants also shared examples concerning how knowledge of potential risks was used in decisions about IT implementation. These examples included: deciding to not install some software because of problems detected ahead of time, disabling some EHR functionality because of the associated risks, and installing major version upgrades only after the subsequent wave of software patches from the vendor had been issued.

One facility was migrating from a custom best-of-breed system (using components from various developers) to an integrated off-the-shelf system in response to the risks posed by continued use of a "fragmented" and "siloed" system. Informants also mentioned instances where investments in new HIT-based safety projects were made only after explicit assessments of risks.

Informants mentioned using anticipated risks to inform decisions about resource investments for safety. One informant stressed how there would still be a need for resources for EHR configuration and evaluation even after the pending migration to an off-the-shelf system. Another mentioned the need to expand the current IT backup infrastructure to keep pace with the rapid growth of HIT in the facility.

Identifying future needs could also involve factors outside the immediate focus of EHR safety. One informant gave an example of this involving the meaningful-use EHR reimbursement requirement (Blumenthal & Tavenner, 2010) concerning greater use of patient portals. Because recent lab results and current medication lists were now more visible to patients, the clinicians became under pressure to make sure those parts of the record were accurate and up-to-date. However, due to workforce distributions and rules regarding scopes of practice, nonphysicians were unable to offload the extra work now required of the primary care physicians.

14.1.3.7 REFLECTING ON RISKS WITH THE SAFETY AND QUALITY CONTROL PROCESS ITSELF

Statements in this category referred to methods by which the facilities recognized and addressed ways the safety and quality control process itself could fail.

> *What the [information systems] leader thinks is happening, in fact, isn't necessarily what's happening.*

14.1.3.7.1 LIMITATIONS OF MONITORING METHODS

Informants emphasized the importance of the quality and accuracy of the information in the system, one stating that "the flow of information about potential errors needs to be very high quality." The limitations of incident reports and verbal feedback were mentioned by two informants, who described them as incomplete sources of information.

Informants mentioned the limitations of testing and how problems have been encountered in live EHR systems despite thorough prior testing. They suggested many reasons. Problems related to specific and infrequently encountered interactions or other circumstances would be more likely to appear only during widespread regular use, not during limited, pre-live testing. Automated testing tools would not necessarily work for a facility's particular customizations, nor would testing environments necessarily capture all the relevant aspects of the real world system. Testing may not have encompassed the range of upstream and downstream components that could be involved in a safety issue.

Informants referred to alternative methods used to mitigate the limitations of current monitoring methods. Focus groups and surveys with end users were used to solicit feedback on patient transfer and handoff tools. After scripts and configurations of IT products were set up, a second IT person would review them before they were implemented.

The way users performed tasks with the software was monitored to see if they had to perform work-arounds due to shortcomings with the software and its fit with the users' workflows. Developers used a similar approach to evaluate a handoff tool designed to provide all the necessary information for clinical staff taking over responsibility of patients. Clinicians were asked if any information was missing and also what problems arose or additional tasks were required as a result of information not being available.

14.1.3.7.2 SENSITIVITY TO FAILURE IN THE QUALITY CONTROL PROCESS

One informant mentioned some limitations with the quality control process itself related to software and workflow validation. One aspect of this was how the introduction of new software functionality almost always occurred in the context of other concurrent changes, meaning that confounding factors made it difficult to establish the particular role of the EHR intervention in affecting outcomes. To improve the quality control process itself, facilities monitored the implementation of EHR interventions more closely and implemented tools to support collaboration across departments.

Some informants raised questions about the underlying assumptions regarding the functioning and scope of the EHR system. One pointed out that many patients move about the country but still need continuity of care, thus requiring EHRs to support real-time health information exchange over a much wider geographic range of clinical partners than currently supported. Addressing the need for continuity and coordination of care across different facilities, another informant suggested that EHRs could and should do more to support coordination beyond the current function of simply exchanging minimal clinical data.

One informant mentioned an effort illustrating how IT managers were willing to acknowledge the risk of problematic inaccuracies in their own interpretations of the current state of the EHR. Work on mitigating these limitations included plans for software to automatically capture additional data on the current state of the EHR.

14.1.3.8 CREDIBILITY ASSESSMENT

Our findings were evaluated against the independently generated codes on HIT safety in general and the independently authored detailed summaries of the interviews for each of the facilities, all generated as part of the parent project (Singh et al., 2013). There were no conflicts between our findings and those coding results and detailed summaries.

To confirm that the categories reflect practices at both facilities, we assessed the distribution of the topics of the comments within and across the two facilities. A chi-square test of the number of comments from Partners and Geisinger for each of the five levels indicates no significant difference in proportions across facilities and category levels (p = .310).

14.1.4 DISCUSSION

14.1.4.1 CATEGORIZATION OF STATEMENTS ABOUT RESILIENT SAFETY PRACTICES

We conducted interviews with 56 key informants from two large health care systems recognized as leaders in use of HIT and EHRs. The interviews focused on HIT and EHR safety issues related to the various roles of the informants, which included IT, informatics, clinical providers, and safety and quality managers. Forty-one participants mentioned practices that reflected some element of the resilience approach to safety. Overall, there were 156 references to resilient practices. The practices covered a wide range of activities related to resilience in health care systems. For instance, almost all of Carthey et al.'s (2001) 20 indicators of institutional resilience in health care systems are reflected in the set of practices. The results also show how these facilities engaged in practices that support different levels of HIT safety: the functioning of the HIT systems themselves, the co-evolution of workflow and HIT systems to enhance performance, and the application of IT to new ways for facilitating safety (Sittig & Singh, 2012).

Our analysis generated a categorization of five levels, each with two or more subcategories. The levels reflect some primary resilience functions: sensitivity to risks (Costella et al., 2009; Nemeth, 2008; Woods et al., 2010), monitoring risks (Hollnagel, 2009; Wreathall, 2011), using control systems to track and modify performance of safety-related operations (Hollnagel, 2008b; Leveson, 2012), maintaining the capability to respond (Hollnagel, 2009; Pariès, 2011), anticipating changes and being proactive (Hollnagel, 2009; Klein, Snowden, & Pin, 2010; Woods, 2011), and mindfulness and reflection on the risks related to the safety management process itself (Reason, 2008; Woods, 2006; Woods et al., 2010; Woods, Schenk, & Allen, 2009).

The practices that reflect resilience are certainly not the only practices important for HIT safety. The importance of best practices in software design, usability testing, and implementation have been stressed (Middleton et al., 2013; Office of the National Coordinator, 2012). Other developments in systems safety, such as high reliability theory (Roberts, 1990), can also offer some contributions. However, the results here emphasize the need for facilities to practice active monitoring and management beyond the initial development and implementation stages. They include practices regarding managing resource constraints and other trade-offs, which is not addressed by high reliability theory.

14.1.4.2 GENERIC PATTERNS THAT FACILITATE RESILIENCE

Because we interviewed informants who were engaged in the real-world work of managing HIT systems in large health care systems, we have made certain that our findings are ecologically valid and grounded in real-world practice. In accordance with the research agenda of cognitive systems engineering (Woods & Hollnagel, 2006), it is important to complement this focus on a specific context by addressing generic, domain-independent patterns in how work is accomplished in complex systems (i.e., macrocognitive functions; Cacciabue & Hollnagel, 1995). We look beyond the specific categories by exploring the commonalities among them, establishing an interpretation of the findings in terms of underlying functions that facilitate resilience.

14.1.4.2.1 REFLECTION

The relationship between the five category levels can be seen in terms of how a critical and reflective view of one level is necessary for the implementation of the subsequent level.

- Level 1 (sensitivity to fundamental risks) is the critical recognition of dynamic risks to which the system is vulnerable.
- Even though the risks are ensconced in uncertainty, in Level 2 (basic monitoring and responding practices), the risks are seen as subject to prediction and management via systematic measurement and intervention.
- In Level 3 (management of monitoring and responding practices), these systematic measurements and interventions are acknowledged as operations that require resources and oversight.

- This ongoing need—for resources and oversight to manage risk—presents pressures for efficiency and effectiveness. Thus in Level 4 (sensitivity to risks beyond the horizon), there is the recognition of the need for planning, anticipation of future demands, and proactive management.
- Because of the anticipation of future demands and planning for them, there is in Level 5 (reflecting on risks with the safety and quality control process itself) recognition of the potential problems owing to the limits of the current safety management approach itself.

Reflection is related to a basic pattern in how cognitive systems manage complex work. The capacity to shift and contrast perspectives is essential in exploring complex situations, generating alternate courses of action, and coordinating work with others (Woods et al., 2010; Woods & Hollnagel, 2006;). The presence of multiple potential points of view encourages reflection and critical evaluation of a given point of view (Hoffman & Woods, 2011). Reflection is also related to one of the basic functions required for resilience: learning (Hollnagel, 2009, 2011). As part of organizational learning (Argyris & Schön, 1996), reflective practice (Schön, 1983) involves being able to detect problems with and make corrections for one's current conceptual model or perspective. This practice is called "double-loop" learning, as it serves as an overarching control loop for improving one's performance at one's normal control loop tasks (Argyris, 1977).

14.1.4.2.2 TRANSCENDING BOUNDARIES

In all of the category levels, there are examples of organizational or technical boundaries being crossed as part of efforts to support resilience. The boundaries of the IT system, as implied by standard HIT use cases (e.g., Cusack et al., 2009), are crossed in the work of HIT systems safety. The informants were sensitive to interactions and risks from various sources, not just IT. They considered impacts to safety from diverse influences, such as external regulations (HIPAA and meaningful use, and also scope of practice) and the effect of patients' direct access to their records. As a means of learning about risks and responding to them, it was a regular practice to communicate about problems with both the vendor and other health care facilities external to their own health care system. Furthermore, as presented in Level 5, Part B (sensitivity to failure in quality control process), a few informants explicitly questioned the scope of the HIT system itself.

One definition of resilience is "stretching at and beyond boundaries" (Woods, Chan, & Wreathall, 2013). This definition emphasizes how fixation (De Keyser & Woods, 1990) on predefined boundaries or scopes of influence can hamper safety management (Reiman & Rollenhagen, 2011). Important risks and opportunities may be overlooked if there is too narrow a view taken on the scope of activities to be monitored or leverage points to be utilized (Woods et al., 2010). In contrast, the function of "seeing the bigger picture" can facilitate resilience.

14.1.4.2.3 SHARP-END STAKEHOLDERS

Across the category levels, there are examples of sharp-end practitioners (e.g., physicians, nurses) serving as active stakeholders in the safety management process, serving to establish a degree of distributed control of the operations of the HIT systems. This role is in contrast to the idea that sharp-end practitioners under automated supervisory control can respond to the technology only through compliance or resistance (see Greenhalgh et al., 2009). Resilience is enhanced by facilitating influence up the chain of command (Carthey et al., 2001) and by sharing some control with sharp-end practitioners (Woods, 2006; Woods & Branlat, 2010). Thus, the function of distributing and coordinating the control of safety across the blunt-end/sharp-end spectrum may facilitate resilience.

These two health care systems established and maintained practices and tools for the purpose of obtaining input from the front-line providers. Informatics and risk management staff were moved to front-line clinical organizations to facilitate problem detection and solving. Tools were developed for end users to report issues and track the organization's response. Evaluations of the IT system included identifying which parts were helpful and not helpful for end users (e.g., drug-drug interaction alerts, fields in patient transfer templates). Of course, these practices constitute only a small range of the possible ways control of the HIT system can be distributed. However, they do show the value of involving sharp-end practitioners as a part of ongoing system management, versus only during an initial requirements elicitation or usability evaluation phase.

14.1.4.3 RECOMMENDATIONS AND FUTURE WORK

The U.S. health care system is undergoing a significant transformation as more and more health care facilities, including smaller clinics and private practices, adopt EHRs and other HIT. Proposals for enhancing EHR safety have emphasized user-centered design approaches and usability testing, and the use of incident reporting, collection, and analysis systems at a large scale (Middleton et al., 2013; Office of the National Coordinator, 2012). Although these are valuable and necessary methods for improving safety, the results of this study suggest that EHR safety also depends on persistent testing and monitoring (Sittig & Classen, 2010; Walker et al., 2008), especially in terms of ongoing appraisal of sociotechnical factors that affect the use and maintenance of the EHR.

Although smaller clinics and private practices have less overall complexity than large health care systems, they are still subject to dynamics and interdependencies that affect risk. Vast numbers of the facilities now adopting HIT lack experience with and/or resources for safety assessment and management of HIT systems (Walker et al., 2008). There is a high demand for workers with HIT operation skills (Furukawa, Vibbert, & Swain, 2012; Hersh & Wright, 2008). These conditions will exacerbate the risks related to dynamics and interdependencies.

By identifying the resilient practices used in a domain, requirements for training and tools to support those practices can be developed (Hale, Guldenmund, & Goossens, 2006; Smith, Davis Giardina, Murphy, Laxmisan, & Singh, 2013). The training for HIT workforce development should address post-implementation quality and safety control, including practices for resilience in complex sociotechnical systems.

The tools available for IT system administrators are poor at supporting the tasks involved in ongoing supervision and management of IT systems (Barrett et al., 2004). Such deficiencies are exacerbated in complex sociotechnical systems, like health care. However, by designing the tools and processes to support collaboration and sensemaking (e.g., Watts-Perotti &

Woods, 2009), they will better support safety management. Resilience will be further enhanced by incorporating ways for sharp-end practitioners to participate in safety and quality control of IT systems.

Because of the large numbers of new EHR adopters lacking in relevant skills and resources, it is more critical to develop techniques to support awareness of the risks (Level 1) and their monitoring and management (Level 2). One method to support awareness of risks is to identify risk indicators that are easily detectable (Reason, 2008; Sittig & Singh, 2013). Ways to facilitate monitoring include the development of easy-to-use measures (Sittig, Campbell, Guappone, Dykstra, & Ash, 2007) and safety audit tools (Singh et al., 2013).

The utility of the information collected via monitoring can be enhanced through cognitive engineering approaches aimed at facilitating interpretation of and responses to the information (Endsley, 1988; Militello & Klein, 2013; Vicente, 1999). This approach includes designing the display of and interaction with the information such that it does not induce cognitive fixation on system boundaries and instead helps the practitioners understand and manage the safety issues as they manifest across system boundaries.

There are also techniques to support the macrocognitive function of reflection, identified as a facilitator of resilient practices across the categorization levels. Methods include doing "pre-mortem" analyses of proposals (Klein, 2007) and generating scenarios to explore new potential risks (Carroll, 2000). Another method is to make explicit the trade-off decisions between goals that organizations face (e.g., being both efficient and thorough), thereby encouraging stakeholders to reflect on how safety is being managed (Branlat & Woods, 2011; Hoffman & Woods, 2011).

14.1.4.4 LIMITATIONS

In this study, we collected data from only two health care systems. However, because both are leaders in the strategic application of HIT, and the focus of the study is on the practices used by health care systems with the greatest experience and success in HIT use, this limitation should not be seen as a threat to appropriate generalizability (Lipshitz, 2010). The two facilities are different: Partners HealthCare is a loosely connected system of hospitals and clinics in a dense metropolitan area, using various EHR tools, many developed internally, whereas Geisinger Health Systems is a tightly integrated system in a nonmetropolitan area, using a leading commercial EHR. This difference in conjunction with the overlap of resilient practices across the two different facilities strengthens the generalizability.

This study did not include analysis of contrasting cases to check if facilities with less experience or success in HIT management also showed evidence of these practices. However, we suspect those with less success are not likely to use the same practices, as there are some indications that differences in HIT safety outcomes are related to HIT implementation practices and management. For example, Han et al. (2005) reported on the unexpected increase in mortality at one facility following implementation of a commercially sold computerized order entry system, whereas Longhurst et al. (2010) found a decrease in their facility's mortality rate following implementation of the same vendor's EHR but with the use of much improved processes and management techniques (Sittig, Ash, Zhang, Osheroff, & Shabot, 2006).

Another limitation is that the information about the practices came from interviews. We did not perform observations or otherwise evaluate for independent evidence of the practices. However, almost all of the practices were mentioned by more than one informant. Some of these facilities' practices have been the subject of scientific publications (Hundt et al., 2013; Paulus et al., 2008; Phansalkar et al., 2010; Teich et al., 1999; Walker et al., 2012). Furthermore, these resilient practices were volunteered during interviews about HIT safety in general (as pertaining to the informant's role); the interviews were not conducted as cognitive task analysis interviews designed to elicit specific types of strategies. This suggests that these practices were strongly associated with general HIT safety in the minds of the informants. A methodology focused on eliciting resilience-specific practices might have uncovered additional ones.

14.1.5 CONCLUSIONS

Our study of quality and safety control processes used for HIT by two leading health care systems shows that resilient safety practices are an important part of safety in complex sociotechnical systems. The practices mentioned were categorized into five areas: (a) awareness of dynamics and interdependencies affecting risk, (b) established monitoring and responding practices, (c) management of resources and methods, (d) anticipating risks, and (e) reflecting on limitations of the safety management process itself. These practices were facilitated by the functions of reflective learning, the capability to revise system boundaries, and systematic sharing of control with sharp-end practitioners.

REFERENCES

1. Aarts J. (2011). Towards safe information technology in health care. Information, Knowledge, Systems Management, 10, 335–344.
2. Amalberti R. (2006). Optimum system safety and optimum system resilience: Agonistic or antagonistic concepts. In Hollnagel E., Woods D., Leveson N. (Eds.), Resilience engineering: Concepts and precepts (pp. 253–274). Farnham, UK: Ashgate.
3. Argyris C. (1977). Organizational learning and management information systems. Accounting, Organizations and Society, 2, 113–123.
4. Argyris C., Schön D. A. (1996). Organizational learning 2. Boston, MA: Addison-Wesley.
5. Ash J. S., Sittig D. F., Campbell E. M., Guappone K. P., Dykstra R. H. (2007). Some unintended consequences of clinical decision support systems. In AMIA 2007 Annual Symposium Proceedings (pp. 26–30). Retrieved from http://www.ncbi.nlm.nih.gov/pmc/issues/177326/
6. Bainbridge L. (1983). Ironies of automation. Automatica, 19, 775–780.
7. Barrett R., Kandogan E., Maglio P. P., Haber E. M., Takayama L. A., Prabaker M. (2004). Field studies of computer system administrators: Analysis of system management tools and practices. In Proceedings of the 2004 ACM Conference on Computer Supported Cooperative Work (pp. 388–395). New York, NY: ACM.
8. Belmont E., Chao S., Chestler A., Fox S., Lamar M., Rosati K., . . . Valenti A. (2013). Minimizing EHR-related serious safety events. Washington, DC: American Health Lawyers Association. Retrieved from http://www.healthlawyers.org/hlresources/PI/InfoSeries/Documents/For%20the%20Healthcare%20Executive/Minimizing%20EHRSSE.pdf
9. Blumenthal D., Tavenner M. (2010). The "meaningful use" regulation for electronic health records. New England Journal of Medicine, 363, 501–504.
10. Branlat M., Woods D. (2011, June). How human adaptive systems balance fundamental trade-offs: Implications for polycentric governance architectures. Paper presented at the 4th Resilience Engineering International Symposium, Sophia Antipolis, France.

11. Cacciabue P. C., Hollnagel E. (1995). Simulation of cognition: Applications. In Hoc J. M., Cacciabue P. C., Hollnagel E. (Eds.), Expertise and technology: Cognition and human-computer cooperation (pp. 55–73). Hillsdale, NJ: Lawrence Erlbaum.
12. Campbell E. M., Guappone K. P., Sittig D. F., Dykstra R. H., Ash J. S. (2009). Computerized provider order entry adoption: Implications for clinical workflow. Journal of General Internal Medicine, 24, 21–26.
13. Carayon P., Schoofs Hundt A., Karsh B. T., Gurses A. P., Alvarado C. J., Smith M., Flatley Brennan P. (2006). Work system design for patient safety: The SEIPS model. Quality & Safety in Health Care, 15(Suppl. 1), i50–i58.
14. Carroll J. M. (2000). Five reasons for scenario-based design. Interacting With Computers, 13, 43–60. doi:10.1016/S0953-5438(00)00023-0
15. Carthey J., De Leval M. R., Reason J. T. (2001). Institutional resilience in healthcare systems. Quality in Health Care, 10, 29–32.
16. Classen D. C., Avery A. J., Bates D. W. (2007). Evaluation and certification of computerized provider order entry systems. Journal of the American Medical Informatics Association, 14, 48–55. doi:10.1197/jamia.M2248
17. Coiera E., Aarts J., Kulikowski C. (2012). The dangerous decade. Journal of the American Medical Informatics Association, 19, 2–5.
18. Committee on Data Standards for Patient Safety. (2003). Key capabilities of an electronic health record system: Letter report. Washington, DC: National Academies Press. Retrieved from http://www.nap.edu/openbook.php?record_id=10781
19. Cook R., Rasmussen J. (2005). "Going solid": A model of system dynamics and consequences for patient safety. Quality and Safety in Health Care, 14, 130–134. doi:10.1136/qshc.2003.009530
20. Cook R. I., Wood D. D. (1994). Operating at the Sharp End: The Complexity of Human Error. In Bogner M. S. (Ed.), Human Error in Medicine (pp. 255–310). Erlbaum.
21. Cook R.I., Nemeth C. (2006). Taking things in one's stride: Cognitive features of two resilient performances. In Hollnagel E., Woods D., Leveson N. (Eds.), Resilience engineering: Concepts and precepts (pp. 205–221). Farnham, UK: Ashgate.
22. Costella M., Saurin T., de Macedo Guimarães L. (2009). A method for assessing health and safety management systems from the resilience engineering perspective. Safety Science, 47, 1056–1067.
23. Cusack C., Byrne C., Hook J., McGowan J., Poon E., Zafar A. (2009). Health information technology evaluation toolkit: 2009 update (No. AHRQ Publication No. 09-0083- EF). Washington, DC: Agency for Healthcare Research and Quality. Retrieved from http://healthit.ahrq.gov/sites/default/files/docs/page/Evaluation%20Toolkit%20Revised%20Version.pdf
24. De Keyser V., Woods D. D. (1990). Fixation errors: Failures to revise situation assessment in dynamic and risky systems. In Colombo A. G., Saiz A., Bustamante de (Eds.), Systems reliability assessment (pp. 231–251). Dordrecht, Netherlands: Springer.
25. Dekker S. (2006). The field guide to understanding human error (1st ed.). Farnham, UK: Ashgate.
26. Department of Health and Human Services. (n.d.). Summary of the HIPAA Security Rule. Retrieved from http://www.hhs.gov/ocr/privacy/hipaa/understanding/srsummary.html

27. Endsley M. R. (1988). Design and evaluation for situation awareness enhancement. In Proceedings of the Human Factors and Ergonomics Society 32nd Annual Meeting (pp. 97–101). Santa Monica: CA: Sage.
28. Furukawa M., Vibbert D., Swain M. (2012). Hitech and Health IT jobs: Evidence from online job postings (No. ONC Data Brief No. 2). Washington, DC: Office of the National Coordinator for Health Information Technology. Retrieved from http://www.healthit.gov/sites/default/files/pdf/0512_ONCDataBrief2_JobPostings.pdf
29. Greenhalgh T., Potts H., Wong G., Bark P., Swinglehurst D. (2009). Tensions and paradoxes in electronic patient record research: A systematic literature review using the meta-narrative method. Milbank Quarterly, 87, 729–788.
30. Grol R., Grimshaw J. (2003). From best evidence to best practice: Effective implementation of change in patients' care. The Lancet, 362, 1225–1230.
31. Hale A., Guldenmund F., Goossens L. (2006). Auditing resilience in risk control and safety management systems. In Hollnagel E., Woods D., Leveson N. (Eds.), Resilience engineering: Concepts and precepts (pp. 289–314). Farnham, UK: Ashgate.
32. Hale A. R., Heming B. H. J., Carthey J., Kirwan B. (1997). Modelling of safety management systems. Safety Science, 26, 121–140.
33. Hammond K., Helbig S., Benson C., Brathwaite-Sketoe B. (2003). Are electronic medical records trustworthy? Observations on copying, pasting and duplication. In AMIA 2003 Annual Symposium Proceedings (pp. 269–273). Retrieved from http://www.ncbi.nlm.nih.gov/pmc/issues/131751/
34. Han Y., Carcillo J., Venkataraman S., Clark R., Watson S., Nguyen T., . . . Orr R. (2005). Unexpected increased mortality after implementation of a commercially sold computerized physician order entry system. Pediatrics, 116, 1506–1512.
35. Hersh W., Wright A. (2008). What workforce is needed to implement the health information technology agenda? Analysis from the HIMSS Analytics™ database. In AMIA 2008 Annual Symposium Proceedings (pp. 303–307). Retrieved from http://www.ncbi.nlm.nih.gov/pmc/issues/177327/
36. Hoffman R. R., Woods D. D. (2011, May/June). Simon's slice: Five fundamental tradeoffs that bound the performance of human work systems. Paper presented at the 10th International Conference on Naturalistic Decision Making, Orlando, FL.
37. Hollnagel E. (2008a). The changing nature of risk. Ergonomics Australia Journal, 22, 33–46.
38. Hollnagel E. (2008b). Safety management: Looking back or looking forward. In Hollnagel E., Nemeth C., Dekker S. (Eds.), Resilience engineering perspectives (Vol. 1, pp. 63–77). Farnham, UK: Ashgate.
39. Hollnagel E. (2009). The four cornerstones of resilience engineering. In Nemeth C., Hollnagel E., Dekker S. (Eds.), Resilience engineering perspectives (Vol. 2, pp. 117–134). Farnham, UK: Ashgate.
40. Hollnagel E. (2011). To learn or not to learn, that is the question. In Hollnagel E., Pariès J., Woods D., Wreathall J. (Eds.), Resilience engineering in practice (pp. 193–198). Farnham, UK: Ashgate.
41. Hollnagel E., Woods D. D., Leveson N. (2006). Resilience engineering: Concepts and precepts. Farnham, UK: Ashgate.
42. Hsieh T. C., Kuperman G. J., Jaggi T., Hojnowski-Diaz P., Fiskio J., Williams D. H., . . . Gandhi T. K. (2004). Characteristics and consequences of drug allergy alert

overrides in a computerized physician order entry system. Journal of the American Medical Informatics Association, 11, 482–491.
43. Hundt A. S., Adams J. A., Schmid J. A., Musser L. M., Walker J. M., Wetterneck T. B., . . . Carayon P. (2013). Conducting an efficient proactive risk assessment prior to CPOE implementation in an intensive care unit. International Journal of Medical Informatics, 82, 25–38. doi:10.1016/j.ijmedinf.2012.04.005
44. Hunte G. S., Wears R. L., Schubert C. C. (2013, June). Structure, agency, and resilience. Paper presented at the 5th Resilience Engineering International Symposium, Soesterberg, Netherlands.
45. Institute of Medicine. (2012). Health IT and patient safety: Building safer systems for better care. Washington, DC: National Academies Press. Retrieved from http://www.nap.edu/openbook.php?record_id=13269
46. Jha A. K., Classen D. C. (2011). Getting moving on patient safety: Harnessing electronic data for safer care. New England Journal of Medicine, 365, 1756–1758.
47. Kannampallil T., Schauer G., Cohen T., Patel V. (2011). Considering complexity in healthcare systems. Journal of Biomedical Informatics, 44, 943–947.
48. Karsh B. T., Weinger M., Abbott P., Wears R. (2010). Health information technology: Fallacies and sober realities. Journal of the American Medical Informatics Association, 17, 617–623.
49. Klein G. (2007). Performing a project premortem. Harvard Business Review, 85(9), 18–19.
50. Klein G., Snowden D., Pin C. L. (2010). Anticipatory thinking. In Mosier K., Fischer U. (Eds.) Informed by knowledge: Expert performance in complex situations (pp. 235–246). New York, NY: Psychology Press.
51. Leveson N. (2004). A new accident model for engineering safer systems. Safety Science, 42, 237–270.
52. Leveson N., Dulac N., Zipkin D., Cutcher-Gershenfeld J., Carroll J., Barrett B. (2006). Engineering resilience into safety-critical systems. In Hollnagel E., Woods D. D., Leveson N. (Eds.), Resilience engineering: Concepts and precepts (pp. 95–123). Farnham, UK: Ashgate.
53. Leveson N. G. (2012). Engineering a safer world: Systems thinking applied to safety. Cambridge, MA: MIT Press.
54. Lipshitz R. (2010). Rigor and relevance in NDM: How to study decision making rigorously with small ns and without controls and (inferential) statistics. Journal of Cognitive Engineering and Decision Making, 4, 99–112.
55. Longhurst C. A., Parast L., Sandborg C. I., Widen E., Sullivan J., Hahn J. S., . . . Sharek P. J. (2010). Decrease in hospital-wide mortality rate after implementation of a commercially sold computerized physician order entry system. Pediatrics, 126, 14–21.
56. Magrabi F., Ong M.-S., Runciman W., Coiera E. (2012). Using FDA reports to inform a classification for health information technology safety problems. Journal of the American Medical Informatics Association, 19, 45–53.
57. Middleton B., Bloomrosen M., Dente M. A., Hashmat B., Koppel R., Overhage J. M., . . . Zhang J. (2013). Enhancing patient safety and quality of care by improving the usability of electronic health record systems: Recommendations from AMIA. Journal of the American Medical Informatics Association, 20(e1), e2–e8.

58. Militello L. G., Klein G. (2013). Decision-centered design. In Lee J., Kirlik A. (Eds.), The Oxford handbook of cognitive engineering (pp. 261–271). Oxford, UK: Oxford University Press.
59. Myers R. B., Jones S. L., Sittig D. F. (2011). Review of reported clinical information system adverse events in US Food and Drug Administration databases. Applied Clinical Informatics, 2, 63.
60. Nemeth C. (2008). Resilience engineering: The birth of a notion. In Hollnagel E., Nemeth C., Dekker S. (Eds.), Resilience engineering perspectives (Vol. 1, pp. 3–9). Farnham, UK: Ashgate.
61. Nemeth C., Cook R. (2007). Healthcare IT as a source of resilience. In IEEE International Conference on Systems, Man and Cybernetics (pp. 3408–3412). doi:10.1109/ICSMC.2007.4413721
62. Office of the National Coordinator. (2012). Health information technology patient safety action & surveillance plan for public comment. Retrieved from http://www.healthit.gov/sites/default/files/safetyplanhhspubliccomment.pdf
63. Pariès J. (2011). Resilience and the ability to respond. In Hollnagel E., Pariès J., Woods D., Wreathall J. (Eds.), Resilience engineering in practice (pp. 3–8). Farnham, UK: Ashgate.
64. Patton M. (2002). Qualitative research and evaluation methods. Thousand Oaks, CA: Sage.
65. Paulus R. A., Davis K., Steele G. D. (2008). Continuous innovation in health care: Implications of the Geisinger experience. Health Affairs, 27, 1235–1245.
66. Pejtersen A. M., Rasmussen J. (1997). Ecological information systems and support of learning: Coupling work domain information to user characteristics. In Helander M. G., Landauer T. K., Prabhu P. V. (Eds.), Handbook of human-computer interaction (pp. 315–346). Amsterdam: Elsevier.
67. Phansalkar S., Edworthy J., Hellier E., Seger D., Schedlbauer A., Avery A., Bates D. (2010). A review of human factors principles for the design and implementation of medication safety alerts in clinical information systems. Journal of the American Medical Informatics Association: JAMIA, 17, 493–501.
68. Qureshi Z. H., Ashraf M. A., Amer Y. (2007). Modeling industrial safety: A sociotechnical systems perspective. In 2007 IEEE International Conference on Industrial Engineering and Engineering Management (pp. 1883–1887). New York, NY: IEEE.
69. Rankin A., Lundberg J., Woltjer R., Rollenhagen C., Hollnagel E. (2013). Resilience in everyday operations a framework for analyzing adaptations in high-risk work. Journal of Cognitive Engineering and Decision Making. Advance online publication. doi:10.1177/1555343413498753
70. Rasmussen J. (1997). Risk management in a dynamic society: A modelling problem. Safety Science, 27, 183–213.
71. Reason J. (1997). Managing the risks of organizational accidents. Farnham, UK: Ashgate.
72. Reason J. T. (2008). The human contribution: Unsafe acts, accidents and heroic recoveries. Farnham, UK: Ashgate.
73. Reiman T. (2011). Understanding maintenance work in safety-critical organisations: Managing the performance variability. Theoretical Issues in Ergonomics Science, 12, 339–366. doi:10.1080/14639221003725449

74. Reiman T., Rollenhagen C. (2011). Human and organizational biases affecting the management of safety. Reliability Engineering & System Safety, 96, 1263–1274.
75. Ritchie J., Spencer L. (2002). Qualitative data analysis for applied policy research. In Huberman M., Miles M. (Eds.), The qualitative researcher's companion (pp. 305–329). Thousand Oaks, CA: Sage.
76. Roberts K. H. (1990). Some characteristics of one type of high reliability organization. Organization Science, 1, 160–176.
77. Roth E., Kilgore R., Burns C., Wears R., Lee J. D., Jamieson G., Bisantz A. (2013). Cognitive engineering across domains: What the wide-angle view can provide. In Proceedings of the Human Factors and Ergonomics Society 57th Annual Meeting (pp. 139–143). Santa Monica, CA: Human Factors and Ergonomics Society.
78. Saleh J. H., Marais K. B., Bakolas E., Cowlagi R. V. (2010). Highlights from the literature on accident causation and system safety: Review of major ideas, recent contributions, and challenges. Reliability Engineering & System Safety, 95, 1105–1116.
79. Schön D. A. (1983). The reflective practitioner: How professionals think in action. New York, NY: Basic Books.
80. Singh H., Ash J. S., Sittig D. F. (2013). Safety Assurance Factors for Electronic Health Record Resilience (SAFER): Study protocol. BMC Medical Informatics and Decision Making, 13, 46.
81. Sittig D. F., Ash J. S., Zhang J., Osheroff J. A., Shabot M. M. (2006). Lessons from "Unexpected Increased Mortality After Implementation of a Commercially Sold Computerized Physician Order Entry System." Pediatrics, 118, 797–801.
82. Sittig D. F., Campbell E., Guappone K., Dykstra R., Ash J. S. (2007). Recommendations for monitoring and evaluation of in-patient computer-based provider order entry systems: Results of a Delphi survey. In AMIA 2007 Annual Symposium Proceedings (p. 671). Bethesda, MD: American Medical Informatics Association.
83. Sittig D. F., Classen D. C. (2010). Safe electronic health record use requires a comprehensive monitoring and evaluation framework. JAMA: The Journal of the American Medical Association, 303, 450-451.
84. Sittig D. F., Singh H. (2009). Eight rights of safe electronic health record use. JAMA, 302, 1111–1113. doi:10.1001/jama.2009.1311
85. Sittig D. F., Singh H. (2010). A new sociotechnical model for studying health information technology in complex adaptive healthcare systems. Quality and Safety in Health Care, 19(Suppl. 3), i68–i74.
86. Sittig D. F., Singh H. (2012). Electronic health records and national patient-safety goals. New England Journal of Medicine, 367, 1854–1860.
87. Sittig D. F., Singh H. (2013). A red-flag-based approach to risk management of EHR-related safety concerns. Journal of Healthcare Risk Management, 33(2), 21–26.
88. Skorve E. (2010). Patient safety, resilience and ICT. A reason for concern? Studies in Health Technology and Informatics, 157, 199–205.
89. Smith M., Davis Giardina T., Murphy D., Laxmisan A., Singh H. (2013). Resilient actions in the diagnostic process and system performance. BMJ Quality & Safety, 22, 1006–1013. doi:10.1136/bmjqs-2012-001661
90. Teich J. M., Glaser J. P., Beckley R. F., Aranow M., Bates D. W., Kuperman G. J., ... Spurr C. D. (1999). The Brigham Integrated Computing System (BICS): Advanced

clinical systems in an academic hospital environment. International Journal of Medical Informatics, 54, 197–208.
91. Teich J. M., Merchia P. R., Schmiz J. L., Kuperman G. J., Spurr C. D., Bates D. W. (2000). Effects of computerized physician order entry on prescribing practices. Archives of Internal Medicine, 160, 2741.
92. Vicente K. (1999). Cognitive work analysis: Toward safe, productive, and healthy computer-based work. Boca Raton, FL: CRC Press.
93. Vicente K. J. (2002). From patients to politicians: A cognitive engineering view of patient safety. Quality and Safety in Health Care, 11, 302–304.
94. Walker J. M., Carayon P., Leveson N., Paulus R. A., Tooker J., Chin H., . . . Stewart W. F. (2008). EHR safety: The way forward to safe and effective systems. Journal of the American Medical Informatics Association, 15, 272–277.
95. Walker J. M., Hassol A., Bradshaw B., Rezaee M. (2012). Health IT hazard manager beta-test: Final report (No. AHRQ Publication No. 12-0058-EF). Washington, DC: Agency for Health Care Research and Quality.
96. Watts-Perotti J., Woods D. (2009). Cooperative advocacy: An approach for integrating diverse perspectives in anomaly response. Computer Supported Cooperative Work (CSCW), 18, 175–198.
97. Wears R. L. (2012). Rethinking healthcare as a safety-critical industry. Work: A Journal of Prevention, Assessment and Rehabilitation, 41, 4560–4563. doi:10.3233/WOR-2012-0037-4560
98. Wears R. L., Hollnagel E., Braithwaite J. (2013). Resilient Health Care. Ashgate Publishing.
99. Woods D. (2011). Resilience and the ability to anticipate. In Hollnagel E., Pariès J., Woods D., Wreathall J. (Eds.), Resilience engineering in practice (pp. 121–126). Farnham, UK: Ashgate Publishing.
100. Woods D. D. (2006). Essential characteristics of resilience. In Hollnagel E., Woods D., Leveson N. (Eds.), Resilience engineering: Concepts and precepts (pp. 21–34). Farnham, UK: Ashgate.
101. Woods D. D., Branlat M. (2010). Hollnagel's test: Being "in control" of highly interdependent multi-layered networked systems. Cognition, Technology & Work, 12, 95–101.
102. Woods D. D., Chan Y. J., Wreathall J. (2013, June). The stress-strain model of resilience operationalizes the four cornerstones of resilience engineering. Paper presented at the 5th Resilience Engineering International Symposium, Soesterberg, Netherlands.
103. Woods D. D., Dekker S., Cook R., Johannesen L., Sarter N. (2010). Behind human error. Farnham, UK: Ashgate.
104. Woods D. D., Hollnagel E. (2006). Joint cognitive systems: Patterns in cognitive systems engineering. Boca Raton, FL: CRC Press.
105. Woods D. D., Schenk J., Allen T. (2009). An initial comparison of selected models of system resilience. In Nemeth C., Hollnagel E., Dekker S. (Eds.), Resilience engineering perspectives (Vol. 2, pp. 73–94). Farnham, UK: Ashgate.
106. Woods D. D., Wreathall J. (2008). Stress-strain plots as a basis for assessing system resilience. In Hollnagel E., Nemeth C., Dekker S. (Eds.), Resilience engineering perspectives (Vol. 1, pp. 143–158). Farnham, UK: Ashgate.

107. Wreathall J. (2011). Monitoring: A critical ability in resilience engineering. In Hollnagel E., Pariés J., Woods D., Wreathall J. (Eds.), Resilience engineering in practice (pp. 61–68). Farnham, UK: Ashgate.
108. Wright A., Henkin S., Feblowitz J., McCoy A. B., Bates D. W., Sittig D. F. (2013). Early results of the meaningful use program for electronic health records. New England Journal of Medicine, 368, 779–780.

SAFER SELF-ASSESSMENT GUIDE: ORGANIZATIONAL ACTIVITIES AND RESPONSIBILITIES FOR ELECTRONIC HEALTH RECORD (EHR) SAFETY

SAFER Guides

Legend	Sources of Input
C	Clinicians, support staff, and/or clinical administration (e.g., medical records and risk managers)
Dx	Diagnostic services, such as laboratory or radiology—could be local or remote
Ev	EHR vendor and/or other IT or HIT vendors
IT	IT support staff, could be local or contracted. Responsible for maintaining the infrastructure
Rx	Pharmacy – could be local or remote
L	Leadership Team – (e.g. Board of Directors, executive team, clinical leadership, operational leadership)
M	Multi-professional Team – (e.g. clinicians, IT, patient safety/quality, informatics)
HI	Health Informatics Team (e.g. content specialists, clinical analysts, nursing/medical informatics, informatics consultants)

ORGANIZATIONAL ACTIVITIES AND EHR SAFETY

Recommended Practices and Responsibilities*	Rationale for Practice or Risk Addressed	Examples of Potentially Useful Practices/Scenarios
Principle 1: Defined decision making activities assure EHR safety.		

Increasing Resilience in an EHR-Enabled Healthcare Organization

Recommended Practices and Responsibilities*	Rationale for Practice or Risk Addressed	Examples of Potentially Useful Practices/Scenarios
1. The highest-level decision makers (e.g., boards of directors or owners of physician practices) are committed to promoting a culture of safety that incorporates the safety and safe use of EHRs.	• Leadership can provide motivation for all staff to pay attention to EHR safety. • Those in authority can provide resources for ensuring EHR safety. • Without leadership involvement, EHR safety efforts will likely fail.	• Highest-level decision makers recognize that EHR safety is integral to patient safety. They ensure that EHR safety is integrated into organizational policies and procedures and risk management practices. • Highest-level decision makers ensure that adequate staffing and resources exist so that safety issues associated with adoption and use of EHRs can be addressed. • Highest-level decision makers review the results of assessments of EHR safety, such as those from SAFER Guide use. • Highest-level decision makers identify EHR-related patient safety goals, assess whether those goals are being reached, and address any shortcomings.
2. An effective decision-making structure exists for managing and optimizing the safety and safe use of the EHR. Responsibility Large organization: Board Responsibility Small organization: Owners Input Source: L,M	• Clarifies responsibility • Maximizes involvement of disciplines • Ensures that important EHR safety issues are addressed	• For larger organizations, all of the following are represented in decision making about EHR safety: clinicians, administrators, patients, Health IT/informatics, board of directors and CEOs, and quality and legal staff. • For smaller ambulatory practices and small hospitals, both clinical and administrative staff members are represented in decision making about EHR safety, with assistance from outside experts. • An EHR safety officer or someone assigned that responsibility part time in a small organization plays a key role in assuring safety. • EHR safety is appropriately included in job performance appraisals. • For a larger organization, an EHR safety oversight committee is in place [1, 2[or these functions are assumed by an EHR or Safety and Quality oversight committee.

Recommended Practices and Responsibilities*	Rationale for Practice or Risk Addressed	Examples of Potentially Useful Practices/Scenarios
3. Staff members are assigned responsibility for the management of clinical decision support (CDS) content. Responsibility (L): Informatics type department Responsibility (S): Providers Input Source: HI, C, M, EV, Rx	• Facilitates decision making about clinical decision support and other content • Provides accountability for decisions • Avoids hazardous wrong or outdated content in EHR	• A decision-making structure exists for making decisions about clinical content.3-6 • Responsibility for management of content, from selection to maintenance, is clear. • Committees or other collaboration mechanisms are in place to approve order sets and documentation templates. • There is clear responsibility for the review of a new decision support that becomes available from developers and other sources (e.g., professional organizations). • Developers provide clear documentation of decision support content and the evidence-base to support that content. • Developers routinely review and update decision support content they provide. • Personnel are available either internally or externally to ensure that decision support is tailored to the workflows of professional roles and specialties.7-1
4. Practicing clinicians are involved in all levels of EHR safety-related decision making that impact clinical use. Responsibility (L): Administration Responsibility (S): Providers Input Source: C, M	• Facilitates wise decision making about clinically relevant issues • Assures focus on patient care • Increases acceptance of decisions	• Clinicians, including physicians, nurses, pharmacists, and others, are included on the EHR safety oversight committee of a large organization. • Clinicians are involved in decision making about all proposed changes to the EHR.

Increasing Resilience in an EHR-Enabled Healthcare Organization 419

Recommended Practices and Responsibilities*	Rationale for Practice or Risk Addressed	Examples of Potentially Useful Practices/Scenarios
5. Clear clinician oversight is maintained when clinicians delegate aspects of order entry, medication reconciliation, or documentation tasks. Responsibility (L): Hospital departments Responsibility (S): providers Input Source: C, M	• Assures that the safety risks of assigning these tasks to medical assistants or scribes are carefully weighed • Assures that responsible providers take the time to review delegated work	• For teaching hospitals and clinics, attending physicians are diligent about reviewing the work of trainees. (Koshy; Santell) • In community non-teaching settings, responsible providers oversee and are diligent about reviewing the delegated work.
Principle 2: Activities to maximize EHR quality and data quality assure EHR safety		
6. Staff members are assigned to regularly monitor EHR hardware, software, and network/ Internet service provider (ISP) performance and safety. Responsibility (L): Safety officer, informatics-type department, IT Responsibility (S): Office management, IT staff or contractor, providers Input Source: L, HI, C, M, IT	• Problems can be caught before harm is done • Providers and others can learn from their mistakes • The impact of changes to the EHR or CDS is transparent	• A plan outlining responsibility for EHR safety monitoring is in place. (Singh; Sittig and Classen; Strom) • Errors involving system-to-system interfaces are routinely monitored. • Providers and others including leadership in large organizations are encouraged to use tools to monitor EHR safety and care quality. • A plan exists for learning from incidents to improve EHR safety. • The review and communication of lab results are monitored. • The test results reporting loop is closed. • Selected post-implementation care outcomes are monitored. • Alert and reminder responses are monitored. • Alert and reminder specificity and sensitivity are appropriately adjusted.

Recommended Practices and Responsibilities*	Rationale for Practice or Risk Addressed	Examples of Potentially Useful Practices/Scenarios
7. Staff members are assigned to regularly test for and promptly correct problems with EHR hardware, software, and network/ISP performance and safety. Responsibility (L): Safety officer, informatics-type department, IT Responsibility (S): Office management, IT staff/contractor, providers Input Source: L, HI, C, IT, Ev	• Customization of either the EHR or content must be skillfully done or upgrades to the EHR can produce unique hazards • Inadequate or unprepared staff members can cause problems to go unaddressed	• The organization has adequate numbers of trained staff members available either on site or elsewhere to modify software. • Adequate technical staff members are available to fix hardware problems during operating hours. • Staff members are available to catch and correct errors such as registration, order entry, or test results communication errors in a timely manner. • When errors occur, a multidisciplinary review and discussion takes place. • The organization has a rigorous process in place for testing new software. (Walker) • The organization has a rigorous process in place for testing new hardware. • Workflow analysis to map the way work is actually done is conducted prior to any system upgrade. • Risk assessments are conducted prior to go live. • The potential impact of any EHR upgrade is carefully assessed.
8. Staff members are assigned responsibility for selecting, testing, monitoring, and maintaining CDS for performance and safety. Responsibility (L): Safety officer, informatics-type department, IT Responsibility (S): Office management, IT staff/contractor, providers Input Source: L, HI, C, IT, Ev	• Untested CDS can lead to patient care errors • Lessons from testing can prevent implementation of error prone CDS	• The organization has a rigorous process in place for testing new CDS. (Walker) • Risk assessments are conducted prior to go live with new CDS. • Clinical content is developed or modified by a multidisciplinary group including clinical specialists when appropriate.

Increasing Resilience in an EHR-Enabled Healthcare Organization

Recommended Practices and Responsibilities*	Rationale for Practice or Risk Addressed	Examples of Potentially Useful Practices/Scenarios
Principle 3: Activities to assure safe use of the EHR can prevent EHR safety hazards		
9. EHR training and support are sufficient for the needs of EHR users and readily available. Responsibility (L): Informatics-type department, IT, vendor Responsibility (S): Office management, vendor Input Source: L, HI, C, IT, Ev	• If the EHR is not used or is poorly used, patient harm can result • Training and support staff must be well trained to maximize effectiveness	• All users are trained prior to their using the system, supported while they are first using the system, and trained again before each change to the system. (Singh) • Different modalities for training are offered to accommodate user schedules and learning styles. • EHR safety is covered in EHR training. • Users are trained on how to proceed during system unavailability (downtimes). • Providers must demonstrate competency via testing in using the system before using order entry. • In larger organizations, IT and informatics staff take vendor training and are certified as appropriate. • A process is in place so users can get help immediately whenever and wherever they need it. (Ash b)
10. EHR training and support are of high quality provided by qualified trainers, and appropriately tailored to specific types of users' needs. Responsibility (L): Informatics-type organization, IT, vendor Responsibility (S): Office management, vendor Input Source: L, HI, C, IT, Ev	• Suboptimal training and support lead to wasted time for users • Lack of diligence can cause EHR safety hazards	• Whether done by dedicated internal trainers or those hired from outside, pre-implementation training prepares users for go-live. • Training and support are provided by individuals who can fill the gap between the clinical and IT languages and understand clinical workflow. (Ash a) • Support is available on site at least during the first week after go-live of the EHR. • A protocol exists so that all users know where to go for technical, software, and connectivity support. • Initial training includes running through scenarios that mirror the tasks users will need to accomplish. • Training stresses that users must be diligent about entering accurate data. (Singh; Thompson; Hogan; Chuo; Magrabi)

Recommended Practices and Responsibilities*	Rationale for Practice or Risk Addressed	Examples of Potentially Useful Practices/Scenarios
		• User skills are monitored and upgraded when needed.
11. EHR training and support are assessed regularly to optimize complete and safe use of the EHR. Responsibility (L): Informatics-type organization, IT, vendor Responsibility (S): Office management, vendor Input Source: L, HI, C, IT, Ev	• Since training and support are ongoing and expensive, feedback for continuous improvement is important	• A training plan outlines regular ongoing training opportunities so that users can optimize their use of the EHR. • Training and support must be tailored to the needs of EHR users. • A plan exists for ongoing assessment of training and support. • Feedback about training and support is responded to effectively.
12. Workflow analysis to map how work is actually done is conducted regularly. Responsibility (L): Informatics-type department Responsibility (S): Office management and vendor or consultant Input Source: L, HI, Ev, M	• Inattention to how the EHR fits workflow can result in wasted time and money. • Workarounds that result from workflow-related problems can lead to errors that affect patients.	• Workflow analysis is conducted prior to implementation of the EHR. (Campbell) • Workflow analysis is conducted prior to any major change to the EHR system. • An effective change management approach guides needed workflow changes based on the workflow analysis.
13. Clinical staff is assigned responsibility for ensuring that CDS content, such as alerts and protocols, supports effective clinical workflow in all practice settings. Responsibility (L): Informatics-type department Responsibility (S): Providers Input Source: C, HI, M, Rx	• Without customization, generic CDS that is not useful to the recipient's role or specialty may create hazards.	• A process exists for the review and modification of any locally-developed, commercial, or freely available CDS so that it is appropriate for a particular setting. (Bates) • A clinical rules committee has a defined process for evaluating and overseeing the testing and monitoring of the CDS. • The unique needs of the pediatric population are taken into account when reviewing and modifying CDS. (Walsh)

Recommended Practices and Responsibilities*	Rationale for Practice or Risk Addressed	Examples of Potentially Useful Practices/Scenarios
14. Organizational policy facilitates reporting of EHR-related hazards and errors and ensures that reports are promptly investigated and addressed. Responsibility (L): Safety officer and all those involved in safety initiatives, informatics-type department Responsibility (S): Office management, providers Input Source: L, HI, C	• A culture of safety relies upon reporting and follow up. If hazards exist but remain unreported they could cause harm.	• The mechanism for reporting EHR-related safety hazards internally is clear to all users. • Those who manage EHR and patient safety initiatives for the organization have a clear protocol for addressing reported problems and for reporting problems externally to the vendor and/or a patient safety organization when appropriate. (Walker; Chuo)
15. Records of reported and addressed EHR-related hazards and errors are maintained. Responsibility (L): Safety officer, informatics-type department Responsibility (S): Office management, providers Input Source: L, HI, C	• If records of these hazards are not maintained, the same problems might arise at a future time without access to prior solutions and mitigation strategies. • There could be some liability risk if the history is undocumented • If users cannot learn the disposition of their reports, they may not bother submitting future reports	• Larger organizations often use help desk software to keep track of internal reports and disposition. The user who reported the issue is notified of the outcome when appropriate. • Smaller organizations develop databases of reports and assign responsibility for maintenance of the database, usually to the health IT person.
Principle 4: Activities to assure the availability of information in the EHR can prevent EHR safety hazards		
16. Staff members are assigned responsibility for the maintenance of the EHR-related hardware, software, CDS, and network/ISP performance. Responsibility (L): IT HI (for CDS) Responsibility (S): IT contractoror internal IT-oriented person Input Source: IT	• Without maintenance, components of the EHR may impede use • Inadequate maintenance could cause the EHR to be unavailable, creating safety risks	• Regular maintenance of hardware, software, network/ISP/CDS is organized and funded.

Recommended Practices and Responsibilities*	Rationale for Practice or Risk Addressed	Examples of Potentially Useful Practices/Scenarios
17. Staff members regularly monitor maintenance of the EHR-related hardware, software, CDS, and network/ISP performance and safety. Responsibility (L): IT, informatics-type department Responsibility (S): Office management Input Source: L, C, IT, HI	• Inadequate maintenance may result in unplanned downtime • Inadequate maintenance may cause the EHR to be unavailable, causing safety risks	• When maintenance for these components is provided from outside the organization, oversight is provided by an internal staff member to assure the competence and performance of the contractors. • When maintenance is provided internally, regular schedules exist for it. • Assessments are conducted on a regular basis to assure adequate maintenance.
18. Organizational procedures ensure that EHR users are able to get timely help when there are EHR-related hardware, software, CDS, or network/ISP problems. Responsibility (L): IT, informatics-type department Responsibility (S): Office management Input Source: L, C, IT, HI	• Without knowing where to go for help, users will develop workarounds, which can be dangerous • Time can be wasted when users and staff members have difficulty finding help	• In small practices, guidelines exist for figuring out whom to seek help outside the organization. • In larger organizations, guidelines exist for users to know how to get help, and for Health IT staff members to know when and how to get outside assistance.
Principle 5: Activities to help the organization learn from EHR safety efforts can prevent EHR safety hazards		
19. Communication mechanisms ensure that EHR users learn of EHR changes promptly, and users are able to give feedback on related safety concerns. Responsibility (L): Vendor, Informatics-type department, IT Responsibility (S): Office management (S) Input Source: L, C, IT, HI, Ev	• If observed errors are not Reported, they will generally not be fixed • If the developer does not receive feedback, he or she will generally not address the issues. • Patient harm can result if hazards are not addressed	• Responsibility is clear for reporting EHR safety errors and getting feedback. • Someone is responsible for being the liaison to the vendor for reporting problems and getting feedback. • Communication channels are in place for including health information management staff in patient registration error correction and feedback. • Software errors or desired changes for safety reasons are routinely reported to the vendor.

Recommended Practices and Responsibilities*	Rationale for Practice or Risk Addressed	Examples of Potentially Useful Practices/Scenarios
		• Reports about EHR safety reach the highest level in the organization on a routine basis and feedback is given.
		• Users know who the go-to person is for reporting EHR safety problems.
20. Staff members with job responsibilities for EHR safety are encouraged to participate in relevant professional activities and communicate with others in similar positions. Responsibility (L): Vendor, Informatics-type department, IT Responsibility (S): Office management Input Source: L, C, IT, HI, Ev	• If key internal people do not network with outsiders, up to date knowledge may not reach them	• Organizations support professional development of staff assigned responsibility for any aspect of EHR safety, by budgeting for and encouraging training. • Staff members with responsibility for EHR safety establish routine mechanisms for discussing problems they encounter as they optimize the safety and safe use of EHRs. This may include participation in specific EHR computer user groups or in professional association activity. • Professional organizations, including those for clinicians and office administration, often provide information about issues that might affect EHR safety.
21. Self-assessments, including use of the SAFER guides, are conducted routinely by a team, and the risks of foregoing or delaying any recommended practices are assessed. Responsibility (L): Safety officer and those involved in safety initiatives, informatics-type department Responsibility (S): Office management, providers Input Source: L, HI, C	• Without learning through use of available self-assessment tools, organizations risk overlooking critical hazards	• Self-assessments related to EHRs and patient safety are done routinely. • The self-assessment process includes setting targets for addressing items the organizational team identifies.

* *Explanation of responsibilities assignments: Organizational structures vary greatly even when they are of the same type and size. The responsible parties listed here are ideal examples and possibilities. We denote large organizations with an L and small organizations such as independent ambulatory clinics with an S. Groups of clinics or hospitals with centralized IT and informatics services are considered large. The EHR safety activities in these large organizations are often included in more general safety and quality initiatives rather than separately.*

REFERENCES

1. Ash J.S., Stavri P.Z., Dykstra R., Fournier L. Implementing Computerized Physician Order Entry: The Importance of Special People. International Journal of Medical Informatics, 2003;69:235-250.
2. Ash JS, Stavri PZ, Kuperman GJ. A Consensus Statement on Considerations for a Successful CPOE Implementation. Journal of the American Medical Informatics Association, 2003; 10(3):229-234.
3. Ash, J.S., M. Berg, and E. Coiera, Some Unintended Consequences of Information Technology in Health Care: The Nature of Patient Care Information System-related Errors. Journal of the American Medical Informatics Association, 2004. 11(2): p. 104-112.
4. Ash J.S., McCormack J.L., Sittig D.F., Wright A., McMullen C., Bates D.W. Standard Practices for Computerized Clinical Decision Support in Community Hospitals: A National Survey. Journal of the American Medical Informatics Association, 2012;19(6):980-987.
5. Ash J.S., Sittig D.F., Guappone K.P., Dykstra R.H., Richardson J., Wright A., Carpenter J., McMullen C., Shapiro M., Bunce A., Middleton B. Recommended Practices for Computerized Cinical Decision Support and Knowledge Management in Community Settings: A Qualitative Study. BMC Medical Informatics and Decision Making, 2012;12:6.
6. Bates, D.W., et al., Reducing the Frequency of Errors in Medicine Using Information Technology. Journal of the American Medical Informatics Association, 2001. 8(4): p. 299-308.
7. Campbell E.M., Guappone K.P., Sittig D.F., Dykstra R.H., Ash J.S. Computerized Provider Order Entry Adoption: Implications for Clinical Workflow. Journal of General Internal Medicine, 2009. 24(1):21-6.
8. Chuo, J. and R.W. Hicks, Computer-Related Medication Errors in Neonatal Intensive Care Units. Clinics in Perinatology, 2008. 35: p. 119-139.
9. Greenes R.A., editor. Clinical Decision Support: The Road Ahead. New York, Elsevier, 2006.
10. Hogan, W.R. and M.M. Wagner, Accuracy of Data in Computer-based Patient Records. Journal of the American Medical Informatics Association, 1997. 4(5): p. 342-355.

11. Koshy S., Feustel P.J., Hong M., Kogan B.A. Scribes in an Ambulatory Urology Practice: Patient and Physician Satisfaction. Journal of Urology, 2010. 184(1): p. 258-62.
12. Magrabi, F., et al., An Analysis of Computer-Related Patient Safety Incidents to Inform the Development of a Classification. Journal of the American Medical Informatics Association, 2010. 17: p. 663-670.
13. Miller, R.A. and R.M. Gardner, Recommendations for Responsible Monitoring and Regulation of Clinical Software Systems. Journal of the American Medical Informatics Association, 1997. 4(6): p. 442-457.
14. Miller, R.A. and R.M. Gardner, Summary Recommendations for Responsible Monitoring and Regulation of Clinical Software Systems. Annals of Internal Medicine, 1997. 127(9): p. 842-845.
15. Nebeker, J.R., et al., High Rates of Adverse Drug Events in a Highly Computerized Hospital. Archives of Internal Medicine, 2005. 165: p. 1111-1116.
16. Osheroff J.A., Pifer E.A., Teich J.M., Levick, S.L, Velasco F., Sittig D.F., Rogers K.M., Jenders R.A. Improving Outcomes With Clinical Decision Support: An Implementer's Guide, 2nd ed. HIMSS, 2012.
17. Santell, J.P., et al., Medication Errors Resulting from Computer Entry by Nonprescribers. American Journal of Health-System Pharmacists, 2009. 66: p. 843-853.
18. Singh, H., et al., Ten Strategies to Improve Management of Abnormal Test Result Alerts in the Electronic Health Record. Journal of Patient Safety, 2010. 6(2): p. 1-3.
19. Sittig, D.F. and D.C. Classen, Safe Electronic Health Record Use Requires a Comprehensive Monitoring and Evaluation Framework. Journal of the American Medical Association, 2010. 303(5): p. 450-451.
20. Strom, B.L., et al., Unintended Effects of a Computerized Physician Order Entry Nearly Hard-Stop Alert to Prevent a Drug Interaction: A Randomized Controlled Trial. Archives of Internal Medicine, 2010. 170(17): p. 1578-1583.
21. Thompson D.A., Duling L., Holzmueller C.G., Dorman T., Lubomski L.H., Dickman F., Fahey M., Morlock L.L., Wu A.W., Pronovost P.J. Computerized Physician Order Entry, a Factor in Medication Errors: Descriptive Analysis of Events in the Intensive Care Unit Safety Reporting System. Journal of Clinical Outcomes Management, 2005. 12(8): p. 407-412.
22. van der Sijs, H., et al., Overriding of Drug Safety Alerts in Computerized Physician Order Entry. Journal of the American Medical Informatics Association, 2006. 13(2): p. 138-147.
23. van der Sijs, H., et al., Time-dependent Drug–Drug Interaction Alerts in Care Provider Order Entry: Software May Inhibit Medication Error Reductions. Journal of the American Medical Informatics Association, 2009. 16(6): p. 864-868.
24. Walker, J.M., et al., EHR Safety: The Way Forward to Safe and Effective Systems. Journal of the American Medical Informatics Association, 2008. 15: p. 272-277.
25. Walsh, K.E., et al., Medication Errors Related to Computerized Order Entry for Children. Pediatrics, 2006. 118(5): p. 1872-1879.

CHAPTER 15

CREATING AN OVERSIGHT INFRASTRUCTURE FOR EHR SAFETY

CREATING AN OVERSIGHT INFRASTRUCTURE FOR ELECTRONIC HEALTH RECORD-RELATED PATIENT SAFETY HAZARDS

Hardeep Singh, David C. Classen, and Dean F. Sittig

15.1.1 INTRODUCTION

Recent passage of the American Reinvestment and Recovery Act (ARRA) incentivizes health care providers and organizations to implement electronic health records (EHRs) at an unprecedented pace to meet reimbursement timelines. EHR implementation is difficult, costly, time-consuming, and might lead to unintended consequences. [1] Post-implementation evaluation often reveals that EHR implementations do not meet minimum safety guidelines, [2-4] a concern that is even more pressing now. The aggressive timeline proposed in the ARRA bill does not allow adequate customization of EHR systems to local workflows. [5,6] Furthermore, cli-

---

Hardeep Singh H, Classen DC, and Sittig DF. Creating an Oversight Infrastructure for Electronic Health Record-Related Patient Safety Hazards. Journal of Patient Safety 7,4 (2011). Reprinted with permission from Wolters Kluwer Health.

nicians increasingly share control of complex processes with computers; in some instances, they assume a higher-level oversight role and allow computers to make routine decisions and carry out appropriate actions (e.g., the computer automatically generates a lab order when certain medications are being ordered). [7,8] As more advanced clinical decision support (CDS) is embedded into existing EHRs, clinicians and their patients are increasingly reliant upon decisions generated by these systems. [9-11] The increasing scope and complexity of tasks that clinicians can perform using EHRs, combined with unprecedented pressure to rapidly adopt these systems, can create a potentially hazardous environment for patient safety.

Reports on EHR-related hazards are now emerging. [12-22] Koppel et al. identified 22 types of errors in the computerized provider entry system within their EHR. [12] Many EHR-related hazards occur at the "blunt end" of the healthcare system, [23] with potential to affect large numbers of patients if not corrected. For instance, we identified a single software configuration error in the EHR that resulted in a lack of timely notification of abnormal test results to several providers, thus affecting a large number of patients. [15] In light of these types of reports, the Office of the National Coordinator for Health Information Technology (ONC) recently sponsored an Institute of Medicine committee to "review the available evidence and the experience from the field on how the use of health information technology (HIT) affects the safety of patient care." [24] Two national databases have also been recently created to facilitate reporting of EHR-related incidents. [25,26] These early EHR hazard reporting initiatives, though important, are insufficient by themselves to address the multitude of complex EHR- related safety concerns. [21,27,27] Furthermore, EHR certification by the ONC Authorized Testing and Certification Bodies (ONC- ATCBs) [28] does not guarantee that EHRs will actually be implemented and work as planned, therefore, ongoing system evaluations and modifications are necessary. At present, it is unclear which single agency is responsible for EHR oversight. Thus, we believe it necessary to establish an independent organized national infrastructure to actively monitor and improve the safety of EHR systems. In this paper, we propose the creation of a national EHR oversight program to provide dedicated surveillance of EHR-related safety events and to promote learning from identified hazards, close calls, and adverse events. We provide an overview of a proposed program and its rationale and discuss its potential organizational components and functions. Although our recommendations might not cover all aspects of EHR-related hazards, they are a starting point to stimulate the much needed discussion and debate in this area.

15.1.2 ORGANIZATIONAL INFRASTRUCTURE OF A NATIONAL EHR SAFETY PROGRAM

Because EHR-related safety issues are an emerging area of knowledge, an EHR safety oversight program should involve robust data gathering and data analysis components. Both mechanisms should be overseen by a new independent board specifically charged with ensuring safety of EHRs nationally. Many useful lessons have been learned from the success of the National Transportation Safety Board (NTSB), [29] which could provide a model for a Congressionally-funded independent EHR safety board. Data gathering activities would necessitate an infrastructure for collecting adverse events and near-misses from various sources, whereas the analytic component would be charged with analyzing the collected information and developing and disseminating preventive strategies on a national scale. The national program should be supported by close collaboration with institutional-level safety initiatives, such as EHR safety committees. [30] This organizational scheme is reminiscent of the unified and cohesive safety program used by the Department of Veterans Affairs National Center for Patient Safety (NCPS), which is charged to lead the VA's patient safety improvement initiatives. [31]

At the institutional level, multidisciplinary EHR safety committees, including a designated EHR patient safety officer, could perform two essential functions: 1) investigate all known EHR-related adverse events and near-misses [32] and report them to the national board using standardized methods; [33] and 2) perform routine safety self-assessments, including tests of all EHR components and applications and "prospective risk assessments" to proactively identify new risks. [34,35] Because providers in smaller practices might not have resources for these functions, their respective local health information exchanges, [36] independent physicians' associations, [37] regional extension centers, [38] quality improvement organizations, [39] or accountable care organizations could house the needed technical resources. Many institutions and practices also have existing legal and risk management infrastructure that can be leveraged to perform these functions. These locally housed investigational and risk assessment initiatives will likely reveal many site-specific, contextual issues that affect EHR safety and these issues could then be addressed at the institutional level without the need for national interference.

At the national level, the proposed board would be charged with analyzing event reports from institutions and investigating major EHR-related incidents (such as those associated with harm to a large number of patients). Aggregate analysis of reports could be used to identify common unsafe conditions for specific EHRs and to inform widely released recommendations to mitigate risks. The board would have both investigative and regulatory components and functions. Potential other roles for the board may include the development of newer error surveillance methods, [40] validation and oversight of EHR safety self-assessment procedures and on-site EHR safety inspections (perhaps in collaboration with hospital certification/accreditation organizations), and dissemination of safety guidelines [41] and benchmarks. The board would work closely with EHR certifying organizations [42] (and thus indirectly with EHR vendors) to improve EHR design and implementation and with other governmental agencies (such as National Institute of Standards and Technology [NIST] and ONC) to coordinate EHR-related rules and regulations. Some of the key functions of such an oversight program are described in the sections below.

15.1.3 DATA GATHERING OPERATIONS OF THE PROGRAM

Voluntary reporting could continue to be an important source of data about EHR patient safety risks. [18] However, existing FDA databases for medical device errors [43-46] appear to be seldom used for reporting EHR-related incidents [47] and a multi-pronged approach to improve reporting is needed. First and foremost, vendor contracts need to be free of non-disclosure or gag clauses [48,49] in order to encourage users and institutions to report errors and adverse events. Second, at the institutional level, standardized models of data collection [50] should be used (e.g., through systems such as the HIT Hazards Reporting Model [Hazard Tracker]). [51] Third, error reporting initiatives should be integrated nationally. For instance, medical professional insurance carriers and the not-for-profit iHealth Alliance [52] recently created PDR Secure™, a website that allows users to report EHR-related safety issues. Patient safety organizations are also working to develop a common format for reporting EHR problems. [33] Additional safety events should be aggressively solicited from EHR-related sentinel events reported to Joint Commission, reports to the vendors, and reports from users, the media, and EHR certifying organizations. Lastly, "major EHR adverse events" need to be defined and their reporting should be mandated.

Although we emphasize reporting for initial discovery, we recognize that it is limited and often neglects minor or latent errors and near-misses; these minor incidents might not be directly implicated in an adverse event but are rich sources of information to prevent future events. [29] Therefore, we propose development of additional methods to collect data. [53] One possible example is automated reporting (such as that of software programs that prompt the user to automatically report an error to the developer when a software glitch is detected). Another area for development is non-voluntary surveillance approaches that rely on electronic triggers to initiate error reporting. [54] For example, Koppel et al. found that orders entered by a physician and subsequently cancelled within 45 minutes were likely errors. [54] We are actively investigating the creation of algorithms that could be run against a wide variety of EHR implementations to establish rates of errors and of other higher-risk scenarios (e.g., counting the number of duplicate patient records in the database or the percentage of orders entered via free text rather than via the structured data entry fields [55]). Methods to query large electronic repositories for safety events in near "real time" should also be developed. For example, through the newly created Sentinel System, the FDA can now query electronic health information of more than 60 million people to monitor the safety of approved medical products. [56]

15.1.4 INVESTIGATIONAL AND ANALYTIC ROLES OF THE PROGRAM

A major adverse event related to an EHR could affect thousands of patients if not corrected rapidly. For example, in 2006 the UK National Health Service was forced to notify over 900 clinicians that their patients who were prescribed Zyban may have mistakenly been given Viagra due to an error in the dispensing pharmacy's medication mapping table. [57] Although there was no reported harm, future similar events might not be harmless. We propose that the board charged with EHR oversight should create a special investigation team dedicated to "major EHR adverse events". Whereas most adverse event investigations could be conducted at the institutional level and reported appropriately, events with particularly broad or catastrophic impact should be investigated by the team at the national board. This strategy will ensure rapid action and wide dissemination of any significant findings. Pre-defined criteria for triggering this level of investigation could specify, for instance, the number of patients at risk for harm (or harmed) in a single incident. For example, criteria for investigation of an unplanned EHR system downtime that adversely affects patient care could include combinations of parameters such as the following: [58]

1. Lasts for more than 24 hours
2. Harms or has potential to harm more than 100 [58] patients [59]
3. Is not the direct result of a natural disaster [60]
4. Occurs in organizations that have implemented the following components of an EHR: admission/discharge/transfer; clinical results review; provider order entry, communication, verification; barcode medication verification; picture archiving and communication; clinical documentation; alert notification; and participation in local health information exchange [61,62]
5. Simultaneously affects at least two of the EHR components identified above

A dedicated multidisciplinary team of core investigators with expertise in safety, informatics, human factors, computer science, and clinical medicine should conduct these major EHR adverse event investigations. The team could also leverage the expertise of an external group of pre-screened, independent consultants composed of clinicians, statisticians, informaticians, lawyers, and human factors engineers who work on an as-needed basis. All board investigators should be given legal authority to examine all records confidentially (including system logs and patient records), to access all relevant computer systems, and to interview all personnel or patients whom they believe are essential to their investigation. If needed, error scenarios should be recreated with test patients in simulated settings using a full range of human factors engineering-based techniques for analysis. [63] Rapid dissemination of identified issues and corrective actions may resemble the processes used by the Joint Commission and VA NCPS [64] to issue advisories. The findings of these investigations should also be publically reported so that other vendors, healthcare organizations, and users can learn from them. To prioritize its efforts, the board could create an annual "Most Wanted List" of improvements, similar to that issued by NTSB [65]. Research and evaluation should also be a key component of this learning process. For instance, centralized, de-identified databases could be created to enable researchers to advance the science and evidence of EHR safety. Because many EHRs share common features and will be used nearly universally, this centralized and standardized approach to high-stakes investigations is likely be beneficial.

15.1.5 REGULATORY AUTHORITY

In addition to providing a safety event reporting clearinghouse and a mechanism for major adverse event investigations, the new national EHR safety board must be charged with appropriate regulatory and legal authority to carry out effective oversight. No government agency is currently fully equipped to perform this function. Because of the increasing complexity and coupling of often unrelated HIT-enabled systems (e.g., from different vendors and/or organizations), the oversight board should coordinate with other agencies (such as ONC and NIST) and industrial trade associations (e.g., the Electronic Health Record Association [62]) in a cooperative fashion to investigate the causes and potential mitigating solutions to the problem(s) identified. This collaboration could be modeled after the aviation industry oversight provided by the Commercial Aviation Safety Team (CAST). [66] Another important aspect of EHR safety oversight will be close collaboration between various state and federal agencies responsible for making and enforcing EHR-related rules, regulations, and certification standards. For example, the new "meaningful use" statute requires e-prescribing in the outpatient setting, but the federal government only recently modified laws restricting the electronic transfer of prescriptions for controlled substances. [67]

To ensure safety and compliance in certain high-risk areas, we recommend strategies to monitor uptake of board recommendations that are of critical importance. One way to do this will be to implement unannounced, randomly scheduled, on-site EHR safety inspections, much like the Line Operations Safety Audit (LOSA) used in aviation. [68] These audits would involve interviews with key stakeholders (e.g., Chief Information Officer, Patient Safety Officer, etc.), observations of important clinical operations, and inspection of the human-computer interface as configured by the organization. [69] Inspectors should be equipped with comprehensive EHR safety inspection guides consisting of a set of "red flags" (e.g., lack of evidence of computer system backups, outdated clinical decision support logic or no evidence of CDS testing) and "best practices" (e.g., pre-printed order and documentation forms for use when the EHR is unavailable, strict policies to address breaches of patient confidentiality and a robust clinician EHR training program) to help guide interviews and observations related to the respective high-risk area. All inspectors should be certified and pre-screened for financial and other conflicts of interest to ensure their impartiality. Inspectors might also build an audit approach modeled on the Leapfrog CPOE Flight Simulator. [70] This simulator evaluates the decision support functionality of implemented EHR systems to measure hospital compliance with the National Quality Forum Safe Practice for CPOE. [71]

Although the approach should initially be non-punitive, eventually the oversight board should have authority to invoke penalties if needed to protect patients. Furthermore, penalties for responsible stakeholders for non-compliance with certain pre-set expectations (such as fines for non-reporting of major EHR adverse events, not fixing a major software bug) could only be ensured using this approach.

15.1.6 NEXT STEPS

The proposal we present herein is ambitious, but a multifaceted and centralized approach to address EHR-related safety is perhaps the best strategy. However, this approach is challenging and any attempt to implement such an important oversight function should proceed using a carefully orchestrated, staged implementation pathway. Given the current state of affairs, some might argue that the scope and cost of the national oversight program might be beyond what is possible now. To jumpstart the creation of this program, we propose that local, institutional-level initiatives to collect and analyze data must be bolstered immediately. This would help characterize the various types and frequencies of EHR-related errors and adverse events. These initiatives must be synergistic with rigorous research to improve the "basic science" of EHR-related safety and simultaneously inform the creation of the national oversight program. Taking this approach would ensure that there is adequate strength of the evidence to justify the scope and cost of implementation of the independent national board, which clearly will take longer to get established. While it might be premature for us to lay out the precise implementation strategy of the entire proposed program, we propose two immediate next steps to advance this agenda:

Establish a standardized reporting infrastructure to facilitate event reporting and investigation of EHR-related safety concerns. The infrastructure should specifically protect reporters and maintain incentives to report. One successful model for this is the Aviation Safety Reporting System (ASRS), a confidential reporting system in which all reports are de-identified before being entered into the incident database. [72] In qualified cases, fines and penalties are waived for unintentional violations of federal aviation statutes and regulations which are reported to ASRS.

Bring together a group of experts and federal stakeholders to explore and define investigational and analytic roles of a national oversight program. This group would need to tackle the following tasks in order to facilitate the creation of a national program:

Define criteria for triggering an investigation and outline a methodology for ensuring timely action and wide dissemination of significant findings. They will also need to develop the investigation methodology and inspection guides in order to facilitate the process.

Help establish the legal and regulatory infrastructure to create the new board and facilitate its work processes. Key issues they will need to address include how to interface with complex technical and organizational governance structures involved with EHR implementation and use at the institutional level, the technical feasibility of gaining access to and examining system logs 'on demand' during investigations and unannounced inspections, and maintaining the confidentiality of the discovery process.

Explore how the board would establish rules and regulations that EHR vendors will need to follow in order to comply with the oversight program; none currently exist in the Office of the National Coordinator for Health Information Technology's Authorized Testing and Certification Body (ONC-ATCB) process. [73]

15.1.7 CONCLUSION

Technological advances give rise to increasingly complex and multifaceted errors in healthcare. EHR-related errors must be conceptualized, analyzed, and mitigated using a robust oversight infrastructure. We propose the creation of a national oversight program which relies on local institution-level EHR safety data collection and analysis, and ultimately leads to the formation of a centralized, non-partisan, multi-disciplinary board specifically charged with ensuring EHR safety. If implemented, the proposed infrastructure could help to identify and reduce EHR-related adverse events and errors and create a safer and more effective EHR-based healthcare delivery system.

REFERENCES

1. Sittig D, Ash J. Clinical information Systems: Overcoming adverse consequences. Jones and Bartlett Publishers, LLC; Sudbury, MA: 2009.
2. Sittig DF, Singh H. Eight Rights of Safe Electronic Health Record Use. JAMA. 2009;302:1111–1113.

3. Kilbridge PM, Classen DC. The informatics opportunities at the intersection of patient safety and clinical informatics. J Am Med Inform Assoc. 2008;15:397–407.
4. Metzger J, Welebob E, Bates DW, Lipsitz S, Classen DC. Mixed results in the safety performance of computerized physician order entry. Health Aff (Millwood) 2010;29:655–663.
5. Sittig DF, Ash JS, Zhang J, Osheroff JA, Shabot MM. Lessons from "Unexpected increased mortality after implementation of a commercially sold computerized physician order entry system" Pediatrics. 2006;118:797–801.
6. Blumenthal D, Tavenner M. The "Meaningful Use" Regulation for Electronic Health Records. New England Journal of Medicine. 2010;363:501–504.
7. Overhage JM, Tierney WM, Zhou XH, McDonald CJ. A randomized trial of "corollary orders" to prevent errors of omission. J Am Med Inform Assoc. 1997;4:364–375.
8. East TD, Wallace CJ, Morris AH, Gardner RM, Westenskow DR. Computers in critical care. Crit Care Nurs Clin North Am. 1995;7:203–217.
9. Campbell EM, Sittig DF, Guappone KP, Dykstra RH, Ash JS. Overdependence on technology: an unintended adverse consequence of computerized provider order entry. AMIA Annu Symp Proc. 2007;94-98
10. McKinley BA, Moore LJ, Sucher JF, et al. Computer protocol facilitates evidence-based care of sepsis in the surgical intensive care unit. J Trauma. 2011;70:1153–1166.
11. Morris AH, Orme J, Jr., Truwit JD, et al. A replicable method for blood glucose control in critically Ill patients. Crit Care Med. 2008;36:1787–1795.
12. Koppel R, Metlay JP, Cohen A, et al. Role of computerized physician order entry systems in facilitating medication errors. JAMA. 2005;293:1197–1203.
13. Horsky J, Kuperman GJ, Patel VL. Comprehensive analysis of a medication dosing error related to CPOE. Journal Of The American Medical Informatics Association: JAMIA. 2005;12:377–382.
14. McDonald CJ. Computerization can create safety hazards: a bar-coding near miss. Ann Intern Med. 2006;144:510–516.
15. Singh H, Wilson L, Petersen L, et al. Improving follow-up of abnormal cancer screens using electronic health records: trust but verify test result communication. BMC Medical Informatics and Decision Making. 2009;9
16. Nerich V, Limat S, Demarchi M, et al. Computerized physician order entry of injectable antineoplastic drugs: an epidemiologic study of prescribing medication errors. Int J Med Inform. 2010;79:699–706.
17. Schulte F, Schwartz E. As Doctors Shift to Electronic Health Systems, Signs of Harm Emerge. The Huffington Post. 2010
18. Magrabi F, Ong MS, Runciman W, Coiera E. An analysis of computer-related patient safety incidents to inform the development of a classification. J Am Med Inform Assoc. 2010;17:663–670.
19. Magrabi F, Ong MS, Runciman W, Coiera E. Using FDA reports to inform a classification for health information technology safety problems. J Am Med Inform Assoc. 2011
20. Harrington L, Kennerly D, Johnson C. Safety issues related to the electronic medical record (EMR): synthesis of the literature from the last decade, 2000-2009. J Healthc Manag. 2011;56:31–43.

21. Sittig DF, Singh H. Defining Health Information Technology-Related Errors: New Developments Since To Err Is Human. Arch Intern Med. 2011;171:1281–1284.
22. Leviss J. H.i.t. Or Miss: Lessons Learned from Health Information Technology Implementation. American Health Information Management Association; 2010.
23. Reason J. Human error: models and management. BMJ. 2000;320:768–770.
24. Institute of Medicine [3-2-2011. 5-23-2011];Activity - Patient Safety and Health Information Technology. http://www.iom.edu/Activities/Quality/PatientSafety-HIT.aspx.
25. PDR Secure [2010. 5-13-2011];EHR Safety Event Reporting Service. http://ehrevent.org/
26. Mosquera Mary. [9-20-2011. 10-7-2011];AHRQ tests tool identify, report health IT hazards. http://govhealthit.com/news/ahrq-tests-tool-identify-and-report-health-it-hazards.
27. Koppel R. Monitoring and evaluating the use of electronic health records. JAMA. 2010;303:1918–1919.
28. Federal Register. 2011 http://edocket.access.gpo.gov/2010/pdf/2010-25683.pdf.
29. Sittig DF, Classen DC. Safe electronic health record use requires a comprehensive monitoring and evaluation framework. JAMA. 2010;303:450–451.
30. Miller RA, Gardner RM. Recommendations for responsible monitoring and regulation of clinical software systems. American Medical Informatics Association, Computer-based Patient Record Institute, Medical Library Association, Association of Academic Health Science Libraries, American Health Information Management Association, American Nurses Association. J Am Med Inform Assoc. 1997;4:442–457.
31. VA National Center for Patient Safety [6-1-2011. 10-7-2011]; http://www.patientsafety.gov/
32. Yackel TR, Embi PJ. Unintended errors with EHR-based result management: a case series. J Am Med Inform Assoc. 2010;17:104–107.
33. Agency for Healthcare Research and Quality Common Formats for Patient Safety Data Collection and Event Reporting. Federal Register. 2010;75:65359–65360.
34. Bonnabry P, spont-Gros C, Grauser D, et al. A risk analysis method to evaluate the impact of a computerized provider order entry system on patient safety. J Am Med Inform Assoc. 2008;15:453–460.
35. DeRosier J, Stalhandske E, Bagian JP, Nudell T. Using health care Failure Mode and Effect Analysis: the VA National Center for Patient Safety's prospective risk analysis system. Jt Comm J Qual Improv. 2002;28:248–67. 209.
36. Sittig DF, Joe JC. Toward a statewide health information technology center (abbreviated version) South Med J. 2010;103:1111–1114.
37. Ash JS, Sittig DF, Wright A, et al. Clinical decision support in small community practice settings: a case study. J Am Med Inform Assoc. 2011
38. Maxson E, Jain S, Kendall M, Mostashari F, Blumenthal D. The regional extension center program: helping physicians meaningfully use health information technology. Ann Intern Med. 2010;153:666–670.
39. Baier R, Gardner R, Gravenstein S, Besdine R. Partnering to improve hospital-physician office communication through implementing care transitions best practices. Med Health R I. 2011;94:178–182.

Creating an Oversight Infrastructure for EHR Safety

40. Dunn AG, Ong MS, Westbrook JI, Magrabi F, Coiera E, Wobcke W. A simulation framework for mapping risks in clinical processes: the case of in-patient transfers. J Am Med Inform Assoc. 2011;18:259–266.
41. Nelson NC. Downtime procedures for a clinical information system: a critical issue. J Crit Care. 2007;22:45–50.
42. Office of the National Coordinator for Health Information Technology. Department of Health and Human Services Establishment of the permanent certification program for health information technology. Final rule. Fed Regist. 2011;76:1261–1331.
43. U.S Food and Drug Administration MedSun: Medical Product Safety Network. 2011 http://www.fda.gov/MedicalDevices/Safety/MedSunMedicalProductSafetyNetwork/defa ult.htm.
44. U.S Food and Drug Administration [2011. 5-23-2011];Medsun Reports. http://www.accessdata.fda.gov/scripts/cdrh/cfdocs/medsun/searchReport.cfm.
45. U.S Food and Drug Administration MAUDE - Manufacturer and User Facility Device Experience. 2011 http://www.accessdata.fda.gov/scripts/cdrh/cfdocs/cf-MAUDE/search.CFM.
46. U.S Food and Drug Administration [2011. 5-23-2011];MDR Database Search. http://www.accessdata.fda.gov/scripts/cdrh/cfdocs/cfmdr/search.cfm?searchoptions=1.
47. Myers RB, jones SL, Sittig DF. Review of reported clinical information system adverse events in US Food and Drug Administration databases. Appl Clin Inf. 2011;2:63–74.
48. Goodman KW, Berner ES, Dente MA, et al. Challenges in ethics, safety, best practices, and oversight regarding HIT vendors, their customers, and patients: a report of an AMIA special task force. J Am Med Inform Assoc. 2011;18:77–81.
49. Koppel R, Kreda D. Health Care Information Technology Vendors' "Hold Harmless" Clause. JAMA. 2009;301:1276–1278.
50. PSO Privacy Center [4-14-2011. 10-7-2011];AHRQ Common Formats Device or Medical/Surgical Supply, including HIT. https://www.psoppc.org/web/patientsafety/ahrq-common-formats-device-or-medical/surgical-supply-including-hit-device.
51. Walker JM, Carayon P, Leveson N, et al. EHR safety: the way forward to safe and effective systems. J Am Med Inform Assoc. 2008;15:272–277.
52. [2011. 5-23-2011];EHR Safety Event Reporting System. http://www.ehrevent.org/
53. Classen DC, Resar R, Griffin F, et al. Global Trigger Tool Shows That Adverse Events In Hospitals May Be Ten Times Greater Than Previously Measured. Health Aff. 2011;30:581–589.
54. Koppel R, Leonard CE, Localio AR, Cohen A, Auten R, Strom BL. Identifying and quantifying medication errors: evaluation of rapidly discontinued medication orders submitted to a computerized physician order entry system. J Am Med Inform Assoc. 2008;15:461–465.
55. Sittig DF, Campbell E, Guappone K, Dykstra R, Ash JS. Recommendations for monitoring and evaluation of in-patient Computer-based Provider Order Entry systems: results of a Delphi survey. AMIA Annu Symp Proc. 2007;671-675
56. Behrman RE, Benner JS, Brown JS, McClellan M, Woodcock J, Platt R. Developing the Sentinel System--a national resource for evidence development. N Engl J Med. 2011;364:498–499.
57. Lister David. [12-14-2006. 5-23-2011];The Latest Smoking Cure: Viagra. http://www.timesonline.co.uk/tol/news/uk/health/article753765.ece.

58. Sittig DF, Classen DC. Monitoring and Evaluating the Use of Electronic Health Records—Reply. JAMA. 2010;303:1918–1919.
59. Kilbridge P. Computer Crash – Lessons from a System Failure. New England Journal of Medicine. 2003;348:881–882.
60. Travis J. Scientist's fears come true as hurricane floods New Orleans. Science. 2005;309:1656–1659.
61. Fischetti L, Mon D, Ritter R, Rowlands D. HL7 HER System Functional Model, Release 1. Health Level Seven ®, Inc.; Ann Arbor, Mich: 2007. Direct Care Functions.
62. HIMSS Analytics [2009. 5-20-2011];U.S. EMR Adoption ModelSM Trends. http://www.himssanalytics.org/docs/HA_EMRAM_Overview_ENG.pdf.
63. Kushniruk AW, Patel VL. Cognitive and usability engineering methods for the evaluation of clinical information systems. J Biomed Inform. 2004;37:56–76.
64. VA National Center for Patient Safety [5-20-2011. 5-23-2011];VHA Patient Safety Alerts and Advisories. http://www.patientsafety.gov/alerts.html.
65. National Transportation Safety Board [2011. 10-7-2011];NTSB Most Wanted List. http://www.ntsb.gov/safety/mwl.html.
66. Pronovost PJ, Goeschel CA, Olsen KL, et al. Reducing health care hazards: lessons from the commercial aviation safety team. Health Aff (Millwood) 2009;28:w479–w489.
67. Department of Justice. Drug Enforcement Administration [3-31-2010];Electronic Prescriptions for Controlled Substances; Final Rule. http://www.deadiversion.usdoj.gov/fed_regs/rules/2010/fr0331.pdf.
68. The University of Texas Human Factors Research Project [2002. 6-13-2011];Line Operations Safety Audit and Threat and Error Management. http://homepage.psy.utexas.edu/homepage/group/helmreichlab/aviation/LOSA/LOSA.ht ml.
69. Schumacher RM, Patterson EM, North R, Zhang J, Lowry SZ, Quinn MT, Ramaiah M, U.S.Department of Commerce. National Institute of Standards and Technology [2011. 10-7-2011];Technical Evaluation, Testing and Validation of the Usability of Electronic Health Records, NISTIR 7804 Draft. http://www.nist.gov/healthcare/usability/upload/Draft_EUP_09_28_11.pdf.
70. Kilbridge PM, Welebob EM, Classen DC. Development of the Leapfrog methodology for evaluating hospital implemented inpatient computerized physician order entry systems. Qual Saf Health Care. 2006;15:81–84.
71. National Quality Forum (NQF) Safe Practice 16: Safe Adoption of Computerized Prescriber Order Entry: 209-215 in Safe Practices for Better Healthcare - 2010 Update: A Consensus Report. National Quality Forum; Washington, DC: 2010.
72. NASA Aviation Safety Reporting System: Confidentiality and Incentives to Report. 2011 http://asrs.arc.nasa.gov/overview/confidentiality.html.
73. Department of Health and Human Services Health Information Technology: Initial Set of Standards, Implementation Specifications, and Certification Criteria for Electronic Health Record Technology, 45 CRF Parts 170. Federal Register. 2010;75:44590–44654.

PATIENT SAFETY GOALS FOR THE PROPOSED FEDERAL HEALTH INFORMATION TECHNOLOGY SAFETY CENTER

Dean F. Sittig, David C. Classen, and Hardeep Singh

The Institute of Medicine's 2012 report on Health IT and Patient Safety called for the establishment of an independent federal entity for monitoring and analyzing patient safety data and investigating serious incidents related to health IT. [1] In an attempt to address this recommendation, President Obama requested $5 million in his 2015 Federal budget for the Office of the National Coordinator for Health Information Technology (ONC) to create a roadmap for a Health Information Technology Safety Center (HIT Safety Center) [2]. This was followed a week later by an influential and much awaited report that responded to the U.S. Food and Drug Administration Safety and Innovation Act (FDASIA) [3]. Briefly, this Act required ONC, Food and Drug Administration (FDA), and Federal Communication Commission (FCC) to describe "strategy and recommendations on an appropriate, risk-based regulatory framework pertaining to health information technology, including mobile medical applications, that promotes innovation, protects patient safety, and avoids regulatory duplication." This report, a culmination of deliberations of the FDASIA workgroup chartered by the FDA, ONC, and FCC [4], reinforced the call for an ONC-based HIT Safety Center [5]. The HIT Safety Center is envisioned as a public-private entity that will serve as "a trusted convener of health IT stakeholders in order to focus on activities that promote health IT as an integral part of patient safety with the ultimate goal of assisting in the creation of a sustainable, integrated health IT learning system that avoids regulatory duplication and leverages and complements existing and ongoing efforts." [5]

Sittig DF, Classen DC, Singh H. Patient safety goals for the proposed Federal Health Information Technology Safety Center. J Am Med Inform Assoc. 2014 Oct 20. pii: amiajnl-2014-002988

The initial funding to establish this new center represents only a fraction of what will be required to put its infrastructure in place and to maintain its functionality. Assuming that the US Congress provides the necessary funding and oversight authority, the HIT Safety Center has the potential to play a key operational role for major national initiatives related to health information technology and patient safety [6]. This also assumes that recent questions regarding the authority of ONC to even create it are answered satisfactorily [7]. More recently, ONC issued a 2-year, task order entitled, "Health IT Safety Center Road Map" that asks contractors to develop a diversified plan including federal funding options, public-private collaboration and potential private sector funding of activities. In this paper, we assume the best case scenario and propose several specific patient safety goals that the HIT Safety Center could adopt to deliver on the promise of creating safe and effective HIT-enabled healthcare systems [8].

As noted in a recent endorsement by the HIT Policy committee [6], the time is ripe for the Health IT Safety Center. The FDASIA report's high-level vision created momentum for its development given the increasing recognition by both frontline clinicians and health care organizations (HCOs) of both the benefits and unintended consequences of the rapidly increasing use of health information technology (HIT), including electronic health records (EHRs). For example, safety concerns have arisen from the design and functioning of HIT and from the disruptions in clinicians' workflow in settings where EHRs have been implemented [9]. Emerging evidence from the scientific literature [10,11,12] as well as anecdotal reports [13] suggest that "HIT-related safety events" (i.e., events arising from unsafe technology or unsafe use of technology) are occurring. Given that neither the FDA nor any other agency will be regulating most forms of HIT, the HIT Safety Center could be instrumental in uniting key "frontline" stakeholders (i.e., clinicians, HCOs, quality and safety personnel, and HIT vendors) with key administrative and policy stakeholders to develop the necessary methods and infrastructure to ensure a cohesive national approach to HIT safety.

To facilitate rapid cycle improvements related to patient safety and to benefit the maximum number of patients, we posit that the HIT Safety Center must lead the coordination of activities to achieve four goals:

- Facilitate the creation of a nationwide "post-marketing" surveillance system to monitor HIT-related patient safety events, including events that lead to patient harm and "near misses" [14];
- Develop the methods and governance structure to support the investigation of major HIT-related safety events;
- Create the infrastructure and methods needed to carry out random assessments of large, complex, HIT-enabled healthcare organizations; and
- Advocate for HIT safety with various government (e.g., U.S. Congress, Centers for Medicare and Medicaid Services (CMS), Office of Civil Rights, Department of Defense, or state departments of health) and private entities (e.g., EHR vendors, healthcare provider organizations).

The following sections provide a brief description of the rationale for these goals and specific actions that could be undertaken.

15.2.1 FACILITATE CREATION OF A NATIONWIDE HIT-RELATED PATIENT SAFETY SURVEILLANCE SYSTEM

Currently, we are unable to quantify the rate of HIT-related patient safety events with any precision using the existing patient safety reporting and analysis infrastructure, [13] which consists of a small number of reports within very large public databases that are not specific to HIT (e.g., FDA_MAUDE [15], Pennsylvania Patient Safety Authority [16], MEDMARX [17]). Moreover, there is still no clear consensus on taxonomy and measurement methods for HIT-related safety events. Thus, the HIT Safety Center could create a robust foundation for improving future measurement and surveillance of patient safety at a national level. For example, ONC could partner with not-for-profit entities (e.g., ECRI's recently formed "Partnership for Promoting Health IT Patient Safety" [18]) to create a federally-funded research and development center for event reporting, analysis, and information sharing, similar in concept to the Veterans Affairs' Informatics Patient Safety office's case tracking database [12]. These centers, in conjunction with local and national PSOs, could play pivotal roles in establishing key safety benchmarks EHR developers and HCOs could use to assess safety performance. This surveillance system should gather data to help HIT developers and clinicians better understand and mitigate risks associated with HIT implementation and use.

A major limitation of existing HIT-related safety event reporting systems is that most clinicians either do not understand what should be reported or cannot recognize that near misses or events have occurred. To facilitate measurement and monitoring of HIT safety, we propose the term "health information technology (HIT) related safety concern" to broadly describe patient safety events that reached the patient (regardless of whether harm occurred), near misses, and unsafe conditions. Although this terminology of "safety concern" is consistent with AHRQ common format reporting standards [19], these standards do not adequately capture the breadth of health IT-related safety concerns defined below and thus need to be broadened. We propose that the AHRQ common format should address five major types of HIT-related safety concerns (Table 15.2.1), including instances in which:

- HIT fails during use or is otherwise not working as designed [20]. The safety concern is directly attributable to the HIT.
- HIT is working as designed, but the design does not meet the user's needs or expectations (i.e., bad design) [21]. HIT is a contributing factor to the safety concern.
- HIT is well-designed and working correctly, but was not configured, implemented, or used in a way anticipated or planned for by system designers and developers [22]. These events are related to use of HIT (i.e., rather than HIT itself) and may be referred to as configuration errors, "work-arounds" or incorrect usage.
- HIT is working as designed, and was configured and used correctly, but interacts with external systems (e.g., via hardware or software interfaces) so that data is lost or incorrectly transmitted or displayed [23]. These events are inevitable due to the interactive complexity of tightly coupled systems. They are often referred to as HIT system interface safety concerns [24].
- Specific HIT safety features or functions were not implemented or not available [25].

At a minimum, event types 1-4 should be subjected to reporting and surveillance. To standardize this process, we propose development of a small set of safety concerns that HCOs and EHR vendors should be required to report at regular intervals to the HIT Safety Center via a Patient Safety Organization (PSO) [26]. Voluntary event reporting by clinicians should be incentivized by providing Continuing Medical Education or Maintenance of Certification credits. In addition, automated reporting mechanisms could greatly advance these surveillance efforts. For instance, the capability to generate and report HIT safety eMeasures (i.e., standardized performance measures in an electronic format) directly from EHRs could be added to future EHR certification requirements [27]. These eMeasures could be modeled after the "near misses" within the airline safety reporting system [28], the types of events and threats reported to the United States Department of Homeland Security's Computer Emergency Readiness Team (US-CERT) [29], or events tracked in mandatory public health reporting systems maintained by the FDA and CDC. Some examples of potential HIT safety eMeasures, which would be a good place to start, are listed in Table 15.2.2.

TABLE 15.2.1: Definitions and Examples of Different Types of HIT-related Safety Concerns

Type of HIT-related safety concern	Examples
1. Instances in which HIT fails during use or is otherwise not working as designed.	Broken hardware or software "bugs"
2. Instances in which HIT is working as designed, but the design does not meet the user's needs or expectations.	Usability issues
3. Instances in which HIT is well-designed and working correctly, but was not configured, implemented, or used in a way anticipated or planned for by system designers and developers.	Duplicate order alerts that fire on alternative PRN pain medications
4. Instances in which HIT is working as designed, and was configured and used correctly, but interacts with external systems (e.g., via hardware or software interfaces) so that data is lost or incorrectly transmitted or displayed.	Medication order for extended release morphine inadvertently changed to immediate release morphine by error in interface translation table
5. Instances in which specific safety features or functions were not implemented or not available (i.e., HIT could have prevented a safety concern).	Hospitalized patient inadvertently receives 5 grams of acetaminophen in 24 hours because maximum daily dose alerting was not available

15.2.2 DEVELOP A FRAMEWORK TO SUPPORT INVESTIGATION OF MAJOR HIT-RELATED SAFETY EVENTS

The HIT Safety Center can also address the problem of slow progress in learning from HIT-related safety events by creating criteria and methods needed to conduct investigations of major HIT-related safety events, defined as those causing severe patient harm or placing more than 100 patients at risk for harm or an HIT-related "sentinel event" reported to the Joint Commission [30]. As more organizations rush to implement comprehensive EHRs, we expect more serious EHR-related safety events. These events would need to be investigated under the auspices of PSOs to identify causes and prevention strategies; most likely similar events will occur at other institutions. Alternatively, Congress could create a new independent agency within the ONC, similar to the National Transportation Safety Board within the Department of Transportation that is authorized to conduct investigations and make recommendations [31]. While the creation of such a new agency may currently appear doubtful given the current socio-political climate, the increasing reliance on the use of HIT within all aspects of healthcare may justify the cause.

For example, over the last several years, several reports have documented long-term (>4 hours) or widespread (i.e., affecting multiple organizations or sites of care) periods of EHR unavailability [32,33]. As the consolidation of HCOs continues, coupled with increasing numbers of large-scale, remotely hosted EHR implementations, similar events are certain to occur. An example of a major HIT-related safety event, that might warrant further investigation to identify generalizable lessons, is a widespread HIT downtime that lasts for more than 24 hours, is unrelated to a natural disaster, and affects at least two of the following EHR functions simultaneously: admission/discharge/transfer; clinical results review; provider order entry, communication, verification; barcode medication verification; picture archiving and communication; clinical documentation; alert notification; or participation in local health information exchange [34].

The types of safety events that need investigation could be further refined by the HIT Safety Center. The investigation format and approach will also depend on the type and severity of the event but in general, analysis should be conducted by independent investigators with deep technical knowledge of the underlying hardware and software systems and extensive clinical knowledge of various healthcare work processes, in conjunction with patient safety experts from PSOs. Investigations should produce comprehensive, publically available reports that outline how similar events can be prevented at other institutions. This HIT Safety Center-

Creating an Oversight Infrastructure for EHR Safety

created investigative framework would be essential for transformation to a "learning" HIT-enabled health care system as the IOM suggests [35].

TABLE 15.2.2: Candidate HIT Safety eMeasures that could be Reported to the HIT Safety Center on a Quarterly Basis.

Proposed HIT Safety EMeasures or Events	Rationale
Unexpected EHR-related downtimes lasting more than 8 hours	After 8 hours it is likely that the downtime event will increase the risk of "change-of-shift".
Mean EHR response time as measured from the end-users viewpoint	As response time increases (e.g., past 10 seconds) the likelihood of "functional downtime" increases.
Interruptive alerts that have fired more than 100 times with 100% override rate	Frequent, synchronous alerts that are repeatedly overridden increase the risk of alert fatigue and clinicians missing potentially life-threatening events.
Erroneous displays of laboratory test results or medications	Incorrect result or medication displays increase the risk of erroneous diagnosis or treatment.
Percent of EHR users trained and passing a competency test before getting a login [35]	Allowing untrained users to login to the EHR can lead to missing key data, erroneous data entry, or failed communication and affect patient care.
Rate of Computer-based provider order entry use	Incomplete CPOE usage, results in duplicative order entry systems which greatly increases risk of errors
Percentage of "order-retract-reorder" events recorded	Order-retract-reorder events are correlated with orders entered on the wrong patient.
Percentage of potential duplicate patients in the live clinical database (i.e., same First name, Last name, and date of birth)	Duplicate patients increase the risk of clinicians missing key information.
Software bugs reported to the EHR vendor	A large quantity of serious software errors increases the risk that data is incorrectly entered, transmitted, stored, or lost.

15.2.3 FACILITATE SAFETY ASSESSMENTS OF LARGE, COMPLEX, HIT-ENABLED HEALTHCARE ORGANIZATIONS

We recently developed self-assessment tools, referred to as Safety Assurance Factors for EHR Resilience (SAFER) guides [36] to help clinicians and HCOs proactively assess the safety and effectiveness of their EHRs [37]. These, freely-available guides help identify areas of vulnerability and enable creation of solutions and culture change to reduce EHR-related concerns [38]. During their development, we learned that even the most highly regarded HIT-enabled healthcare organizations often had significant gaps in their EHR features, functions, or usage [39]. For example, one organization noted for its longstanding, highly successful computer-based provider order entry (CPOE) system did not have an interface between the EHR system used by physicians to enter orders and the laboratory system used to generate and report results. Similarly, another organization noted for its effective use of advanced clinical decision support never implemented CPOE.

Over the last 15 years, education and outreach alone have been insufficient to improve the safety of the healthcare system [40]. Therefore, we believe that more rigorous assessments are needed to improve the safety of EHR-enabled health care. We propose that the HIT Safety Center should work with an independent entity to refine the SAFER methodology and become a coordinating hub (i.e., establish the assessment criteria and aggregate the results) for random, preferably unannounced, on-site assessments of large, complex organizations that have received meaningful use incentives. These assessments could be carried out as part of current CMS site visits or by independent entities such as existing CMS deeming authorities (e.g., The Joint Commission) as a part of their accreditation process site visits. Assessment activities could include interviews with stakeholders, live EHR demonstrations, observations of clinicians as they interact with the EHR, tours of key clinical and technical sites, and reviews of EHR-related policies and procedures [41]. Reports of these visits could be submitted to regulatory organizations such as FDA, U.S. Inspector General, Office of Civil Rights, or CMS for their review and follow-up and made available on public websites [42]. While this might require additional resources, we believe some form of an EHR assessment strategy is key for organizations to reduce health IT safety issues [43].

15.2.4 ADVOCACY FOR HIT SAFETY, EVIDENCE GENERATION AND KNOWLEDGE DISSEMINATION

The HIT Safety Center must work with leading organizations that represent the broad range of "users" of HIT systems and the resulting data including patients, clinicians, ancillary service providers, policy makers, and payers, for example, to inform policy decisions and regulation related to HIT-related safety issues. This will ensure that future mandates take into account complex socio-technical and clinical implications of these decisions [44]. In addition, it must work with private entities involved in design, development, and use of these systems to help them understand why certain, safety-critical mandates were enacted and perhaps suggest potential technical solutions to address them. For example, EHR vendors may be reluctant to implement the eMeasures previously described [45]. The HIT Safety Center should coordinate, along with AHRQ, research required to generate and disseminate best evidence regarding the intricacies of designing, developing, implementing, and overseeing HIT within complex adaptive healthcare organizations [46]. Initially, the focus could be key research topics that need to be quickly resolved, such as development and validation of methods to measure, monitor, and improve EHR usability [47,48] and methods to achieve widespread interoperability [49]. Immediate deliverables could include acceleration of the long-standing work by the National Library of Medicine and the ONC on the standardization of clinical vocabularies [50] and technical data interchange standards [51]. For instance, to resolve the persistent and widespread problem of patient identification across healthcare organizations [52,53,54], the HIT Safety Center could encourage research, development, and implementation of innovative approaches to patient identification and matching [55]. Such solutions may not only reduce the burden of incorrect diagnosis and treatment but also improve the efficiency of healthcare processes by reducing duplicate testing and manpower required to merge and validate duplicate patient records.

15.2.5 CONCLUSIONS

We applaud FDASIA's recommendation to create a federally-supported HIT Safety Center. Although the initial funding request is insufficient to establish and maintain such a Center, we are optimistic about its development and future funding decisions. The convening ability of such a center could be critically important to our transformation to safe and effective HIT-enabled healthcare systems. To ensure progress and to avoid failure of this transformation, we need to move this recently created vision to reality, in keeping with the rapid pace of HIT implementation. A HIT Safety Center focused on the exemplary goals and activities we outline will more likely realize the transformative benefits of state-of-the-art health information technology and enable patients to receive HIT-facilitated, safe, and high value healthcare that they deserve.

REFERENCES

1. Institute of Medicine. Health IT and Patient Safety: Building Safer Systems for Better Care. The National Academies Press, Washington DC. (2012).
2. Advancing the health, safety, and well-being of the nation: FY 2015 President's Budget for HHS. 128-129. 2014. U.S. Department of Health & Human Services. Available at: http://www.hhs.gov/budget/fy2015/fy-2015-budget-in-brief.pdf
3. Section 618 of the Food and Drug Administration Safety and Innovation Act (FDASIA), Public Law 112-144. Available at: http://www.gpo.gov/fdsys/pkg/PLAW-112publ144/html/PLAW-112publ144.htm
4. HIT Policy Committee FDASIA Workgroup Transcript; April 29, 2013. Available at: http://www.healthit.gov/facas/sites/faca/files/2013-04-29_fdasia_final_transcript.pdf
5. FDASIA Health IT Report: Proposed Strategy and Recommendations for a Risk-Based Framework. April 2014. Available at: http://www.fda.gov/downloads/AboutFDA/CentersOffices/OfficeofMedicalProductsandTobacco/CDRH/CDRHReports/UCM391521.pdf
6. Tang P, for the Health Information Technology Policy Committee. Letter to Dr. Karen DeSalvo. August 6, 2014. Available at: http://www.healthit.gov/facas/sites/faca/files/STF__Safety_Center_Transmittal_2014-08-05.pdf
7. Upton F, Pitts JR, Blackburn M, Walden G. For US Congress Committee on Energy and Commerce. Letter to ONC Director, Karen DeSalvo. June 3, 2014. Available at: http://energycommerce.house.gov/sites/republicans.energycommerce.house.gov/files/letters/201406003ONC.pdf
8. Singh H, Classen DC, Sittig DF. Creating an oversight infrastructure for electronic health record-related patient safety hazards. J Patient Saf. 2011 Dec;7(4):169-74. doi: 10.1097/PTS.0b013e31823d8df0.
9. Sittig DF, Singh H. Electronic health records and national patient-safety goals. N Engl J Med. 2012 Nov 8;367(19):1854-60. doi: 10.1056/NEJMsb1205420.
10. Koppel R, Metlay JP, Cohen A, Abaluck B, Localio AR, Kimmel SE, Strom BL. Role of computerized physician order entry systems in facilitating medication errors. JAMA. 2005 Mar 9;293(10):1197-203.
11. Magrabi F, Ong MS, Runciman W, Coiera E. Using FDA reports to inform a classification for health information technology safety problems. J Am Med Inform Assoc. 2012 Jan-Feb;19(1):45-53. doi: 10.1136/amiajnl-2011-000369.

12. Meeks DW, Smith MW, Taylor L, Sittig DF, Scott J, Singh H. An Analysis of Electronic Health Record-Related Patient Safety Concerns. J Amer Med Inform Assoc 2014 Jun 20. pii: amiajnl-2013-002578. doi: 10.1136/amiajnl-2013-002578.
13. Bad Informatics can Kill. Available at: http://iig.umit.at/efmi/badinformatics.htm
14. Institute of Medicine. Patient Safety: Achieving a New Standard for Care. Washington, D.C.: National Academies Press, 2004.
15. Magrabi F, Ong MS, Runciman W, Coiera E. An analysis of computer-related patient safety incidents to inform the development of a classification. J Am Med Inform Assoc 2010;17:663-670.
16. ECRI Institute PSO Deep Dive: Health Information Technology. ECRI Institute PSO [serial online] 2012; Accessed March 14, 2014.
17. Santell JP, Kowiatek JG, Weber RJ, Hicks RW, Sirio CA. Medication errors resulting from computer entry by nonprescribers. Am J Health Syst Pharm. 2009 May 1;66(9):843-53. doi: 10.2146/ajhp080208.
18. Possanza L. ECRI's Partnership for Promoting Health IT Patient Safety. Available at:https://www.ecri.org/Products/PatientSafetyQualityRiskManagement/Pages/Partnership-for-Promoting-Health-IT-Patient-Safety.aspx
19. Clancy CM. Common formats allow uniform collection and reporting of patient safety data by patient safety organizations. Am J Med Qual. 2010 Jan-Feb;25(1):73-5. doi: 10.1177/1062860609352438.
20. Kilbridge P. Computer crash--lessons from a system failure. N Engl J Med. 2003 Mar 6;348(10):881-2.
21. Horsky J1, Kuperman GJ, Patel VL Comprehensive analysis of a medication dosing error related to CPOE. J Am Med Inform Assoc. 2005 Jul-Aug;12(4):377-82. Epub 2005 Mar 31.
22. Koppel R, Wetterneck T, Telles JL, Karsh BT. Workarounds to barcode medication administration systems: their occurrences, causes, and threats to patient safety. J Am Med Inform Assoc. 2008 Jul-Aug;15(4):408-23. doi: 10.1197/jamia.M2616
23. Spencer DC, Leininger A, Daniels R, Granko RP, Coeytaux RR. Effect of a computerized prescriber-order-entry system on reported medication errors. Am J Health Syst Pharm. 2005 Feb 15;62(4):416-9.
24. Perrow C. Normal Accidents: Living with High-Risk Technologies. Basic Books, 1984.
25. Bobb A, Gleason K, Husch M, Feinglass J, Yarnold PR, Noskin GA. The epidemiology of prescribing errors: the potential impact of computerized prescriber order entry. Arch Intern Med. 2004 Apr 12;164(7):785-92.
26. Clancy CM. New patient safety organizations can help health providers learn from and reduce patient safety events. J Patient Saf. 2009 Mar;5(1):1-2. doi: 10.1097/PTS.0b013e318198dca3.
27. Health Information Technology: Standards, Implementation Specifications, and Certification Criteria for Electronic Health Record Technology, 2014 Edition; Revisions to the Permanent Certification Program for Health Information Technology. Federal Register Vol. 77, No. 171; September 4, 2012; pgs. 54163 – 54292. Available at: http://www.gpo.gov/fdsys/pkg/FR-2012-09-04/pdf/2012-20982.pdf

28. Aviation Safety Information Analysis and Sharing. "Wrong Runway Departures". July 2007. Available at: http://www.cast-safety.org/pdf/asias_wrong_rwy_report_2007.pdf
29. United States Computer Emergency Readiness Team (US-CERT). Available at: http://www.us-cert.gov/
30. The Joint Commission. Sentinel Event Alert: Safely implementing health information and converging technologies. Issue 42, December 11, 2008. Available at: http://www.jointcommission.org/assets/1/18/SEA_42.pdf
31. History of the National Transportation Safety Board. Available at: https://www.ntsb.gov/about/history.html
32. Terhune C. Patient Data Outage Exposes Risks of Electronic Medical Records. Los Angeles Times, August 3 2012. Available at: http://articles.latimes.com/2012/aug/03/business/la-fi-hospital-data-outage-20120803
33. Robertson K. Sutter electronic records system crashed. Sacramento Business Journal. Aug 27, 2013. Available at: http://www.bizjournals.com/sacramento/news/2013/08/27/sutter-electronic-records-system-down.html?page=all
34. Sittig DF, Classen DC. author reply 1918-9 to: Koppel R. Monitoring and evaluating the use of electronic health records. JAMA. 2010 May 19;303(19):1918.
35. Smith M, Saunders R, Stuckhardt L, and McGinnis JM. (eds.). BEST CARE AT LOWER COST: The Path to Continuously Learning Health Care in America. Institute of Medicine, The National Academies Press, Washington, DC; 2013.
36. The Safety Assurance Factors for EHR Resilience (SAFER) Guides. Available at: http://www.healthit.gov/safer/
37. Sittig DF, Ash JS, Singh H. ONC issues guides for SAFER EHRs. J AHIMA. 2014 Apr;85(4):50-2.
38. Sittig DF, Ash JS, Singh H. The SAFER Guides: Empowering Organizations to Improve the Safety and Effectiveness of Electronic Health Records. Am J Managed Care, 6/2014 (in press).
39. Smith MW, Ash JS, Sittig DF, Singh H. Resilient Practices in Maintaining Safety of Health Information Technologies. J Cognitive Engineering; 2014 (in press).
40. Kuehn BM. Patient Safety Still Lagging: Advocates Call for National Patient Safety Monitoring Board. August 20, 2014. Available at: http://jama.jamanetwork.com/article.aspx?articleid=1899762
41. McMullen CK, Ash JS, Sittig DF, Bunce A, Guappone K, Dykstra R, Carpenter J, Richardson J, Wright A. Rapid assessment of clinical information systems in the healthcare setting: an efficient method for time-pressed evaluation. Methods Inf Med. 2011;50(4):299-307. doi: 10.3414/ME10-01-0042.
42. VA Office of the Inspector General. Review of Defects in VA's Computerized Patient Record System Version 27 and Associated Quality of Care Issues. Report No. 09-01033-155 June 29, 2009. Washington, DC Available at: http://www4.va.gov/oig/54/reports/VAOIG-09-01033-155.pdf
43. Schneider EC, Ridgely MS, Meeker D, Hunter LE, Khodyakov D, Rudin R. Promoting Patient Safety through Effective Health Information Technology Risk Management. RAND Health Report RR-654-DHHSNCH, May 2014. Available at: http://www.healthit.gov/sites/default/files/rr654_final_report_5-27-14.pdf

44. Sittig DF, Singh H. A new sociotechnical model for studying health information technology in complex adaptive healthcare systems. Qual Saf Health Care. 2010 Oct;19 Suppl 3:i68-74. doi: 10.1136/qshc.2010.042085.
45. Electronic Health Record Association. Comments on SAFER guides. Available at: http://www.himssehra.org/docs/SAFER%20Guides%20Comments%20Final.pdf
46. Rouse WB. Health care as a complex adaptive system: implications for design and management. The Bridge, National Academy of Engineering, Washington, DC. 2008:17e25.
47. Middleton B, Bloomrosen M, Dente MA, Hashmat B, Koppel R, Overhage JM, Payne TH, Rosenbloom ST, Weaver C, Zhang J; American Medical Informatics Association. Enhancing patient safety and quality of care by improving the usability of electronic health record systems: recommendations from AMIA. J Am Med Inform Assoc. 2013 Jun;20(e1):e2-8. doi: 10.1136/amiajnl-2012-001458.
48. Testimony of the American Medical Association for Health IT Policy Committee's Workgroups on Certification/Adoption and Implementation. Implementation and Usability of Certified Electronic Health Records; July 23, 2013. Available at: http://www.healthit.gov/sites/default/files/archive/FACA%20Hearings/2013-07-23%20Standards%3A%20Implementation,%20Meaningful%20Use,%20and%20Certification%20%26%20Adoption%20WGs,%20%20Implementation%20%26%20Usability%20Hearing/ama_usabilitytestimony_0.pdf
49. Furukawa MF1, Patel V, Charles D, Swain M, Mostashari F. Hospital electronic health information exchange grew substantially in 2008-12. Health Aff (Millwood). 2013 Aug;32(8):1346-54. doi: 10.1377/hlthaff.2013.0010.
50. Humphreys BL, Lindberg DA, Schoolman HM, Barnett GO. The Unified Medical Language System: an informatics research collaboration. J Am Med Inform Assoc. 1998 Jan-Feb;5(1):1-11.
51. MedVirginia. Nationwide Health Information Network: Trial Implementation. NHIN Evaluation Deliverable #15; 11/14/08. Available at:
52. Grannis SJ, Overhage JM, Hui S, McDonald CJ. Analysis of a probabilistic record linkage technique without human review. AMIA Annu Symp Proc. 2003:259-63.
53. McDonald CJ. Computerization can create safety hazards: a bar-coding near miss. Ann Intern Med. 2006 Apr 4;144(7):510-6.
54. Mannos D. NCPS Patient Misidentification Study: A Summary of Root Cause Analyses. Topics In Patient Safety 3(1); June/July 2003. Available at: http://www.patientsafety.va.gov/docs/TIPS/TIPS_Jul03.pdf
55. Weber GM, Mandl KD, Kohane IS. Finding the Missing Link for Big Biomedical Data. JAMA. 2014 May 22. doi: 10.1001/jama.2014.4228.
56. Sittig DF, Singh H. Defining health information technology-related errors: new developments since to err is human. Arch Intern Med. 2011 Jul 25;171(14):1281-4. doi: 10.1001/archinternmed.2011.327.
57. Sittig DF, Gonzalez D, Singh H. Contingency Planning for Electronic Health Record-based Care Continuity: A Survey of Recommended Practices. Int J Med Informatics 2014 (under review).
58. Belmont E, Chao S, Chestler AL, Fox SJ, Lamar M, Rosati KB, Shay EF, Sittig DF, Valenti AJ. EHR-related Metrics to Promote Quality of Care and Patient Safety. American Health Lawyers Association, Washington DC, 2013. Available at: http://

www.healthlawyers.org/hlresources/PI/InfoSeries/Documents/For%20the%20 Healthcare%20Executive/Minimizing%20EHRSSE.pdf
59. McCoy AB, Waitman LR, Lewis JB, Wright JA, Choma DP, Miller RA, Peterson JF. A framework for evaluating the appropriateness of clinical decision support alerts and responses. J Am Med Inform Assoc. 2012 May-Jun;19(3):346-52. doi: 10.1136/amiajnl-2011-000185
60. Sittig DF, Campbell E, Guappone K, Dykstra R, Ash JS. Recommendations for monitoring and evaluation of in-patient Computer-based Provider Order Entry systems: results of a Delphi survey. AMIA Annu Symp Proc. 2007 Oct 11:671-5.
61. Adelman JS, Kalkut GE, Schechter CB et al. Understanding and preventing wrong-patient electronic orders: a randomized controlled trial. J Am Med Inform Assoc 2013;20:305-310.
62. McCoy AB, Wright A, Kahn MG, Shapiro JS, Bernstam EV, Sittig DF. Matching identifiers in electronic health records: implications for duplicate records and patient safety. BMJ Qual Saf. 2013 Jan 29.

AUTHOR NOTES

CHAPTER 1.1

Acknowledgements

Dr. Sittig is supported in part by a grant from the National Library of Medicine R01- LM006942 and by a SHARP contract from the Office of the National Coordinator for Health Information Technology (ONC #10510592). Dr. Singh is supported by an NIH K23 career development award (K23CA125585), the VA National Center of Patient Safety, Agency for Health Care Research and Quality, a SHARP contract from the Office of the National Coordinator for Health Information Technology (ONC #10510592), and in part by the Houston VA HSR&D Center of Excellence (HFP90-020). These sources had no role in the preparation, review, or approval of the manuscript. We thank Laura A. Petersen, MD, MPH, VAHSR&D Center of Excellence, Michael E. DeBakey Veterans Affairs Medical Center, and Baylor College of Medicine, and Eric J. Thomas, MD, MPH, University of Texas, Houston-Memorial Hermann Center for Healthcare Quality and Safety and Department of Medicine, University of Texas Medical School, Houston, for their guidance in this work and Annie Bradford, PhD, for assistance with medical editing, for which they received no compensation.

CHAPTER 1.2

Funding Source

Dr. Sittig is supported in part by a grant from the National Library of Medicine R01- LM006942. Dr. Singh is supported by an NIH K23 career development award (K23CA125585), the VA National Center of Patient Safety, Agency for Health Care Research and Quality and in part by the Houston VA HSR&D Center of Excellence (HFP90-020). These sources had no role in the preparation, review, or approval of the manuscript.

CHAPTER 1.3

Acknowledgments

Dr. Menon is supported by a training fellowship from the Keck Center AHRQ Training Program in Patient Safety and Quality of the Gulf Coast Consortia (AHRQ Grant No. T32 HS017586-05). Dr. Meyer was a VA Health Services Research postdoctoral fellow supported by the VA Office of Academic Affiliations and in part by the Houston VA Center for Innovations in Quality, Effectiveness and Safety (CIN 13–413).. Dr. Sittig was supported in part by a grant from the American Society for Healthcare Risk Management to the American Health Lawyers Association. Dr. Singh is supported by the VA Health Services Research and Development Service (CRE 12-033; Presidential Early Career Award for Scientists and Engineers USA 14-274), the VA National Center of Patient Safety and the Agency for Health Care Research and Quality (R01HS022087) and in part by the Houston VA Center for Innovations in Quality, Effectiveness and Safety (CIN 13–413). These sources had no role in the preparation, review, or approval of the manuscript. We thank Annie Bradford, PhD, for assistance with medical editing.

CHAPTER 2.1

Conflict of Interest

The authors declare that they have no conflicts of interest in the research.

Protection of Human and Animal Subjects

Neither human nor animal subjects were included in this project.

CHAPTER 2.2

Acknowledgments

The authors are very grateful to the NHS hospitals that participated in their evaluation and to all individuals who kindly gave their time. They would like to thank their colleagues on the NHS CRS evaluation team, led by Professor Aziz Sheikh. They wish to thank the independent project steering committee, which was chaired by Professor David Bates, and also Michael W. Smith, PhD, for his contribution to the systems analysis approach.

Footnotes
Contributors DWM, AT, DFS, HS and NB participated in the conception and design of this study. AT and NB conducted data collection and primary analysis to identify patient safety relevant data from the original UK study. DWM, DFS and HS participated in the data analysis and interpretation of results. DWM wrote the initial draft. All authors performed critical review of drafts and approved the submitted version.

Funding
This paper is independent research commissioned by the NHS Connecting for Health Evaluation Programme (005 08/S0709/97) led by Professor Richard Lilford. The views expressed in this publication are those of the authors and not necessarily those of the NHS, the National Institute for Health Research or the Department of Health. HS is supported by the VA National Center of Patient Safety, Agency for Health Care Research and Quality, and in part by the Houston VA HSR&D Center of Excellence (HFP90-020). DWM is supported by the Baylor College of Medicine Department of Family and Community Medicine post-doctoral fellowship program and the Ruth L. Kirschstein national research service award (T32HP10031). These sources had no role in the preparation, review, or approval of the manuscript.

Ethics Approval
The NHS Connecting for Health Evaluation Programme received approval from a NHS ethics committee.

Data Sharing Statement
The authors of the original study may be contacted for the dataset.

CHAPTER 2.3

Acknowledgments
We acknowledge the VA IPS for their collaboration on this work.

Funding

Dr. Singh is supported by the VA National Center of Patient Safety and the Agency for Health Care Research and Quality(R18HS017820). Dr. Meeks is supported by the Baylor College of Medicine Department of Family and Community Medicine Postdoctoral Fellowship program and the Ruth L. Kirschstein National Research Service Award (T32HP10031). This material is based upon work supported (or supported in part) by the Department of Veterans Affairs, Veterans Health Administration, Office of Research and Development, and the Center for Innovations in Quality, Effectiveness and Safety (CIN 13-413).

Disclaimer

The views expressed in this article are those of the authors and do not necessarily reflect the position or policy of the Department of Veterans Affairs or the United States government.

Contributors Statement

DWM analyzed the data, wrote the manuscript, and reviewed drafts for important intellectual content. MWS analyzed the data and reviewed drafts for important intellectual content. DFS and HS conceptualized the study and reviewed drafts for important intellectual content. LT and JS obtained the data and reviewed drafts for important intellectual content.

CHAPTER 3.1

Acknowledgements:

Dr. Sittig is supported in part by a grant from the National Library of Medicine R01- LM006942 and by a SHARP contract from the Office of the National Coordinator for Health Information Technology (ONC #10510592). Dr. Singh is supported by an NIH K23 career development award (K23CA125585), the VA National Center of Patient Safety, Agency for Health Care Research and Quality, a SHARP contract from the Office of the National Coordinator for Health Information Technology (ONC #10510592), and in part by the Houston VA HSR&D Center of Excellence (HFP90-020). These sources had no role in the preparation, review, or approval of the manuscript.

We thank Ben Shneiderman, PhD for encouraging us to explore this topic. We also thank Elmer V. Bernstam, MD, MS, Gilad J. Kuperman, MD, PhD, Daniel G. Miller, MD, PhD, Laura A. Petersen, MD, MPH, Ryan P. Radecki, MD, Heidi V. Russell, MD, M. Michael Shabot, MD, Ben Shneiderman, PhD, Geeta R. Singhal, MD and Eric J. Thomas, MD, MPH for their helpful comments on early drafts of this manuscript and Annie Bradford, PhD for assistance with medical editing. We also thank several anonymous reviewers who gave us invaluable comments on earlier drafts of this manuscript.

CHAPTER 4.1

Acknowledgments

We thank Donna Espadas and Adol Esquivel, MD, PhD for their help creating the graphical depiction of the model. We also thank the two reviewers of this paper for their constructive criticism. This research was supported in part by the National Library of Medicine R01- LM006942 (DFS), NIH K23 career development award (K23CA125585) to HS, the VA National Center of Patient Safety (DFS, HS), Agency for Health Care Research and Quality (R18 HS17820) to HS and in part by the Houston VA HSR&D Center of Excellence (HFP90-020) (HS). These sources had no role in the preparation, review, or approval of the manuscript. We also thank Andrea Bradford, PhD for editorial assistance.

CHAPTER 5

Authors' Contributions

HS, JSA, and DFS drafted the protocol. All authors made substantial contributions to the design of the study and contributed feedback on the protocol. HS drafted the manuscript, to which all authors provided feedback and final approval.

Acknowledgements

Andrea Bradford, Ph.D. provided medical editing services on behalf of the authors.

Funding Information

The SAFER project is supported through a subcontract from Westat (HHSP-23320095655WC0095655; Anticipating the Unintended Consequences of Health IT) funded by the Office of the National Coordinator for Health Information Technology (ONC) (HHSP23337003T; to Drs. Singh, Ash, and Sittig); and the Houston VA Health Services Research and Development Center of Excellence (HFP90-020) (Dr. Singh). The views expressed in this article are those of the authors and do not necessarily represent the views of the Department of Veterans Affairs or the ONC. Neither ONC nor its funds were involved in manuscript writing process.

CHAPTER 6.1

Acknowledgements

The SAFER project is supported through a subcontract from Westat (HHSP-23320095655WC0095655; Anticipating the Unintended Consequences of Health IT) funded by the Office of the National Coordinator for Health Information Technology (ONC) (HHSP23337003T; to Drs. Sittig, Ash, and Singh). Dr. Singh is supported in part by the Department of Veterans Affairs, Veterans Health Administration, Office of Research and Development and the Center for Innovations in Quality, Effectiveness and Safety. The funders had no role in the design and conduct of the study; collection, management, analysis, and interpretation of the data; and preparation, review, or approval of the manuscript; and decision to submit the manuscript for publication. Andrea Bradford, Ph.D. provided medical editing services on behalf of the authors.

CHAPTER 6.2

Acknowledgements

Dr. Sittig was supported in part by a grant from the American Society for Healthcare Risk Management to the American Health Lawyers Association. Dr. Singh is supported by the VA National Center of Patient Safety, Agency for Health Care Research and Quality, and in part by the Houston VA HSR&D Center of Excellence (HFP90-020). These sources had no role in the preparation, review, or approval of the manuscript. We thank Annie Bradford, Ph.D., for assistance with medical editing.

CHAPTER 7.1

Acknowledgement
We thank Shelli Williamson, Executive Director of the Scottsdale Institute and Ricki Levitan for their help and support of our survey, as well as all the Scottsdale member organizations who participated in the survey.

Funding Statement
Dr. Singh was supported in part by the Houston VA Health Services Research and Development Center of Excellence (HFP90-020). The funders had no role in the design and conduct of the study; collection, management, analysis, and interpretation of the data; and preparation, review, or approval of the manuscript; and decision to submit the manuscript for publication.

CHAPTER 9.1

Acknowledgments
The authors thank Laurence Berg for his contributions to the data collection and analysis. The authors were supported in part by NCATS UL1 TR000371 (ABM, EVB, DFS), RC1 RR028254 (EVB), and UL1 RR025780 (MGK); NLM grant R00 LM009556-06 (JS); NSF grant III 0964613 (EVB); ONC SHARP Grants #10510592 (ABM, AW, EVB, DFS) and #10510924 (EVB); and Westat Contract #HHSP23320095655WC0095655, part of the ONC Task Order HHSP23337003T (DFS).

CHAPTER 10.1

Acknowledgments
The CPOE SAFER guide and survey were developed by all authors. The fieldwork and data analysis were performed by Dr. Vartian in partial fulfillment of the requirements for a master's degree in biomedical informatics at the University of Texas Health Science Center at Houston. Dr. Sittig

was his supervisor. Dr. Vartian wrote the first draft of the manuscript. All authors contributed to and approved the final version of the manuscript. Andrea Bradford, PhD, provided medical editing services on behalf of the authors. The views expressed in this article are those of the authors and do not necessarily represent the views of the U.S. Department of Veterans Affairs or the Office of the National Coordinator for Health Information Technology.

Funding

The SAFER project was supported through a subcontract from Westat (HHSP-23320095655WC0095655; Anticipating the Unintended Consequences of Health IT) funded by the Office of the National Coordinator for Health Information Technology (ONC) (HHSP23337003T; to Drs. Singh and Sittig). Dr. Singh is also supported by the VA Health Services Research and Development Service (CRE 12-033; Presidential Early Career Award for Scientists and Engineers USA 14-274), the VA National Center of Patient Safety and the Agency for Health Care Research and Quality (R01HS022087). This work is supported in part by the Department of Veterans Affairs, Veterans Health Administration, Office of Research and Development, and the Houston VA Center for Innovations in Quality, Effectiveness and Safety (CIN 13–413). Neither the ONC nor its funds were involved in the manuscript writing process.

CHAPTER 11.1

Author Contributions

HS conceived of the study, participated in its design and coordination, and drafted the manuscript. LW participated in the statistical analysis and qualitative data collection. LP participated in drafting and editing the manuscript as well as the design of the study. MS was involved in the qualitative data collection and analysis. BR was involved in qualitative data collection with Information Technology personnel and drafting the results of the manuscript. DE participated in the coordination of the study as well as qualitative data collection. DS participated in data analysis and provided edits to the final manuscript. All authors read and approved the final manuscript.

Author Information

The study was supported by an NIH K23 career develop- ment award (K23CA125585) to Dr. Singh, the VA National Center of Patient Safety, and in part by the Hou- ston VA HSR&D Center of Excellence (HFP90-020). These sources had no role in the design and conduct of the study; collection, management, analysis, and interpreta- tion of the data; and the preparation, review, or approval of the manuscript. The views expressed in this article are those of the authors and do not necessarily represent the views of the Department of Veterans Affairs.

CHAPTER 13

Acknowledgments

Project funded by Baylor College of Medicine Center for Globalization Demonstration Project Grant, and in part by the Houston VA HSR&D Center of Excellence (HFP90-020).

CHAPTER 14.1

Acknowledgements

The SAFER project is supported through a subcontract from Westat (HH-SP-23320095655WC0095655; Anticipating the Unintended Consequences of Health IT) funded by the Office of the National Coordinator for Health Information Technology (ONC) (HHSP23337003T; to Drs. Sittig, Ash, and Singh); and the VA HSR&D Center for Innovations in Quality, Effectiveness and Safety (#CIN 13-413), at the Michael E. DeBakey VA Medical Center, Houston, TX. (HFP90-020) (Dr. Singh). The funders had no role in the design and conduct of the study; collection, management, analysis, and interpretation of the data; and preparation, review, or approval of the manuscript; and decision to submit the manuscript for publication. Andrea Bradford, Ph.D. provided medical editing services on behalf of the authors.

CHAPTER 15.1

Acknowledgment
The authors thank Annie Bradford, PhD, for assistance with medical editing.

CHAPTER 15.2

Funding Statement
This work was supported in part by the Houston VA Center for Innovations in Quality, Effectiveness and Safety (CIN 13-413).

Competing Interests
Two of the authors (DFS & HS) were funded by the ONC and were key contributors to the development of the SAFER guides. These guides are freely available at: http://www.healthit.gov/safer. The authors have no other competing interests to declare.

Contributorship Statement
DFS wrote the first draft of the article. HS & DCC revised and re-revised the article significantly and approved the final version.

Acknowledgement
Andrea Bradford, Ph.D. provided medical editing services on behalf of the authors.

INDEX

A

accountability, 20, 34, 86, 113, 183, 208, 392, 418
    vendor accountability, 34
accreditation, 15, 142, 201, 432, 450
    accreditation test, 15
adverse drug effects (ADEs), 36, 294
    ADE Spontaneous Triggered Event Reporting (ASTER), 36, 44, 46, 50
Affordable Care Act, 51, 342, 354
alert, xxv, xxvii–xxviii, 4–6, 9, 13, 23–25, 27, 39, 41, 65, 69, 77, 92–93, 97, 99, 102, 111, 114, 116–117, 119, 121, 123–124, 128–131, 134, 151, 154, 160–162, 166–167, 172–173, 175, 177, 179, 184–185, 210, 217, 227, 230, 240, 242, 261, 264, 272–273, 276, 280, 287–288, 290, 292–295, 297–300, 302–303, 306–308, 313–326, 334–336, 338–339, 360, 363–366, 391, 395, 397, 405, 411, 413, 419, 422, 427, 434, 442, 447–449, 454, 456
    alert fatigue, xxviii, 128–129, 160, 184, 272, 276, 298, 320, 334, 391, 397, 449
    alert notification, 117, 319, 321, 434, 448
    alert notifications, 314, 322–323, 325, 334
allergies, 3, 39, 41, 77, 79, 128, 130, 135, 175, 230, 272, 279–280, 285, 290–291, 293, 394, 411

ambiguity, 314
ambulatory care, xxviii, 97, 160, 184, 313, 355
American Health Lawyers Association (AHLA), 18–19, 21, 28, 162, 189, 195–196, 409, 455, 458, 462
American Medical Informatics Association (AMIA), xxviii, 13, 16, 32, 34, 49–51, 68–69, 84–87, 103, 121, 133–134, 185–186, 208, 222, 232–233, 255–256, 266, 282–285, 296–299, 339, 354–357, 365, 409–415, 426–427, 439–441, 455–456
American Recovery and Reinvestment Act (ARRA), 14, 50, 113, 123–124, 210, 268, 429
American Society for Healthcare Risk Management (ASHRM), 16, 18–19, 21, 28, 162, 195, 458, 462
auto-population, 348
automation, 9, 99, 103, 133, 221, 299, 343, 348, 383, 387, 395, 409
autonomy, 68, 91

B

backup, 45, 90, 110, 113, 127, 165, 174, 191–195, 197, 199, 202–207, 224, 235, 237–238, 240, 262, 399, 436
biomedicine, 108
blood bank, 36–37, 43, 80, 224
Bluetooth, 369–370

C

cancer, xxv, 9, 67, 86, 124, 134, 151, 163, 183–184, 213, 256, 288, 301–302, 309–313, 338, 340, 355–356, 439
    cancer screening, xxv, 288, 301, 309–312
care delivery, xvii, 105, 113–114, 121, 137, 148, 154, 210, 367, 370, 381, 384–386
children, xx, 35, 98–103, 173, 247, 282, 287, 296, 369, 376, 427
classification, xxviii, 31, 40, 42, 56, 64, 68, 86–87, 96, 132, 150–151, 161, 256, 297, 357, 382, 412, 427, 439, 452–453
clinical
    clinical content, 4, 45, 52, 56–63, 74–75, 78–80, 82, 111, 116, 119, 139–140, 156, 175, 177, 188, 238–239, 286, 289, 315, 343–344, 373–374, 418, 420
    clinical context, 60, 63, 93, 243, 287
    clinical decision support (CDS), xvii, xix, xxiv–xxv, 11, 13, 15, 17, 19, 23, 25, 27, 29, 90, 93–95, 97–98, 102, 109, 111, 114, 116–118, 122, 127–130, 133–135, 141, 151, 153, 172, 175–176, 183–185, 190, 210, 232, 242–243, 245, 267–269, 271, 273–277, 279, 281, 283–289, 291–300, 315, 328, 357, 370, 374, 379, 385, 387, 409, 418–420, 422–424, 426–427, 430, 436, 440, 450, 456
    Clinical Information Systems (CISs), xxii, 18, 29, 32–39, 42–45, 47–48, 123, 145, 150–151, 223, 381, 413, 438, 442, 454

clinical practice, 20, 54, 98–99, 108, 125, 131, 148, 175, 181, 198, 240, 286, 297, 310, 330, 344, 372
close calls, xxvi, 430
cloud, 11, 369
coding, 56, 65, 67, 73–74, 85, 130, 150, 175, 238, 266, 385, 388–389, 392, 401, 439, 455
communication, xvii, xix, xxv–xxvi, xxviii, 1, 5, 9, 13, 16, 18, 36, 41–43, 46, 52, 57–58, 60–61, 63–64, 67, 69, 74–75, 77–79, 82, 86, 94, 96, 106, 112, 115–121, 124, 129, 133–134, 139–141, 144, 151, 153, 156, 167, 179, 183, 189, 193, 203, 206, 211, 224–225, 255, 287, 290, 301–303, 306, 308–317, 319–320, 323–327, 330–331, 336–338, 340–361, 363, 365, 368, 373, 375, 387, 398, 419–420, 424, 434, 439–440, 443, 448–449,
    automated communication, 302, 311
    electronic communication, xxv, 120, 124, 129, 303, 309–311, 314–315, 347, 350, 358, 360, 363, 368
    communication system, 77, 121, 302–303, 306, 308, 311
computer-generated interventions, 92–93, 179
Computerized Patient Record System (CPRS), 133, 319, 321, 323–324, 454
computerized provider order entry (CPOE), xix, xxiv, xxvii–xxviii, 2–3, 5, 8–9, 29, 31–32, 35, 42, 49, 66–68, 85, 98, 103, 115–117, 119–120, 126–127, 129, 131, 133–134, 141, 150–151, 160–161, 172–173, 176, 180, 183–185, 210–211, 227, 237, 239, 241, 243, 266–279, 282–283,

Index

285–290, 293–294, 296–298, 300, 315, 329–330, 354, 365, 382, 385, 410, 412, 426, 436, 439–440, 449–450, 453, 463
confidentiality, 25, 58, 65, 90, 97, 101, 177, 181, 197, 205, 227–228, 436, 438, 442,
    adolescent confidentiality, 101
content, xxi, 4, 11, 19, 39, 41, 43, 45, 49, 52, 54, 56–63, 65, 74–80, 82, 95, 101–102, 106, 108–109, 111, 115–117, 119, 128, 133–135, 139–142, 144, 146, 155–156, 161–162, 175, 177, 184, 188, 192, 204, 214, 216–217, 225, 227–229, 238–239, 271, 276, 278, 286, 289, 299, 315, 343–345, 354, 362, 372–374, 379, 381, 416, 418, 420, 422, 460
contingency, xix, xxi, 159, 161, 187–189, 196–200, 211, 276, 284, 455
    contingency planning, xix, xxi, 159, 161, 187–189, 196–200, 211, 276, 284, 455
coordination of care, 51, 342, 349, 401
copy and paste, 58, 61, 395
cost, xx, 7, 15, 34, 43, 59, 90–91, 103, 135, 150, 221, 268, 283, 346, 355, 365, 391, 437, 454
    cost-effective, xx, 91
credibility, 388, 401

D

data
    data collection, 19, 54, 65, 72, 93, 145, 247, 303, 375, 380, 433, 438, 440, 459, 463–464
    data entry, 4, 13, 58, 78, 94–95, 110, 123, 175, 207, 221, 229, 261, 268, 370, 375, 433, 449
    data management, 56
    data reporting, 23, 27, 47, 370

data sharing, 56, 101, 255, 459
delivery of care, 60
Department of Veterans Affairs (VA), xx, 71–74, 81–83, 86, 98, 133, 200, 302–303, 309–310, 316–317, 319–321, 324–325, 340, 431, 435, 440, 442, 454–455, 457–466
design, xvii, xxiv, 3–4, 6, 8–9, 12–13, 15–16, 36, 43, 51, 53, 60, 62, 65, 67, 70–71, 85–87, 92, 94–95, 102–103, 105–106, 109, 112–113, 116, 118–119, 121, 123, 139, 141, 153, 172–173, 185, 220, 257, 266, 276, 279, 299–300, 307, 312, 325, 339, 343–345, 351, 354, 357, 364, 368, 372, 375, 378–380, 402, 406, 410–411, 413, 432, 444, 446–447, 451, 455, 459, 461–465, –
developers, xviii, xxii, 2, 7, 12, 15, 30, 54, 57, 63, 71–72, 75, 91, 111–112, 138–139, 149, 155, 158, 182, 198, 225–228, 236, 286–289, 277, 347, 399–400, 418, 424, 433, 445–447
diagnostic, xix, xxv, 9, 16, 24–25, 32, 66, 69, 87, 92, 96, 110–111, 119, 124, 128, 130, 135, 157, 161, 168, 170, 175, 180–181, 183, 223, 230, 233, 255, 258, 279, 286–287, 289, 291–292, 299, 301–303, 305, 307–309, 311–313, 315, 317–319, 321, 323, 325–333, 335, 337, 339, 357, 359, 370, 379, 414, 416
disclosure, 14, 18, 432
documentation, 6, 17, 66, 94, 115, 127, 165, 195, 203, 229, 289, 316, 331, 364, 391, 395, 418–419, 434, 436, 448
downtime, xxi–xxii, 17, 20, 23–24, 27–28, 42, 45, 92, 126–127, 141, 149, 159, 165, 172, 174, 187–199, 201–208, 236, 273, 276, 387, 421, 424, 434, 441, 448–449

drug, xx, xxiii, xxvii–xxviii, 8–9, 23, 27, 29, 31–35, 39, 44, 46–47, 49, 67, 77, 85, 97, 99, 102, 121, 124, 128, 130, 132–135, 150, 160, 166, 175–176, 184–185, 230, 272–273, 275, 277, 280, 282, 284–285, 287, 290–292, 295–300, 384, 392, 394, 397, 405, 411, 413, 427, 439, 441–443, 452
    drug dose, 29, 39, 99, 176

E

e-iatrogenesis, 2, 8, 32, 67, 85, 93
efficiency, 10, 15, 48, 51, 69, 90, 94, 96, 114, 138, 150, 221, 264, 267, 270, 293, 322, 342, 346, 348–350, 353, 365, 377, 380, 384, 404, 451
electronic health record (EHR)
    EHR access, 58, 91, 234
    EHR accessibility, 126
    EHR adoption, 15, 21, 43, 51, 90, 135, 155, 210
    EHR design, xvii, 51, 62, 70, 153, 364, 432
    EHR implementation, xvii, xix–xx, 13–15, 17, 19–21, 29, 53–54, 56–57, 62–66, 73, 75, 82, 91, 99, 125, 131, 139, 148, 153–154, 156, 269, 315, 387, 429, 438
epidemiology, xiii, 99, 282, 295, 453
error
    diagnostic errors, v, 16, 66, 96, 223, 255, 312, 319, 325, 337
    error analysis, 7
    error mitigation, 246, 251–252, 258
    human error, 8, 121, 129, 384, 410, 415, 440
    juxtaposition errors, 138, 169
    medical errors, xxv, 10, 18, 84, 87, 185, 209, 221, 298, 337, 359
    medication error, 10, 173, 282, 285, 295, 297, 427
    silent errors, 211, 219
    technology-related errors, xix, xxviii, 1, 7, 32, 67, 86, 132, 151, 172, 210, 300, 351, 357, 440, 455
    user input error, 23, 27–28
    wrong-patient errors, 101

F

feedback, 5, 71, 90, 93, 109, 120, 130, 142–145, 177, 189, 214, 270, 276–277, 289, 292, 299, 320, 343, 345, 349, 379, 393, 400, 422, 424–425, 461
framework analysis, 56, 73, 388
free text, 130, 175, 180, 189–190, 224–225, 287, 289–290, 394, 433
functionality, 4, 15, 39, 41, 43, 58–59, 62–63, 71, 101, 126, 159, 176, 198, 203, 216, 227, 234, 272, 277, 279, 293–295, 329, 335, 347, 363, 370, 372, 379, 381, 399, 401, 436, 444

G

generators, 110, 127, 191, 193, 195, 203
global positioning satellites (GPS), 355–356, 370

H

hard-stop, 6, 9, 92, 97, 184, 298, 427
hardware, xix, xxi–xxiii, 2–4, 10–11, 19, 23, 45, 52, 56–60, 63, 65, 74–75, 78–80, 82, 103, 106, 108–113, 115–116, 119, 125, 127, 138–140, 144, 156, 159, 165, 174, 176, 188, 191–193, 195, 202, 207, 212,

Index

215–216, 224–229, 233–234, 236, 238, 240, 315, 343–344, 371, 374, 385, 419–420, 423–424, 446–448
health care quality, 91, 138, 342, 367
health information
    health information exchange, 14, 96, 101, 245, 401, 434, 448, 455
    health information management, 16, 50, 65, 150, 252, 258, 356, 440
    health information technologies, 89, 112, 187, 201, 383, 454
health information technology (HIT),
    health IT adoption, 367–368
    HIT development, 7, 105
    HIT error, 2, 7
    HIT errors, 2–3, 7
    HIT evaluation, 7, 118
    HIT interventions, 106, 109, 118
    HIT use, 2, 7, 116–118, 188, 404, 407
Health Information Technology for Economic and Clinical Health (HITECH), xvii, xxvii, 17, 113, 153, 159, 187, 210, 268, 342, 352–353, 411
Health Insurance Portability and Accountability Act (HIPAA), 113, 183, 196–198, 200, 205, 208, 352, 392, 404, 410
healthcare law, 18, 21
healthcare quality, 51, 71, 93, 161, 457
helpdesk, 71
Henriksen's model, 107
Hippocratic Oath, 91, 97
Hold Harmless, 34, 44, 48, 441
human factors, 3, 8, 72–74, 81, 87, 100, 103, 120, 122, 129, 142, 144, 155, 185, 270, 293, 326, 411, 413–414, 434–435, 442
human-computer interface, 4, 52, 57–61, 74–75, 78, 106, 111, 119, 139, 156, 178, 241, 320, 322, 333, 343–345, 436

I

identifiers, xxiii, 14, 39, 87, 103, 223, 232, 245–249, 251–254, 260, 266, 456
immunization, 102–103, 255, 265, 288, 298
implementation, xvii, xix–xxv, xxviii, 2–7, 9–10, 12–17, 19–21, 24, 29, 35, 43, 49, 51–54, 56–60, 62–66, 68, 70, 73, 75–76, 78, 82, 85–87, 91, 95, 99, 105–106, 108–109, 112–114, 117–120, 123, 125, 129, 131, 133–135, 137–139, 148–150, 153–158, 183, 185, 197–199, 211, 214–215, 225, 233, 243, 251, 257–258, 266, 268–272, 274–276, 283–284, 293, 295–296, 300, 302, 310, 312, 314–315, 325, 327–328, 338–339, 351–352, 357–359, 367–368, 372–373, 375, 380–381, 387, 391, 395, 398, 401–403, 406, 408, 411–414, 419–422, 426, 429, 432, 437–440, 442, 445, 451–453, 455
incentives, 14, 17, 29, 31, 43, 87, 113, 115, 132, 210, 268, 283, 285, 288, 353, 395, 437, 442, 450
incident reporting, 70, 128, 232, 393, 406
incomplete, xxvi, 3, 6, 20, 23–25, 27–30, 39, 92, 97, 133, 163, 167, 170–171, 176, 240, 263, 276, 347, 400, 449
India, xxvi, 367–369, 372, 380–382
informatics, xviii, xxviii, 8, 13, 15–16, 34, 50–51, 68–69, 71–74, 81, 84–87, 90, 97, 122–123, 134, 137, 142, 149, 155, 157, 160–161, 173, 180, 182–183, 187, 202, 208–209,

220, 222, 239, 243, 270–271, 274, 282–285, 301, 312, 326, 339–341, 354, 356–357, 367, 382, 388–389, 397, 402, 405, 410, 412–427, 434, 439–440, 445, 453, 455, 463,
   Informatics Patient Safety Office (IPS), 71–73, 81, 83, 445, 459
information transfer, 13, 90, 126, 318
infrastructure, xviii–xix, xxi, xxiii, xxvi, 32, 52, 54, 58, 63, 71, 75, 83–84, 97, 106, 110, 123, 125, 127, 130, 135, 137, 139, 157, 165, 191, 193, 195, 201, 203, 206, 211, 232–233, 236, 243, 258, 328, 359, 396, 399, 416, 429–433, 435, 437–439, 441, 443–445, 447, 449, 451–453, 455
innovation, 13–14, 92, 94, 106, 120–121, 268, 388, 413, 443, 452, 458, 460, 462, 464–466
inspection, 4, 15, 353, 436, 438
Institute of Medicine (IOM), xvii, xxvii, 1–2, 8, 16, 18, 31, 47, 50, 64, 67, 84–85, 97, 120, 131–132, 138, 151, 154, 160, 196, 201, 209–211, 269, 283, 412, 430, 440, 443, 449, 452–454
integrated systems, 210, 226, 342, 392, 408
Interactive Socio-technical Analysis (ISTA), 108
interdependencies, 110, 115, 390–392, 406, 409
interface, xviii–xix, xxii–xxiii, 3–5, 7, 12, 16, 39, 41, 43, 45, 52, 56–61, 74–83, 87, 90, 93, 100–101, 106–109, 111–112, 115–117, 119, 122, 127–128, 139–141, 154, 156, 163, 165, 167–168, 170, 172, 176, 178, 185, 188, 191–193, 195, 198, 207, 209, 211–217, 219–233, 235, 237, 239, 241–243, 262, 266, 287, 298–299, 315–316, 320, 322, 324, 329–330, 333, 337, 343–345, 348, 352, 354, 357, 362, 369, 372, 374–375, 378–379, 381, 387, 392, 395, 419, 436, 438, 446–447, 450
internet, 3, 11, 48–50, 95, 123, 174, 191, 195, 203, 238, 241, 345, 355–357, 374, 419
interoperability, 14, 53, 96, 213, 225, 339, 451
interpretation, 6, 23, 27, 56, 66, 73–74, 84, 93, 130, 225–226, 229–230, 329, 331, 335, 403, 407, 459, 462–463, 465

L

liability, 8, 47, 92, 184, 298, 319, 423
limitations, 13, 30, 64–65, 83, 93–94, 106, 109, 115, 130, 148, 199, 210, 231, 254, 277, 299, 309, 324, 381, 390, 400–401, 407–409, 446
literature review, xxvi, 20, 249, 300, 372, 411

M

malfunctioning, 2–3, 19, 34–35, 39, 41, 45, 125–126, 138, 156, 242
mandatory reporting, 34–35, 48
Manufacturer and User Facility Device Experience (MAUDE), xx, 35, 37, 48, 441
mapping, xxiii, 56, 73–74, 228, 234, 434, 441
master patient indexes (MPIs), 9, 246, 252, 258
meaningful use, xxvii, 14, 17, 31, 36, 87, 125, 129, 131, 134, 150, 160–161, 173, 200, 223, 232, 268, 282,

Index 473

284–285, 342, 361, 366, 382, 392, 404, 409, 416, 436, 439, 450
Medical Device Reporting (MDR), 34–35, 37, 49, 441
Medical Product Safety Network, 35, 49, 441
MedSun, 35, 37, 49, 441
medical record numbers (MRNs), 58, 248, 252, 261
mental health, 54, 61, 68
meta-narrative, 384–385, 411
misspellings, 37, 251
misuse, 15, 125
mobile device, 368, 374, 377, 380
mobility, 273
monitoring, 6, 15–16, 28, 30, 32, 37, 48, 50, 52, 57–58, 62, 64–65, 71, 73, 75–76, 78, 82, 106, 109, 112–114, 116, 120, 122–123, 130, 133, 138–140, 156, 175, 181, 183, 186, 188, 192, 195, 197, 201, 207–208, 216–217, 220, 225, 228, 231, 233, 241, 243, 273, 286, 289, 292, 294, 298, 310–311, 317, 336, 343–344, 351–352, 364, 368, 373, 377, 386, 388–397, 400, 402–403, 406–407, 409, 414, 416, 419–420, 422, 427, 440–443, 446, 454, 456
mortality, 16, 24–25, 35, 49, 120, 123, 133, 150, 266, 283, 296, 312, 357, 381, 408, 411–412, 414, 439

N

notification, 5, 9, 14, 16, 32, 67, 69, 71, 86, 96, 117–119, 124, 128, 130, 151, 157, 165, 171, 173, 183, 237, 246, 251, 258, 299, 302–303, 309–310, 312–315, 317–325, 332, 334–335, 337–340, 350, 357, 363, 365, 398, 430, 434, 448,

O

open-access, 11, 255
operating procedure, 14, 302, 377
operating system, 3, 110, 236, 238
order sets, 13, 94, 115, 119, 127, 133, 164, 175, 177, 184–185, 238, 243, 277, 279, 286, 295, 297, 391, 418
outpatient, xxi–xxii, xxiv–xxvi, xxviii, 9, 32, 69, 80, 124, 135, 151, 159, 161, 183, 257, 269, 273, 283–284, 296, 309, 312–313, 318–319, 322, 325–327, 331, 334, 337, 339, 341, 343–345, 347, 349, 351–355, 358–359, 362, 365–366, 436
oversight, xxvi, 7, 11, 14, 20, 32–34, 46, 48, 91, 93, 97, 135, 157, 170, 181, 207, 231, 241, 265, 294, 336, 364, 385, 403–404, 417–419, 424, 429–439, 441, 443–445, 447, 449, 451–453, 455

P

password, 60, 92, 127, 192, 205, 235–236, 240–241, 273, 275, 280, 292, 374
pathology, 33, 61, 134, 173, 266, 329, 338, 396
patient
    patient access, 347
    patient advocacy, 47
    patient communication, 343, 347
    patient compliance, 348
    patient confidentiality, 65, 181, 205, 436
    patient data, 1, 12, 24, 43, 77, 90, 92–93, 95, 110, 127, 163, 166, 168, 177, 183, 192, 194, 200, 204–205, 207–208, 210, 227, 236–240, 245, 369, 454

patient harm, xxv, 2, 61, 77, 83, 93, 128, 176, 178, 181–182, 194–195, 212, 229, 231, 235, 246, 254, 258, 260–261, 329, 337, 359, 361, 421, 424, 445, 448

patient identification, xix, xxiii–xxiv, 14, 19, 24, 141, 162, 164, 172, 177, 205, 209, 245, 247, 249, 251, 253, 255–261, 263, 265–266, 387, 392, 396, 451

patient information, 14, 23, 27–28, 76, 100, 196, 262, 281, 293, 319, 349

patient record, xxiv, 16–17, 29, 50, 71–72, 77, 133, 164, 178, 205, 207, 211–212, 245–246, 249–251, 253, 255, 258–259, 261–265, 319, 359, 411, 426, 433, 435, 440, 451, 454,

patient safety, xvii, xx, xxii, xxv–xxviii, 1–2, 6, 8–12, 14, 16, 18, 21, 28–32, 34, 47, 50–54, 56, 60, 62–74, 79, 81, 83–87, 95–97, 103, 108, 120–121, 123–128, 130–132, 134–135, 138, 141–142, 146–151, 154–155, 157–158, 160–163, 171–172, 177, 180–183, 185–186, 196, 199–201, 203, 207, 209–211, 218, 221–223, 225–226, 229, 231–232, 240–241, 245, 247, 253–254, 256–257, 261, 265–270, 274, 276, 283, 287, 294, 296–297, 299–300, 313, 318–320, 325, 327–328, 336, 339, 355, 357–359, 364–365, 368, 385, 390, 410, 412–417, 423, 425, 427, 429–433, 436, 439–440, 442–446, 448, 452–462, 464–465

patient safety model, 10

patient summaries, 92, 224

patient verification, 33, 101

performance measurement, 93, 97

physical environment, xix, xxiii, 65, 107, 156, 188, 377

point of care (POC), xvii, xix, 93, 101, 121, 129, 131, 176, 179, 192–194, 287, 290, 292, 302, 346, 367, 369–370, 374, 380–381

policies, xviii, xxiii, 5–8, 13–14, 18, 23, 27, 43, 49, 52, 57–58, 65, 70, 76, 78, 83, 91, 93, 95, 106, 113, 116–120, 124, 127, 139–140, 151, 156, 170,173–174, 177, 180, 182, 184, 188–189, 191–193, 195, 205–207, 211, 216, 225–227, 229, 232, 234–236, 238, 240–241, 259–262, 264, 269, 289–290, 293–294, 297–298, 310–311, 314, 316–318, 331, 335, 338, 343, 345, 350–351, 360–361, 364–365, 367, 377, 381, 391–392, 414, 417, 423, 436, 450–452, 455, 460

power supply, 110, 191, 203

primary care provider (PCP), xxviii, 119–120, 172, 178, 303, 306–308, 310, 315, 317, 324, 333, 339, 342–346, 348–351, 353, 360, 365

prioritization, 143, 315, 327

privacy, 14, 50, 69, 92, 101, 135, 200, 237, 243, 410, 441

productivity, 63, 90, 120, 368

provider, xvii–xix, xxiv–xxv, xxvii–xxviii, 2, 4, 10, 17, 29, 31–32, 35, 42–43, 47–49, 52–53, 57, 59, 61–62, 67, 90, 95, 98, 103, 107–108, 115, 117–120, 122, 124–127, 129, 133, 135, 138, 141, 145, 150–151, 153–154, 157, 160–161, 171–172, 183–186, 191, 196–198, 203, 208, 210–211, 225, 233, 243, 255–256, 266–267, 269–273, 275, 277, 279–283, 285, 287, 289, 291, 293, 295–297, 299, 302–303, 306–311, 314–317, 323–327, 329–337, 339,

341–354, 363, 365–366, 368, 370, 377, 382, 384–385, 387, 391, 394–397, 402, 405, 410, 414, 418–423, 425–427, 429–431, 434, 439–441, 445, 448–451, 453, 456,
   provider characteristics, 107
   provider order-entry, 10

R

records
   duplicate records, 87, 103, 164, 223, 245–249, 251–255, 258–259, 261, 265–266, 456
red flags, 163–172, 310, 436
referral, xix, xxv, xxviii, 64, 69, 94, 303, 309, 341–366, 375, 377–378, 382
   electronic referral, 69, 342–343, 345–355, 357, 362, 366
reflection, 402–404, 407
regulatory authority, 435
reliability, xx, 8, 11, 24, 56, 78, 90, 144, 176, 216, 226, 236, 310, 342, 350, 379, 403, 410, 414
reminders, xxviii, 11, 13, 93, 133, 128, 130, 169, 177, 184, 190, 238, 288, 293, 299, 335, 348, 365, 370, 376, 419
reporting, ix, xix–xx, xxv, 6, 14–15, 18, 23–25, 27, 30, 33–37, 40, 43–45, 47–49, 58, 62, 70–71, 83–85, 87, 128, 130–131, 141, 157, 163, 173, 181, 191, 220, 232, 247, 285, 289, 297, 301, 303, 305–307, 309–311, 313–315, 317, 319–321, 323, 325–331, 333, 335–340, 370, 372–373, 379, 387, 393, 397, 406, 419, 423–425, 427, 430, 432–433, 435–437, 440–442, 445–447, 453
   reporting system, xx, 35, 44–45, 71, 87, 128, 311, 379, 427, 437, 441–442, 446–447

automated reporting, 36, 433, 446
voluntary reporting, xx, 35, 71, 432
resilience, xviii, xxvi, xxviii, 59, 83, 87, 137–138, 147–148, 154–155, 160, 195, 197, 201–202, 211, 223, 269–270, 284, 382–383, 385–391, 393, 395, 397, 399, 401–417, 419, 421, 423, 425, 427, 449, 454
responsibility, xix–xx, 33, 44, 89–94, 98–102, 119, 134, 158, 185, 226, 229, 314, 316–317, 324, 330–332, 341, 348, 350, 361, 364, 388, 400, 416–426
Rights of Safe EHR Use, 10
risk
   risk assessment, xix, 1, 3, 5, 7, 9, 11, 13, 15, 17, 19, 21, 23, 25, 27, 29, 31, 66, 71, 85, 146, 149, 154–155, 269, 398, 412, 432
   risk management, 16, 18, 122, 142, 148, 155, 158, 162, 195, 201, 257, 327, 359, 385, 395, 398, 405, 413–414, 417, 432, 454, 458, 462,
   risk managers, 18, 21, 30, 148, 157, 163, 171, 233, 258, 327–328, 359, 388, 416
   risk reporting, 18
rural, xxvi, 368, 372, 375, 380–381

S

safety
   patient safety, xvii, xx, xxii, xxv–xxviii, 1–2, 6, 8–12, 14, 16, 18, 21, 28–32, 34, 47, 50–54, 56, 60, 62–74, 79, 81, 83–87, 95–97, 103, 108, 120–121, 123–128, 130–132, 134–135, 138, 141–142, 146–151, 154–155, 157–158, 160–163, 171–172, 177, 180–183, 185–186, 196, 199–201, 203, 207, 209–211, 218, 221–223, 225–226, 229,

231–232, 240–241, 245, 247, 253–254, 256–257, 261, 265–270, 274, 276, 283, 287, 294, 296–297, 299–300, 313, 318–320, 325, 327–328, 336, 339, 355, 357–359, 364–365, 368, 385, 390, 410, 412–417, 423, 425, 427, 429–433, 436, 439–440, 442–446, 448, 452–462, 464–465, , –
  resilient safety practices, 387–388, 402, 408
  safety benefits, 62, 130, 257, 276, 327, 359
  safety concerns, vii–viii, xvii–xx, 9, 16–18, 24, 30–32, 52–53, 56–60, 62–63, 65–67, 69–76, 79–84, 86, 90, 124–125, 127–129, 138, 148, 153–154, 156, 161–163, 171–172, 183, 201, 221, 269, 274, 313, 325, 338–339, 393, 414, 424, 430, 437, 444, 446–447, 453
  safety events, xx, 17–21, 23–24, 26–32, 47, 52–53, 68, 70, 82, 86–87, 125–126, 128, 130–131, 143, 148–149, 156, 162, 172, 195, 211, 225, 229, 231–232, 241, 257, 274, 327, 359, 409, 430, 433, 435, 440–441, 444–448, 453
  safety hazards, xxvi, 1, 32, 63, 93, 97, 135, 139, 150, 155, 182, 269, 421, 423–424, 429, 439, 452, 455
  safety measures, 20, 24–25, 28, 30–31, 177
scrolling, 117, 168, 293, 316, 323, 333
self-assessment, xviii, xxi–xxv, 138–146, 149, 154, 156, 158, 161, 197, 211, 214–215, 223, 256–257, 267, 269–270, 275–278, 326, 358, 416, 425, 432, 449
SNOMED-CT, 11, 92, 175, 225, 285
social media, 347
sociopolitical, 52, 385

sociotechnical, xviii, xx, 2, 4, 7–9, 51–53, 55–67, 70, 73–75, 78–79, 81, 84–86, 90, 95–96, 105–110, 112, 115–116, 118–119, 122, 134, 138–140, 144, 146, 149, 151, 155–156, 160–161, 188–189, 193, 201, 212, 232, 269, 282, 284, 296, 310, 315, 317–318, 342–344, 346, 351–352, 354, 367–369, 372–373, 380–382, 385, 387, 390–391, 397, 406, 409, 413–414, 451, 455
  sociotechnical analysis, 52, 67, 70, 81, 85, 122
software
  clinical application software, 10
  software configuration, xxv, 118–119, 138, 233, 304, 306–308, 392, 430
  software modification, 72
  software reliability, 379
  software upgrades, 165, 227, 393, 398
Spontaneous Reporting Systems (SRSs), 35
stakeholders, xxi–xxii, xxvi, 6, 143, 54, 66, 72, 91, 93, 95, 99, 130, 141, 143–144, 146, 148, 155, 157–159, 220, 276, 328, 344, 349, 351–352, 368, 380, 397–398, 405, 407, 436–437, 443–444, 450
stealth mode, 395
supervisory control, 383–384, 405
Swasthya Slate, 369–372, 378, 381–382
system interface, xxii, 56, 58, 75, 77, 79–82, 107, 128, 141, 163, 167, 176, 209, 211–217, 219–231, 233, 235, 237, 239, 241, 243, 315, 330, 387, 419, 446
  system-system interface, 77, 79, 82, 141, 387
Systems Engineering Initiative for Patient Safety (SEIPS), 10, 16, 67, 86, 108, 121, 410

Index

T

temporality, 64–65
training, xxiv, 5, 12–13, 17, 20, 24, 29, 35, 58, 72–73, 93–94, 112, 119, 127, 180, 189, 205, 220, 232, 240, 246, 251, 254, 257–258, 262, 268, 272, 276, 288, 320, 325, 361, 372–373, 375, 380, 406, 421–422, 425, 436, 458
typology, 7, 356

U

upgrades, 14, 39, 41, 45, 47, 74, 76, 78–79, 81, 165, 192, 204, 216, 225–227, 237–238, 288, 329, 392–393, 396, 398–399, 420, 422
urgency, 159, 315–316, 322, 335, 344, 347, 351, 360–361, 363–364
usability, xvii, 15, 70, 85, 93, 103, 116, 122, 142, 153, 155, 173, 179, 185, 197, 207, 222, 230, 241, 260, 263, 270, 293, 300, 333, 342, 351, 354, 364, 372–373, 379, 381–382, 402, 406, 412, 442, 447, 451, 455
user
    user familiarity, 24, 29
    user interface, 12, 16, 39, 41, 45, 58, 76, 83, 90, 93, 100–101, 108–109, 111, 115–117, 119, 122, 140, 170, 172, 188, 192–193, 198, 207, 232, 241, 266, 316, 324, 333, 372, 374, 379, 381
    user training, 5, 29, 112, 320

V

vendor, 2, 7, 12, 15, 18, 20, 24, 29, 34, 36–37, 39–41, 43–46, 48–49, 58, 91–95, 98, 124, 127, 131, 138, 143, 149, 157, 159, 171, 197, 215–216, 220, 233, 235, 238, 243, 258, 279, 325, 327–328, 359, 391–393, 398–399, 404, 408, 416, 421–425, 432–433, 435, 438, 441, 444–446, 449, 451
Vincent's framework, 107
vocabularies, xxiii, 11–12, 108–109, 111, 128, 216, 225, 229, 285, 385, 451
vulnerability, xviii, 52, 155, 329, 449

W

wording, 76, 142, 276
workarounds, xxviii, 5, 9, 17, 51, 67, 70, 172, 283, 296, 307–308, 310–311, 313, 365, 390, 422, 424, 453
workflow, xvii–xviii, xxiv, 5, 7, 11, 13, 20, 24, 29, 32, 35, 45, 47, 52–53, 57–61, 63–64, 67, 70, 74–75, 77–80, 82, 90–91, 95, 98, 106, 111–113, 115–117, 119, 121, 139–140, 144, 173, 177, 188–189, 191–193, 214, 221, 263–264, 268, 290, 299, 302–304, 306–307, 309–310, 316, 320, 322–323, 334, 343–344, 348–350, 368, 373, 375, 379, 381, 391–392, 400, 402, 410, 420–422, 426, 444
    workflow disruption, 91
workstations, 41, 63, 144, 234,